High Dimensional Space to Formulate Marriage and Birth Functions

Shuichirou Ike

Professor, Department of Sociology, Faculty of Letters
Chief of Educational Center for Information Processing
Teikyo University
Hachiouji-Shi, Tokyo, Japan

CRC Press
Taylor & Francis Group
Boca Raton London New York

CRC Press is an imprint of the
Taylor & Francis Group, an **informa** business
A SCIENCE PUBLISHERS BOOK

First edition published 2022
by CRC Press
6000 Broken Sound Parkway NW, Suite 300, Boca Raton, FL 33487-2742

and by CRC Press
4 Park Square, Milton Park, Abingdon, Oxon OX14 4RN

Library of Congress Cataloging-in-Publication Data (applied for)

ISBN: 978-0-367-18222-9 (hbk)
ISBN: 978-1-032-13471-0 (pbk)
ISBN: 978-0-429-06018-2 (ebk)

DOI: 10.1201/9780429060182

Typeset in Times New Roman
by Radiant Productions

Foreword

Fertility decline or population decline is now becoming a major problem in all of the developed countries around the world. In order to explain the fertility decline, several demographers, economists, sociologists, feminists and social policy-makers have pointed out many socioeconomic causes and have made proposals to restore the birth rate based on them. However, most of them are based on false assumptions and they cannot explain the real causes of decline in the fertility rate.

This may come as a surprise to some, but this is the startling conclusion that Professor Ike, the author of this book, draws.

According to Ike, marriages occur depending on its neighborhood, and the fertility decline that result from the decline in marriages can be described as an effect of geographic reaction-diffusion system. Ike assumes that the fertility decline is a simple alteration of a three-dimensional surface by its gradients, and it can identify the locations where fertility has historically declined in Europe and reproduce the diffusion of the fertility decline.

Furthermore, Ike argues that the actual number of children is not affected by the ideal number of children or the socioeconomic factors. He concludes that a causal birth model can only be constructed on the time-space dimensional space as a causal chain of the birth behavior.

These conclusions are not only mathematically well-formulated models, but are also falsifiable theories based on empirical data. I hope that this book will intellectually stimulate readers who are interested in the study of fertility decline and to create new academic findings.

Manabu Akagawa
Professor of Tokyo University
https://www.u-tokyo.ac.jp/focus/ja/people/people000194.html#

Preface

In 2016, I published *Fertility Decline and Background Independence—Applying a Reaction-Diffusion System as a Stochastic Process*. At the time of writing it, I intended for readers to understand easily the implications of the proposed reaction-diffusion mathematical model and the impact of the model on demographic theory. I understand now that I was wrong.

The reaction-diffusion model is not just a fine-fitted model of fertility declinc but requires a fundamental revision of the existing behavioral model. Can we scientifically answer the following questions?

"How do marriages occur? Further, how do births occur?"

Individual physiological maturation is not a sufficient answer. Today, the number of mature adults who do not marry (common law) has increased. This phenomenon must be considered as having collective emergent properties. Consequently, I ask you another question:

"How do these occurrences change dynamically?"

Traditionally, we assume that our conscious will and attitudes shift in response to these problems. However, when we question why our will and attitudes shift, we recognize that we have no adequate answers. We typically assume that there is a socioeconomic background causing such shifts. Ordinary social scientists research the relationship between socioeconomic variables and individual fertility behaviors.

Research only implicitly postulates that each individual is affected by socioeconomic variables. We must be skeptical about this postulation. Because the lack of substantial causal behavioral model, many researches cannot help depending on the ready-made statistical methods. We have not achieved clarity for many years under this postulation and these methods. We need to develop comprehensive models of fertility. The concrete model of fertility behaviors which can make quantitative predictions deserves it. The models are the marriage and birth functions, which require us to drastically change our philosophy to analyze phenomena. In addition, models must be built on the stochastic process to help us understand the historical trend of fertility decline.

A.J. Coale recognized that marriages of a cohort occur according with a probability density function for the first time. Namely, he cultivated the object as a stochastic marriage function. He and his colleagues also realized that fertility decline has diffused in space. While this phenomenon has been empirically recognized, stochastic models must be applied to it.

Constructing a stochastic and mathematical model requires us to change our perspective on human behavior. The diffusion process is inevitably the interaction of neighboring spaces (neighboring others). We are perpetually exposed to the affection of neighboring others, whether we are aware of it or not. This is the locomotive power and dynamics of marriage and birth. Thus, marriage and birth functions are constant floating fluids, especially at the micro-scale.

"Are you a utilitarian individual, maximizing your own interests?" No. This is a false image built by a delusive hypothesis. The eclecticism of utilitarian egos and norms has long prevailed in the social sciences. However, this is simply a programmed feeling. The true law that our behaviors obey is different; however, we are incapable of intuitively understanding such behaviors while they occur. We find it difficult to admit the effects of others on our self-consciousness.

The decisive conception is that previous neighboring marriages stochastically decide the succeeding ones. Neither conscious free will nor individual selection is necessary to build a scientific model of the occurrences of marriage and birth. Clearly, we behave (or respond) according to the behavior of those around us; we are programmed to believe that we think and act as individuals.

You need not be disappointed that free will does not exist. In the multiplex variable neighboring state of space, you will still feel free, provided that fascists do not compel you to follow pronatalist policies.

I am grateful to Vijay Primlani, the science publisher of the CRC Press for giving me the opportunity to write this book. I would also like to thank Editage (www.editage.com) for English language editing.

I am indebted to Emeritus Professor E. A. Wrigley of Cambridge University. It is pleasant to express gratitude to him who willingly grant the permission to reprint contents of his book *Population and History* as citations.

I am also grateful to maintainers of "Ubuntu", LaTeX, GNU-Octave. I cannot write any articles without them.

Finally, I benefited from my wife Takako, and my children (Kenichirou, Reiichirou).

Contents

Section II: Marriage Function in High Dimensional Space

Section III: Birth Function in High Dimensional Space

I

HISTORY OF GEOMETRICAL DIFFUSION MODEL

Chapter 1

The Dawn of Reaction-Diffusion Dynamics

> The random-walk problem is adopted as a starting point for the analytical study of dispersal in living organisms.

J.G. Skellam, Random Dispersal in Theoretical Populations, in *Biometrika* [69], p. 196.

In 1937, R.A. Fisher's "The Wave of Advance of Advantageous Genes" [26] first introduced diffusion dynamics into geographical phenomena. Although Torsten Hägerstrand is known as the pioneer of geographical diffusion, Fisher's formulation is more fundamental in the mathematical phase space. This is because, for the first time, Fisher presented the reaction-diffusion equation and the progressive wave, and mathematically deduced the velocity of the wave.

The limitations of Fisher's work were that he assumed that is was "advantageous genes" that diffused and spread; moreover, he developed his theory in single-dimensional space. Like him, his successors Skellam and Hägerstrand supposed solely geographical dimensions. However, higher-dimensional space is indispensable for the development of spatial pattern formation theories. Moreover, a classical reaction-diffusion theory must be realized. Therefore, this chapter explains the fundamentals of reaction-diffusion theories.

1.1 Fisher's Advantageous Genes

Fisher's "The Wave of Advance of Advantageous Genes" [26], marked the beginning of the reaction-diffusion model. His belief in fictive kin genetics made him develop this work. The development of science can be ironic; a false premise sometimes bears a true mathematical model. Fisher attempted to explain how a mutation occurring at a point in one-dimensional (linear) space diffuses. He deduced the velocity of the wave, now known as the Fisher-Kolmogorov-Petrovsky-Piskunov (FKPP) wave. Let us consider the formulation.

1.1.1 Fisher's Equation

Here, it is supposed that Fisher imagined the reaction-diffusion equation from a purely mathematical perspective. This is both the merit and demerit of his theory. Had he developed it based on the observation of real phenomena, he would have realized the true value of the reaction-diffusion model.

> Let p be the frequency of the mutant gene, and q is that of its parent allelomorph, which we shall suppose to be the only allelomorph present. Let m be intensity of selection in favour of the mutant gene, supposed independent of p. Suppose that the rate of diffusion per generation across any boundary may be equated to
>
> $$-k\frac{\partial p}{\partial x} \tag{1.1}$$
>
> at the boundary, x being the co-ordinate measuring position in the linear habitat. Then p must satisfy the differential equation
>
> $$\frac{\partial p}{\partial t} = k\frac{\partial^2 p}{\partial x^2} + mpq \tag{1.2}$$
>
> where t stands for time in generations.
>
> The constant k is a coefficient of diffusion analogous to that used in physics. Its use should be appropriate in many cases.
>
> The Wave of Advance of Advantageous Genes, R.A. Fisher [26], p. 355.

Fisher's formulation (as you see equation (1.2)) is a simple differential equation involving diffusion and logistic terms. The analysis that adopts the reaction-diffusion system investigates the geometric space satisfying this differential (or difference) equation.

1.1.1.1 *Mapping*

$\frac{\partial p}{\partial t}$ is a continuous mapping in continuous time–space. Specifically, the ratio p of some generation is a mapping from the ratio p of a previous generation;

after all, the reaction-diffusion model is a mapping from one time-point to next time-point in geometric space.

Fisher could not recognize this striking property of the equation. It is marvelous that, despite his false religious beliefs in heredity, he mathematically derived the reaction-diffusion equation over ten years before A.M. Turing. He maintained the classical diffusion frame represented by the concept of final stability and smoothing.

1.1.1.2 Autonomy—Background Independence

"Background independence" is another important mapping-related feature that appeared in the initial model. In this context, the following p is decided based exclusively on the present p. Specifically, the frequency of p at a successive time in a space is described exclusively by the previous p in that space. Other plausible factors, such as gamete distribution, can be eliminated from the models. When this model is applied to other research fields, all variables except the object variable in a geometric space can be excluded, given that its performance corresponds to reality or it sufficiently explains the phenomena. For example, the process of fertility decline can be explained quite effectively using the reaction-diffusion equation; specifically, plausible variables, such as socioeconomic conditions, can be neglected when analyzing the decline. This will be explored further in later chapters.

1.1.2 Progressive Wave

Fisher's most significant contribution was that he deduced the velocity of the progressive wave[1] from Equation (1.2). He calculated it in his article postulating wave stationary form.

> If we seek for a solution of (1)[2] representing a wave of stationary form advancing with velocity v, we may put
>
> $$\frac{\partial p}{\partial t} = -v\frac{\partial p}{\partial x} \tag{1.3}$$
>
> and obtain the differential equation (2)[3] involving only one independent variable:
>
> $$k\frac{\partial^2 p}{\partial x^2} + v\frac{\partial p}{\partial x} + mpq = 0 \tag{1.4}$$
>
> Since the variable x does not appear explicitly, we may write, for the frequency gradient,
>
> $$g = -dp/dx,$$

[1] These days, "traveling wave" or FKPP wave are the more commonly used terms. I adopt Fisher's original term "progressive wave."

[2] In this book Equation (1.2) corresponds to (1).

[3] In this book Equation (1.4) corresponds to (2).

whence

$$\frac{\partial^2 p}{\partial x^2} = -\frac{dg}{dx} = g\frac{dg}{dp}$$

and so find the relation between q and p,

$$kg\frac{dg}{dp} - vg + mpq = 0 \tag{1.5}$$

At the point of inflection $dg/dp = 0$, and $vg = mpq$; in advance of this point dq/dp is positive. If g/p tends to a limit u as p tends to zero, then u must satisfy the equation,

$$ku^2 - vu + m = 0, \tag{1.6}$$

a quadratic equation in u, which has real roots only if v^2 is not less than 4 km; but g/p cannot tend to zero for $vg > mpq$, and cannot to infinity because $v > k$, dg/dp. hence solutions only exist for velocity of propagation is equal to, or exceeds, $2\sqrt{(km)}$.

The Wave of Advance of Advantageous Genes, R.A. Fisher [26], p. 356.

Fisher first proposed the concept of the wave running in geometric (geographical) space. He developed the novel idea that the micro-variation gradually proceeds with the wave in space until the transition is completed.

In addition to this, owing to his remarkable mathematical ability, the aforementioned definition of progressive wave velocity was given.

A.M. Turing is generally regarded as the founder of modern reaction-diffusion theory. This is because he was far more radical and revolutionary in his approach than Fisher. Fisher did not recognize the possibility of complex pattern formation in the reaction-diffusion equation.

It is sad, but fortunate in a sense that Fisher's contribution and his prejudice originating from false eugenics genetics have been forgotten. Had he departed from gene theory and observed any real process of reaction-diffusion, he may have recognized the true power of the reaction-diffusion system, such as its autonomy. It is undeniable that the remarkable identity of the reaction-diffusion system is in the complexity originating from its simplicity. From this perspective, his works were purely those of a mathematician and statistician.

1.1.3 Do Genes Really Propagate?

It is generally believed that some entities propagate in space during the diffusion process. However, in the diffusion of advantageous genes, this concept does not correspond to real ecological life. The same gene does not travel from one area to another. Rather, the essential substance of gene diffusion is the frequency of the copies of the gene in neighboring spaces.

The diffusion process is based on a micro random walk model. However, it is not always true that the minute particles travel from one area to another. It is entirely natural that the slight stochastic vibrations of an area cause slight neighboring stochastic vibrations. Therefore, by simply assuming that the alteration of an area results in the alteration of a neighboring area, the reaction-diffusion model can be applied to specific objects.

Fisher was unaware of this rigorous scientific definition. In social science fields, the diffusion process must be treated precisely in this context. Conventional ideas of diffusion must be discarded. It is the most canonical conception that the slight alteration of a space provokes a slight alteration of neighboring spaces. When these processes are observed at a macro scale, they are referred to as "diffusion," "dispersion," or "spread."

Another point to be emphasized is that for the exact description of these phenomena, it is necessary to consciously employ (mathematical) mapping. In sociology, mapping from the present space to the successive one defines the causality of a space concerned with a particular phenomenon. High-dimensional geometrical imagination is utilized for these problems.

1.2 Skellam's Muskrats Diffusion Model

Fisher's work was ahead of its time. As mentioned earlier, he developed an impractical and abstract model. Practical cases were needed to ground the reaction-diffusion model. In 1951, J.G. Skellam published "Random Dispersal in Theoretical Populations" in *Biometrika* [69]. In this article, a real instance of the reaction-diffusion model and evidence of diffusion deduced from the reaction-diffusion equation were exhibited for the first time.

It is fascinating to read the introductory chapter of his article, in which he compressed the English history of diffusion studies. The origin of diffusion studies was *The Origin of British Flora* by Clement Reid (1899). Essentially, the reaction-diffusion system originated from botany. After Fisher, biologists developed the system with mathematical techniques utilized by J.C. Maxwell in developing the kinetic theory of gases.

The word "**dispersal**" signifies that diffusion is not the transfer of the something from A to B; rather, it represents the original dispersal of seeds. Seeds are not original entities, but clones of clones. In this case, the same entities do not move through space.

It is important to understand this slight but crucial difference. In the case of fertility decline, no mysterious entities need to be assumed as ideologies or innovations.

1.2.1 Skellam's Instance

Skellam's greatest contribution to the reaction-diffusion theory was that he obtained an empirical confirmation of diffusion.

Let μ be a coefficient of diffusion (Fisher expressed it as k) and α be the coefficient of the reaction (Fisher expressed it as m). μ and α are applied more

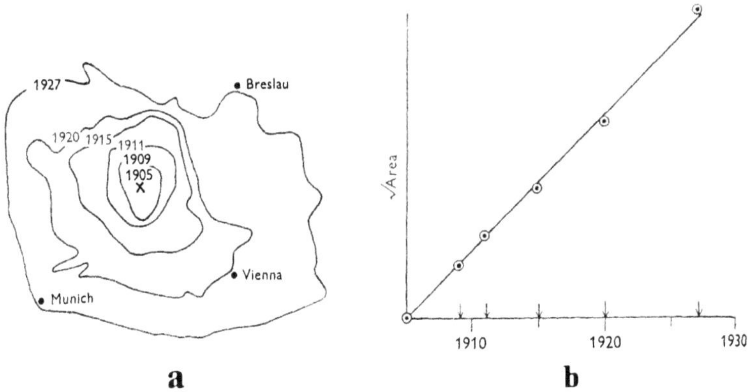

Figure 1.1: Spread of muskrats—(a) Apparent boundaries; (b) Relation between time and $\sqrt{}$area.

Source: J.G. Skellam [69], p. 200.

generally; moreover, km may be mistaken for km, the unit of distance. Thus, the velocity of diffusion is $2\sqrt{(\mu\alpha)}$ in one-dimensional space. The distance of the expansion of muskrats from the center is $t \cdot 2\sqrt{(\mu\alpha)}$. The living area of the muskrats is $\pi(t2\sqrt{(\mu\alpha)})^2$. Essentially, the living area of muskrats expands in proportion to $t^2 4\mu\alpha$. Thus, $\sqrt{Area}$ is proportional to time t.

The same regularity is deduced by the relation between the diffusion equation and its elementary solution. The elementary solution of the diffusion equation is a function of time t.

$$\frac{1}{\sqrt{4\pi t}} \exp(-\frac{x^2}{4t}) \tag{1.7}$$

The probability density function of normal distribution is as follows:

$$\frac{1}{2\sqrt{\pi}\sigma} \exp(-\frac{x^2}{4\sigma^2}) \tag{1.8}$$

The relation $2t \sim \sigma^2$ is easily obtained; σ^2 of the diffusion distribution increases in proportion to $2t$ at the initial stage of reaction-diffusion. It is difficult to precisely formulate this linear relation between time t and variance σ^2, except for the diffusion model.

It was Skellam who first found the real object for this relation between $\sqrt{(Area)}$ and t. His model has been adopted in the fields of botany and biology; in spite of this, it appears difficult to apply it to the human domain. However, parallel to Skellams work, Hägerstrand showed that the diffusion model powerfully explains ordinary human actions.

1.3 Hägerstrand

In 1958, Torsten Hägerstrand published *Innovation Diffusion as a Spatial Process* [30] in Swedish. In 1968, it was translated to English. Until then, the language barrier prevented the popularization of this geography-based understanding of diffusion.

Hägerstrand used the Monte Carlo simulation to demonstrate the geographical diffusion process. He insisted that the spatial stochastic process is a primary factor in the diffusion of innovations and new technologies. He excluded other socio-economic factors from the diffusion process. E.M. Rogers [65], who is the representative researcher on the spread of innovation, criticized Hägerstrand for his excessive emphasis on spatial variables.

However, this emphasis is a monumental achievement. In the social sciences, he was the first investigator to consciously find the significance of the geographical diffusion process and the possibility of excluding other plausible variables. Exclusion of other variables from the diffusion process should be praised rather than criticized. The fact that all variables except the spatial dimension can be omitted is crucial. The diffusion process is trustworthy because of Hägerstrand's empirical Monte Carlo simulation of human actions. The logical consequence is that other variables are excluded from the diffusion process. The belief that socioeconomic variables must always affect the diffusion process is merely a sentiment and not a logical inference.

1.3.1 Concrete Record of the Diffusion Process

Hägerstrand concretely recorded a real diffusion process in a manner characteristic of a geographer. This graphical and explanatory recording is highly valuable. Owing to his years of labor, the idea that innovative services spread diffusively can be intuitively understood. The idea that the diffusion process can be applied to social services, rather than just material products, deserves special consideration. Let us consider the outcome of his work on the diffusion of the postal checking service.

Hägerstrand wrote:

> Chapter 12.1.[4] Postal checking services were conceived of and introduced in Austria during 1882–83. The idea was soon adopted by the postal authorities in countries adjacent to Austria, and later in more distant countries: Hungary (1890), Switzerland (1905), Germany (1909), Belgium (1913). The Netherlands and France (1918), Denmark (1920) (Nylander, 1927; Örne, 1942).
>
> By this time interest was aroused in Sweden as well. From 1917 to 1924 the matter was placed on the agenda of several legislative fact-finding committees. In 1922, a government bill proposing introduction of the Danish system of postal checking services was defeated in parliament, but another bill was passed in 1924 (Wilhelmsson, 1959). The legislation covering postal checking services (SFS, 1924:378) went into effect on January 1, 1925.
>
> Source: T. Hägerstrand [30], p. 105.

[4]These quoted paragraphs are taken from the twelfth chapter of his book. To ensure accuracy, I have quoted them exactly as written.

As a cultural geographer, Hägerstrand explained the sequence of the European spread of postal checking services. This global sequence of adjacent transfer suggested the mechanism of diffusion.

He set his study area in the Swedish location "Kinda-Ydre." He elucidated geographical information about Kinda-Ydre as follows.

> The dimensions of Kinda-Ydre are approximately 60 km from east to west and 50 km from north to south. Its land area is 2,139 square kilometers.
>
> Source: T. Hägerstrand [30], p. 16.

He counted the number of individuals holding postal checking accounts on eight selected dates, and drew maps depicting this number.

Geographical diffusion can be intuitively imagined from the isolate points to all areas on the map. Hägerstrand was criticized for his extreme exaggeration of geographical diffusion. However, from the perspective of a radical diffusionist such as myself, Hägerstrand paid attention to other social factors as well, such as agricultural and non-agricultural residents. Nonetheless, his emphasis on geographical distance was consistent in his book.

> The only universal fact is that, no matter what the variations in such phenomena as telephone traffic, migration, personal meetings, and postal correspondence, data concerning them can always be placed in relation to distance. This relation itself can be expressed in many different ways.
>
> Source: T. Hägerstrand [30], p. 241.

He certainly recognized that the position of shortest distance is the most probable place to diffuse. Therefore, he may be the first person to use the "neighborhood effect" as the mechanical basis of diffusion.

> ... We can speak of the appearance of a *neighborhood* or *proximity* effect. Such a term, however, should not be interpreted too literally. A strict spatial sequence from farm to farm does not exist. If we leave the microstudy area and return to a careful observation of the absolute maps of the agricultural indicators for Kinda-Ydre in its entirely, we can observe the neighborhood effect again and again. In appears as the dominant feature of the innovation process (cf. 9.8.2, 10.8.3, and 11.4).
>
> T. Hägerstrand, *Innovation Diffusion as a Spatial Process* [30], p. 163.

He certainly achieved the insight that the neighborhood effect is the fundamental principle of diffusion.

1.3.2 Limitations of Hägerstrand's Work

Although there is no doubt that Hägerstrand is a great pioneer, his work had certain shortcomings, which can be summarized as follows.

1. Absence of a mathematical formulation.
2. Implicit assumption of what spreads in space.
3. No extension to higher-dimensional space.

1.3.2.1 Absence of a Mathematical Formulation

Hägerstrand's Monte Carlo simulation is a random walk simulation. Random walk transposes to the diffusion equation in terms of limit migration. Thus, he ought to have introduced the diffusion equation. It is odd that he and his successors did not refer to the diffusion equation; almost as though they did not know it.

In the 1960s, in a study of fertility decline in Europe, A.J. Coale and his colleagues inferred that fertility decline in 19th century Europe was a diffusion process led by elimination that deleted other possible variables. They manifestly concluded that the decline was a result of geographical diffusion. However, they also did not mention the diffusion equation.

Hägerstrand often referred to the "normal gradient" in the relation between distance and diffusion. In order to obtain the normal distribution as the fundamental solution of the diffusion equation, it can be expected that one would seek to apply the diffusion equation. However, he did not.

Hägerstrand realized the enduring fact that people telephoned one another predominantly within their neighborhood. This remains true for Internet-based communication today. Now, it is impossible for us to analyze the logs of subscribers in relation to their distance from a destination, because the privacy of respondents must be protected. From this perspective, his findings are all the more valuable.

Although he captured the principle of neighborhood interaction and the normal distribution, Hägerstrand could not advance to the diffusion equation. He estimated the exponential equations as follows:

> When the median points are plotted against a logarithmic scale it becomes evident that they can be very well depicted by the exponential equation:
>
> $$\mathbf{T} = C \cdot D^{-n}, \tag{7}$$
>
> where C is a constant indicating $\mathbf{T}$ when $D = 1$, and n a concentrated expression for the degree of slope. Obviously, as n increases, the slope steepens. C and n are determined on the basis of the empirical data in Table 1.1.
>
> The following fit is obtained after employing the least squares method:

Table 1.1: Distance Zone and corresponding T.

Distance Zone	T → Median	
[3–4 km	263	] eliminated
4–5 "	317	
5–6 "	200	
6–7 "	116	
7–8 "	65	
8–9 "	52	
9–10 "	45	
10–11 "	36	
11–12 "	33	
12–13 "	30	
13–14 "	26	
14–15 "	23	

$$\mathbf{T} = 7,612 \cdot D^{-2.2}, \qquad (8)$$

The corresponding curve is found in Figure 1.2, and hereafter is regarded as the normal gradient of telephone traffic for outgoing exchange located within Kinda-Ydre.

T. Hägerstrand, *Innovation Diffusion as a Spatial Process* [30], pp. 211–212.

Did Hägerstrand, Coale, and his colleagues not know the diffusion equation? The diffusion equation is constructed based on Fick's law. Fick's law is the principle of the dispersal of specific molecules concentrated at a point in some liquid. Even if they had known Fick's law or Fourier's law, they could not have imagined the relationship between human behaviors and these laws.

Defining the mapping from a state of a neighborhood to that of the neighborhood is indispensable for formalizing the innovation diffusion process. The absence of this mathematical formulation brought about the idea that the diffusion hypothesis is an inductive theory merely based on observations, rather than a deductive theory based on the law of human behavior. This impression is biased and premature in terms of considering the basis of human behavior. The idea will be explained in depth later in this book.

Regardless, the absence of a mathematical formulation is a disadvantage of Hägerstrand's model. The reason for this shortcoming may be assumed to be that Hägerstrand was also limited by the available information exchange entities. The diffusion of the automobile, postal checking service, and telephone does not represent the transmission of the idea or information about them among people. The existence of their users affects those who do not own them yet.

1.3.3 Entities that Spread in Space

It must be emphasized that diffusion dynamics are fundamentally the stochastic fluctuation of some spatial state from a space to its neighborhood. In this

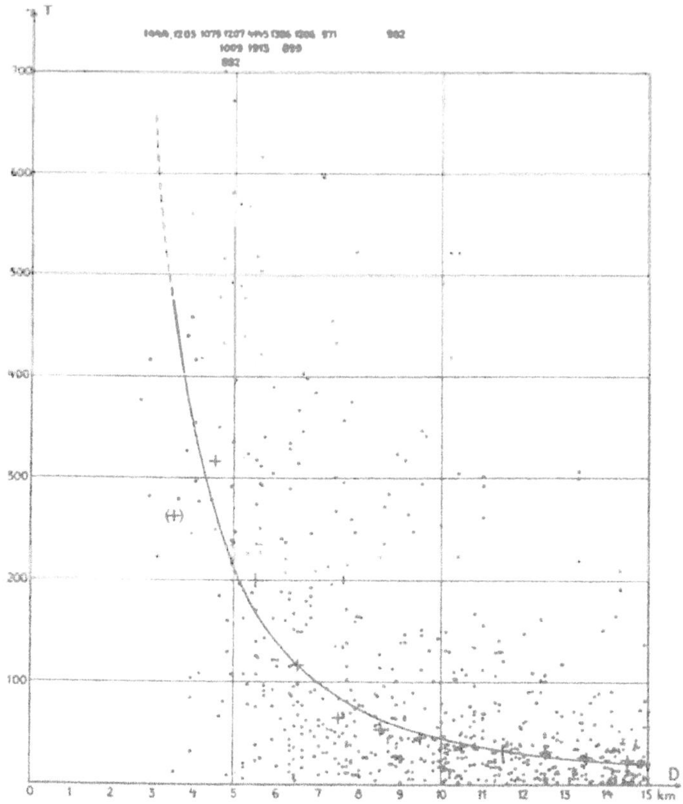

Figure 1.2: The normal gradient of telephone traffic for outgoing exchange located within Kinda-Ydre.

T. Hägerstrand, *Innovation Diffusion as a Spatial Process* [30], p. 210.

fundamental schema, there is no scope for the interchange of information or ideas.

The transmission of information is generally imagined as ideas flying past each other, such as in transmission control protocol/internet protocol (TCP/IP) packets on the Internet. However, this idea is not adequate for explaining sociological objects. True information transmission is a precise alteration of the state of destinations. For example, however enthusiastically teachers address their students, if the state of the students brains does not change, no information is transmitted at all.

It was natural that, in his study of innovation diffusion, Hägerstrand assumed that there exists something that spreads in space. The process of diffusion of phones and automobiles in geographical segments is increasing. Everybody assumes that innovative goods diffuse in space.

However, this postulation is too homespun to effectively analyze human behavior. Is it true that automobiles diffuse in space during the diffusion process? Even in Fisher's theory, advantageous genes do not strictly transfer in space. In terms of social phenomena, no single entity moves from human to human. In any case, social diffusion is not a bucket brigade of ideas, meaning, and information. It is entirely untrue that some infectious information is transmitted from one brain to another; this is only a metaphor.

Nothing spreads precisely in space. Hägerstrand emphasized the "neighborhood effect," but he failed to capture the actual principles of diffusion. The most essential concept is that at the next sequential time point, neighboring spaces either do or do not have the goods in question (such as phones). These probabilities are highly essential and are theoretical objects of the reaction-diffusion theory. They are governed by the previous neighboring state of diffusion. Based on this, a mapping can be used to form a mathematical definition of social diffusion.

Some may reject the aforementioned mathematical concept stating that there are no human subjective motivations. Hereinafter, the notion that conscious subjective motivation drives human behaviors is recognized as a false hypothesis. However strongly conscious subjects appear to reflect human actions, mere reflection cannot be the basis of science.

It is relatively easy to demonstrate that conscious subjective motivation has at best secondary significance in causing human behavior. Whether or not someone purchases a cell phone, they cannot have conscious subjective motivation, if they did not recognize the neighboring cell phone before the next sequential time point. Namely, the neighboring state is primarily significant. If no neighboring people have a cell phone, one cannot have conscious subjective motivation in the first place.

I believe that Hägerstrand was simply a geographer. Hence, his assumptions about human behaviors were too naïve; his contemporaries were also naïve. He should have paid attention to human nature. For instance, T. Veblen described human desire as follows:

> In the nature of the case, the desire for wealth can scarcely be satiated in any individual instance, and evidently a satiation of the average or general desire for wealth is out of the question. However widely, or equally, or "fairly", it may be distributed, no general increase of the community's wealth can make any approach to satiating this need, the ground of which is the desire of every one to excel every one else in the accumulation of goods. If, as is sometimes assumed, the incentive to accumulation were the want of subsistence or of physical comfort, then the aggregate economic wants of a community might conceivably be satisfied at some point in the advance of industrial efficiency; but since the struggle is substantially a race for reputability on the basis of an invidious comparison, no approach to a definitive attainment is possible. What has just been said must not be taken to mean

> that there are no other incentives to acquisition and accumulation than this desire to excel in pecuniary standing and so gain the esteem and envy of one's fellow-men. The desire for added comfort and security from want is present as a motive at every stage of the process of accumulation in a modern industrial community; although the standard of sufficiency in these respects is in turn greatly affected by the habit of pecuniary emulation. To a great extent this emulation shapes the methods and selects the objects of expenditure for personal comfort and decent livelihood.
>
> T. Veblen, *The Theory of the Leisure Class* [77], pp. 21–22.

The generally accepted belief about human individual selection is not a well-studied idea. Humans do not select goods as a result of receiving information about them. We are not the clever consumers that neo-classical economists assume us to be. Rather, our desires are others' desires; we are always gazing upon our neighbors. This is the key to reaction-diffusion theory in the social sciences.

Although Hägerstrand was condemned for his excessive emphasis on spatial factors, his emphasis on geographical distance is reasonable. I believe that his reflections on human nature were far from sufficient. However, he was the first person to introduce the diffusion mechanism into the realm of liberal arts. When Hägerstrand presented the Monte Carlo method-based diffusion process, it became clear that the "neighboring effect" is a primary factor of the diffusion process. The subjective idea that socioeconomic variables must have an effect is only a belief, and not a scientific induction.

1.3.3.1 Information does not Diffuse

Hägerstrand should be criticized for the absence of completeness in his diffusion mechanism. In social phenomena, ideas or meanings do not diffuse spatially. However, the neighboring sequential chain of behaviors is an accurate indicator of diffusion; surprisingly, marriage and birth can be described by it.

Hägerstrand certainly emphasized on the "neighborhood effect." However, he did not make a clear distinction between the "neighborhood effect" and private information.

1.4 Delay in Human Conscious Sensory Awareness

Modern social theories assume that humans consciously and reasonably behave in response to received information. This Cartesian view is completely false. We are always unconsciously recognizing the world. Our brains unconsciously observe other people and evaluate what they do. This is exactly what phenomenology teaches us as "Geworfenheit" (thrownness). We already exist a priori in society. Society is anything but constituted by ourselves as the social contract theory explains.

A prominent physiologist, Libet [51], found a "delay in conscious sensory awareness."

> ... The brain needs a relatively long period of appropriate activations, up to about half a second, to elicit awareness of event! Your conscious experience or awareness of your finger touching the table thus appears only after the brain activities have become adequate to produce the awareness.
>
> B. Libet, *Mind Time: The Temporal Factor in Consciousness* [51], p. 33.

It is the exposure to the possession of something by others that induces a conscious desire. The more we observe accumulation of good by others, the more likely it is that we would desire goods without information transmission. Therefore, the diffusion process can be formulated as a reaction-diffusion-equation with no relation to the information, as long as we effectively describe the spatial state of what diffuses as an equation.

Clearly, this has a firm foundation of physiology and epistemology as compared to neo-classical economics or rational choices theory.

1.4.0.1 Interpretation of Conscious Motivations is Tedious

I wish to elaborate on one additional point. When neo-classical economists or rational choice theorists interpret the conscious motivations behind purchase decisions, it is always a tedious process. They are classified as being subjectively advantageous or utile for the purchaser. These explanations are redundant and unsurprising.

When conscious motivations are postulated in the next chapter, readers of this book will understand that there were no consistent or significant advantages of fertility decline. These explanations based on conscious motivations cannot describe the occurrence of fertility decline according to time–space. It should be the central problem of the theory of fertility decline to calculate the level of fertility at each time–space point, not to understand the motivations of parents. The crucial reality is the probability of alteration in time–space dimensions.

1.4.1 Geographical Dimensions

Being a pioneer, Hägerstrand developed his theory for innovation diffusion on the geographical surface. However, his approach was insufficient. He observed the phenomena of diffusion exclusively in terms of the distance as quantitative values.

The limitation of his exclusive observance of the distance between individuals suggested that he remained invested in the ideas of individualism shared by his contemporaries. However, humans must be considered on the two-dimensional space at the very least, because social relations are always one-to-numbers or numbers-to-numbers. In order to describe these relations, the geometric two-dimensional space is necessary. Mere distances are not sufficient for describing social phenomena.

Moreover, the diffusion is not limited to geographical dimensions. For example, as a geographer, he was not able to imagine an "age dimension." If such a dimension is included, the diffusion of cell phones can be described in a more eloquent and realistic manner.

1.5 A.M. Turing, Hodgkin, and Huxley—Autonomous Pattern Formation

In 1952, in the article "The Chemical Basis of Morphogenesis" [73], A. Turing, an outstandingly talented computer scientist and mathematician, proposed a precise paradoxical possibility of pattern formation. Skellam's "Random Dispersal in Theoretical Populations" previously mentioned in this chapter was presented in 1951. His next article, "Studies in Statistical Ecology" [70] was presented in 1952. Was this a mere coincidence?

Some intellectual fashions of the spatial pattern may have been observed in 1950s England. It can be assumed that A. Turing may have been influenced by this. However, in those days, his ideas were too radical for many biologists to admit the significance of his proposition. The biologists believed that all biomedical issues were simply constructed from the design drawn by DNA. In the days when "DNA"—genes—held sway, his study was ahead of its time. The importance of his model went unacknowledged for about twenty years.

Indeed, in 1952 England, the spatial pattern formation by the diffusion equation saw a surprising, coincidental advancement. Hodgkin and Huxley [34] proposed that the simultaneous partial differential equations of one diffusion equation and three ordinary differential equations causes the wave of pulses.

$$\begin{cases} \dfrac{\partial V}{\partial t} = k\dfrac{\partial^2 V}{\partial x^2} - a_1 m^3 h(V - V_1) - a_2 n^4 (V + V_2) - a_3 (V + V_3), \\ \dfrac{\partial m}{\partial t} = \alpha_m(V)(1-m) - \beta_m(V)m, \\ \dfrac{\partial h}{\partial t} = \alpha_h(V)(1-h) - \beta_h(V)h, \\ \dfrac{\partial n}{\partial t} = \alpha_n(V)(1-n) - \beta_n(V)n, \end{cases} \tag{1.9}$$

Generally, reaction-diffusion equations cannot be solved analytically. Hodgkin and Huxley solved them numerically by employing the then-developed computer, and they found the wave of pulses with a constant velocity.

As mentioned earlier in this chapter, R.A. Fisher had already found the wave with a constant velocity in 1937. Japanese mathematician Kametaka had also presented a proof of it [41].

The reaction-diffusion equation has a wide range of possibilities for describing the spatial pattern formation. In the demographic realm, the possibility of such astonishing pattern formation may not exist; however, an important pattern of fertility decline can be described by the reaction-diffusion equation.

Unfortunately, my mathematical ability is not sufficient to venture into a more profound development of the reaction-diffusion equations in applied mathematics. I have merely adapted the bi-stable reaction-diffusion equation to the phenomenon of fertility decline in modern times.

Chapter 2

Decline of the Number of Children

> There are more things in Heauen and Earth, Horatio,
> Then are dream't of in our Philosophy.

W. Shakespeare, *Hamlet*, https://www.gutenberg.org/ebooks/2265.

2.1 Fertility Decline—Demographic Transition?

Are you aware of the term "demographic transition"? Though it is already collasped, fertility of European countries except France declined in the second half of 19th century. And fertility decline in modern European countries has mainly been caused by the decline in the number of children per family. Although lower fertility rates are also caused by the delay or diminishment of marriage, the fertility decline of 19th century Europe was caused by the decline in the number of children.

2.1.1 Early Diffusionists

In the 1970s, based on the method of eliminating other possible causes [13], A.J. Coale, J. Knodel, and others insisted that fertility decline is an instance of geographical diffusion. They were called "diffusionists" and hence, I refer to them as "early diffusionists".

Let me explain the method of elimination more specifically. We tend to suppose that human fertility declined in accordance with the alteration of socioeconomic conditions. The early diffusionists had supposed so in their initial research hypotheses, and they empirically collected European historical data to verify this. Surprisingly, their hypotheses that industrialization, education, urbanization, and decline of mortality cause fertility decline met with contradictions at every point.

One example of these contradictions is as follows. The theory of demographic transition supposed that surplus children caused by the decline of infant mortality was the primary cause of the fertility decline in 19th century Europe. However, several districts were in fertility decline before the decline in infant mortality.

Tables 2.1 and 2.2 show that there were several districts in Europe that contradicted the demographic transition theory.

Several other contradictions to the demographic transition were found. The idea of demographic transition had already failed in the 1970s, leaving only the diffusion hypothesis. Early diffusionists found the diffusion process to be the cause of fertility decline. They visualized this process in the following maps.

Table 2.1: Order of the onset of infant mortality and marital fertility decline.

Country	Number of Units	Infant Mortality Declines First	Marital Fertility (I_g) Declines First	Both at the Same Time
Switzerland	181	172	8	1
Germany	71	36	34	1
Belgium	9	1	8	-

Source: Francine van de Walle, Infant Mortality and Demographic Transition [76], p. 228.

2.1.1.1 *Diffusion of Lower Fertility*

Figures 2.1–2.4 depict maps of Europe; the shades of colors vary based on the level of I_g. I_g is an index of marital fertility calculated by Coale and his colleagues. It measures marital fertility in comparison with the average marital fertility of Hutterites[1]. Deeper shades of red or blue indicate lower or higher fertility, respectively.

Although the diffusion of lower fertility increased, geographical barriers prevented its spread. The Pyrenees mountain range historically constrained diffusion processes between France and Spain. However, this barrier was circumvented and diffusion occurred in waves, penetrating Spain through a small gap along the Mediterranean Sea. This circumventing wave can be precisely observed.

Other mountain ranges, notably the Alps and the Apennines, also obstructed waves of lower fertility. These barriers constrained contact between people, and therefore, prevented the diffusion of lowered fertility.

It was then and is now evident that fertility decline in Europe expresses the phenomena of diffusion. Unfortunately, early diffusionists did not know how to scientifically and empirically consider the diffusion process.

[1]Hutterites are a group of Christian fundamentalists who prohibit contraception. They reside in the state of Alberta in Canada.

Table 2.2: Demographic and socioeconomic indexes at the onset of fertility decline for selected countries.

	Date of Decline in Marital Fertility by 10 Percent	Marital Fertility Before Decline (I_g)	Index of Proportion Married (I_m)	Overall Fertility (I_f)	Infant Mortality (per thousand)	Percent of Male Labor Force in Agriculture	Percent Rural[d]	Percent in Cities over 20,000 Population	Percent Illiterate[e]
France	ca. 1800	.70	.51[a]	.30	185c	70	81	7	High
Belgium	1882	.82	.44	.35	161	30	56	22	.30
Switzerland	1885	.72	.44	.29	165	33	78	9	Low
Germany	1890	.76	.50	.39	221	38	68	21	Low
Hungary	ca. 1890	.63	.70	.45	250	73	84	11	49[f]
England and Wales	1892	.68	.48	.31	149	15	28	57	Low
Sweden	1892	.71	.42	.31	102	49	81	11	Low
Scotland	1894	.75	.42	.31	124	13	27	49	Low
Netherlands	1897	.85	.45	.35	135[c]	29	26	42	Low
Denmark	1900	.68	.47	.32	131	42	61	23	Low
Norway	1904	.75	.42	.30	76	37	72	18	Low
Austria	1908	.68	.51	.36	205	40	–	19	21
Finland	1910	.70	.46	.31	114	66	85	9	44
Italy	1911	.68	.54	.36	146	46	38	28	39
Bulgaria	1912	ca. .70	ca. .74	ca. .45	159	70	82	7	60
Spain	1918	.64	.51	.30	158	66	45	26	46
Ireland	1929	.71	.35	.23	69	48	73	20	Low
Costa Rica	1962	.89[b]	.50[b]	.57	74	58	66	20	14
Taiwan	1963	.70	.70	.42	49	47	42	31	30
Chile	1964	.65[b]	.50[b]	.39	103	37	29	53	15
Thailand	ca. 1970	ca. .75	.75	ca. .51	77	75	85	12	18

NOTES: Country borders are of the date of decline. All figures refer to the year estimated as the date of a 10 percent decline in marital fertility, except the index of the level of marital fertility before decline. Estimates were obtained by interpolation or extrapolation when data were not directly available for the year indicated.

[a] In 1831.

[b] Excluding consensual unions.

[c] Children dead after registration only.

[d] In communities of less than 5,000 or as per legal definition.

[e] Both sexes, over 10 or 15 years of age; high refers to percentages of young adults unable to sign their name on the marriage certificate or of illiterate army recruits, exceeding 50 percent; low refers to percentages under 10 percent.

[f] 6+.

SOURCE: Adapted with additions and correlations from Etienne van de Walle and John Knodel, "Demographic Transition and Fertility Decline: the European Case," *Contributed papers: Sydney Conference, International Union for the Scientific Study of Population*, 1965, p. 55.

Source: J. Knodel & Etienne van de Walle [44], pp. 221–222

Figure 2.1: Values of I_g by province in Europe, 1870.

Figure 2.2: Values of I_g by province in Europe, 1900.

From the 1960s to the present, ready-made statistical analyses and tests have been thought to enable empirical and scientific verification in the field of social sciences. However, in principle, these are not scientific verifications. It is unfortunate that social scientists[2] have innocently consumed them just

[2]They have been pseudoscientists.

Figure 2.3: Values of I_g by province in Europe, 1930.

Figure 2.4: Values of I_g, by province in Europe, 1960.

Source: Figures 2.1–2.4 from *The Decline of Fertility in Europe* [13].

because every neighboring pseudoscientist consumed them. This was and is human reality. However, the common approach is not necessarily the truth. It is not acceptable that truth relies on statistics and popular belief.

Moreover, the diffusion hypothesis could not be tested by ready-made statistical analyses. Early diffusionists condemned the fact that they frequently could not avail regression analysis or statistical tests. The idea that classic and ready-made statistical methods can analyze diffusion is absurd.

2.1.2 Reaction-Diffusion System

In 2012, I published an article titled "Singularity of Fertility Decline is Lot-et-Garonne [36]". In it, I applied the reaction-diffusion equation to the fertility decline of 18th–19th century Europe and 20th century Japan.

$$\frac{\partial c(x,y,t)}{\partial t} = \mu \frac{\partial^2 c(x,y,t)}{\partial x^2} + \mu \frac{\partial^2 c(x,y,t)}{\partial y^2} + \alpha(\beta - c(x,y,t))(\gamma - c(x,y,t)) \quad (2.1)$$

$c(x,y,t)$ is the mean number of children born to each family at geographical points x,y at time t. β and γ represent the maximum and minimum of the mean number of children.

The gradients of the means of the number of children in the space show slight neighboring declines. These slight declines have been developed by the logistic term. The mean number of children in a space at a time t is the mapping from the mean number of children in the neighboring space of the immediately
preceding t.

Specifically, human fertility behaviors are dependent on each other and react to one other, but contrary to common assumptions, they are essentially independent of socioeconomic "background." Humans are dependent on each other; this situation creates illusions of background dependence.

When postulating human dependence on neighboring individuals, this differential equation can be written using the two-dimensional value c and Laplacian Δ in another format as follows:

$$\frac{\partial c}{\partial t} = \mu \Delta c + \alpha(\beta - c)(\gamma - c) \quad (2.2)$$

Specifically, the spatial undulation of $c(x,y,t)$ and the reaction term decide the number of children in a two-dimensional geographical space. Therefore, the number of children is a three-dimensional surface geometrically. The number of children of parents in a space at the following time t is the mapping from the present neighboring number of children in families in the space. Fundamentally, human actions are exclusively dependent on our actions. From the nineteenth century to the present, industrialization[3] and capitalism have been thought to be the cardinal factors causing the diversity of our societies. These factors have been overestimated.

[3] Informatization is also included.

2.1.2.1 Variances in Fertility Decline are Proportional to Time During the Initial Stage

As long as the effects of the reaction-term remain small, the differential equation (2.1) can be recognized as a normal diffusion equation. An analytical basic solution is obtained in the form of Equation (2.1).

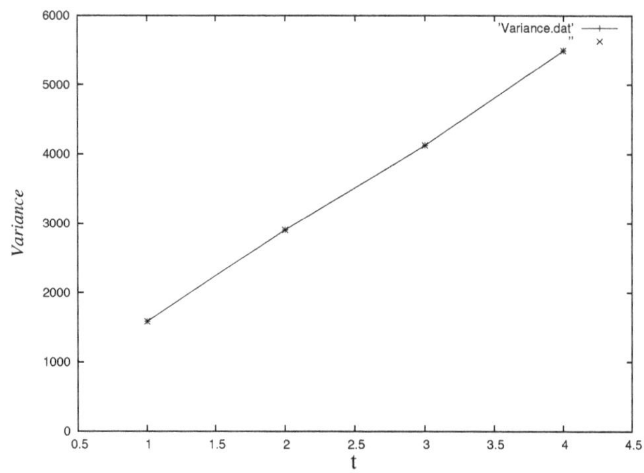

Figure 2.5: Variances of fertility decline with time t in Hokkaido.

I have already observed the fertility decline processes in three areas of Japan: Hokkaido, Tokyo, and Osaka. All three areas have been observed over 50 years, and the data on their fertility decline variances during the initial 20 years were analyzed. All the areas showed a very similar upward tendency; however, the proportionality in Tokyo and Osaka cases appeared slightly imperfect. The linearity in Hokkaido was almost perfect (Figure 2.5).[4]

As mentioned previously, J.G. Skellam [69] reported the same linearity of the development of the area of diffusion of muskrats from Bohemia to mid-Europe. In his article, the space was widened from Munich to Breslau in a circle of radius 300 km.

2.1.2.2 Diffusion is not Imitation

It is known that humans think and act subjectively. This ordinary common knowledge conceals the grave perceptual illusion suggested by Libet. I do not argue this problem here. It is crucial to understand that the diffusion process is not an imitation of the parents in the process. Individuals scarcely have

[4] I believe that this is because of the size of the area covered by the data, with Tokyo and Osaka being much smaller regions than Hokkaido. In order to demonstrate perfect proportionality, in line with the argument by J.G. Skellam, I resorted to using the Hokkaido case. Hokkaido is approximately nine times as large as Tokyo and Osaka.

the subjective consciousness to imitate the number of children of neighboring individuals.

We inevitably face our neighboring environment and recognize and justify our actions depending on it. An observer just observes it as imitation. However, individuals is not conscious to her birth behavior as imitation. It is important to realize that the diffusion mechanism is not imitation at all.

When considering "imitation," the original is contrarily postulated somewhere in our minds. What is the original? All that can be done is replications with little or no difference of previous events. As M. Foucault pointed out, the "original" and "imitation" are both illusions[5].

In particular, the diffusion mechanism causes our ordinary thinking and reason to depend on the people around us. The diffusion hypothesis is based upon human nature.

The hypotheses that assume that fertility behaviors are based on rational choices are more unnatural. "Rational choice" cannot be theoretically and transcendentally defined. What is rational or irrational is an affective evaluation by observers, not a scientific definition. What I would like to emphasize is that to call the same or similar result occurring in succession an "imitation" is an affective evaluation, not an objective recognition.

In this book, marriage and birth functions are formulated as mappings from present to succeeding occurrences in high-dimensional space. The sentimental evaluation of fertility behaviors must be entirely rejected. In other words, it is necessary to concretely map the present to the successive present.

2.2 Search for a Singularity Origin of Fertility Decline in Europe

2.2.1 Progressive Wave

A nontrivial solution of the reaction-diffusion differential equation (2.1) is called a "Kolmogorov-Petrovsky-Piskunov progressive wave" [46][6].

This wave is graphically represented in Figure 2.6. A progressive wave spreads from a singularity to its peripherals. In time, the points where minimum fertility decline occurred had advanced to the peripherals. These points are called the "front" or "edge" of the progressive wave (see "*Front*" in Figure 2.6). In other words, when the front is reached, fertility decline begins.

2.2.1.1 Velocity for One-Dimensional Space

R.A. Fisher [26] showed that the progressive wave has a constant velocity. His mathematical model is designed for one-dimensional space. This low dimensionality is rather convenient for our estimations.

[5] I will clarify that the imitations of articles or parts of it are actual imitations, and are illegal. They are not illusions.

[6] It is also called a "traveling wave" or FKPP wave.

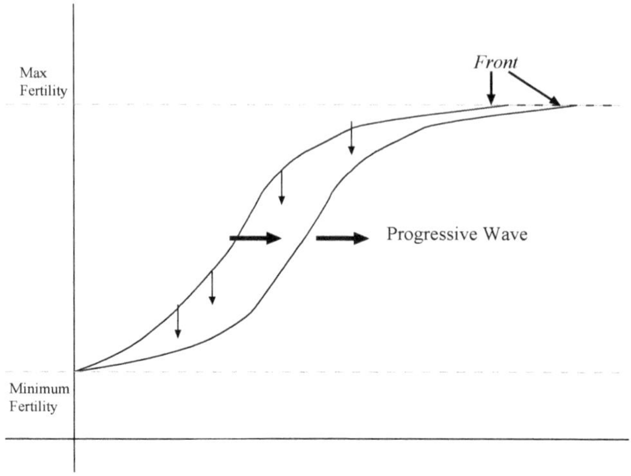

Figure 2.6: Progressive wave of fertility decline.

For each geographical point, the time taken to reach the front of the wave differs in proportion to the distance from a singularity. Thus, by taking the difference in time, when marital fertility decline began, the velocity of the progressive wave can be estimated.

For Japan, I empirically estimated 10 km/y $\leq v \leq$ 14 km/y as the velocity of the front of the wave. This is based on differences in fertility time decline occurrences recorded in census data, shown in Table 2.3.

Table 2.3: Distance and time taken by the progressive wave (front) to arrive in Japan.

From	Arrived Area	Distance	Passed Time	Estimated Velocity
Tokyo	marginal Kanto area	(max) 50 km	<5 year	10 km/year >
Tokyo	Niigata	270 km	from 20 to 25 years	from 10.5 to 13.5 km/year
Sapporo	Hokkaido	(max) 140 km	in 10 years	14 km/y

*Kanto is the name of a region near Tokyo. Tokyo is one of the singularities of fertility decline in Japan. Niigata is the largest of the cities on the coast of the Sea of Japan. Tokyo is the city's front on the Pacific Ocean; the progressive wave ran across the Japanese Island. Sapporo is the center of Hokkaido, which is an isolated island of northern Japan.

Method to estimate μ, α

Conversely, I obtained the estimations of μ, α using the simplex method [17]. Fisher deduced that $v = 2\sqrt{\mu\alpha}$ in a one-dimensional space. I computed μ, α for the following difference equation isodynamic to the differential equation (2.1).

$$\begin{aligned} & c(x_i, t_{k+1}) \\ = \quad & \mu c(x_{i-1}, t_k) + \mu c(x_{i+1}, t_k) + \\ & (1-2\mu)\left[(c(x_i, t_k) + \alpha(\beta - c(x_i, t_k))(\gamma - c(x_i, t_k))\right], \end{aligned} \tag{2.3}$$

Hereinafter, this difference equation is referred to as the difference equation of spatial distribution of children (DESDC).

In this work, "direct search methods" [58] were used to estimate μ, α. Direct search methods are those that obtain a minimizer (or maximizer) using only the values of functions. They do not depend on Taylor expansion, but on convex analysis. The "Nelder–Mead simplex method" [17], [82] is the most frequently applied direct search method. It is robust against noises in function values and is useful for problems with low-dimensional variables. However, this may not converge to a true minimizer, and there is no adequate convergence analysis.

Here, the Nelder–Mead simplex method and subsidiary original algorithm (simulated annealing method [58]) were applied to search for the minimizer using procedures to branch the separated local space and narrow the search area iteratively. This algorithm lacks analytical basis. Unfortunately, the Nelder–Mead simplex method depends upon the choice of the initial simplex, and thus, several local minimizers are found with very slight differences in the sum of squares of residuals.

Nonetheless, the simplex method can obtain a plausible minimizer for DESDC. Meanwhile, the simplex method and simulated annealing method can find highly approximate minimizers for μ, α (observed values are set for β, γ).

In order to accord the observations with a unit of velocity (km/y), it is necessary to multiply the intervals of the ratio of the square of space (h^2) and divide the ratio of time (k) for each coefficient. The data employed to estimate them are composed of the means of the number of children in each segment (5 km long) over five years. For this, $\mu\alpha$ is multiplied by $\frac{h^2}{k}$. Thus, the velocity of the progressive wave of fertility decline (km/y) is calculated by the following expression:

$$2\sqrt{\mu\alpha}\,\frac{h^2}{k} \tag{2.4}$$

Previously obtained estimations (Table 2.4) were substituted for each μ, α in the expression (2.4). Thus, the velocity (km/y) of the progressive wave can be estimated by the following expression:

$$11.45 < 2\sqrt{\mu\alpha}\,\frac{5^2}{5} < 11.68$$

These estimations resemble the values obtained by differences of time and distances between two points at which fertility decline occurred. This resemblance supports the validity of the DESDC.

Table 2.4: Estimated μ, α of Tokyo area.

Method of Estimation	μ	α	$\sum e^2$
Nelder–Mead simplex method(2)	0.47815	2.80000	3.73783
Simulated annealing method	0.47773	2.74658	3.74058

2.2.1.2 Velocity in Two-Dimensional Space

The velocity of a wave in two-dimensional space is $\sqrt{2\mu\alpha}$. Namely, the front of the wave moves with velocity $\sqrt{2\mu\alpha}$ on this surface. I obtained estimations of μ, α for two-dimensional space composed of 1 km^2 rectangles over one month. In this case, DESDC developed into a two-dimensional random walk and reaction term. Thus, $\mu < \frac{1}{4}$.

Therefore, the velocity of the progressive wave of fertility decline (km/y) for two-dimensional space is calculated by the following expression:

$$\sqrt{2\mu\alpha\frac{h^4}{k^2}} \tag{2.5}$$

The estimations are somewhat smaller than those for one-dimensional space.

$$10.04 < \sqrt{2\mu\alpha}\,\frac{1^2}{\frac{1}{12}} < 10.67.$$

However, this similarity between the estimations for one and two dimensions suggests the validity of two difference equations.

2.2.2 The Year a Singularity Appeared

In the differential equation (2.2), on returning to time $t = 0$,

$$\lim t \to 0, \qquad \frac{\partial c}{\partial t} = \mu\Delta c + \alpha(\beta - c)(\gamma - c) \to 0. \tag{2.6}$$

It can be observed that the partial differentiation is infinitely small. A quantum of the decline of c at time $t = 0$ must be postulated. This point is a singularity, that is, the origin of fertility decline.

By means of the estimated velocity of the progressive wave of fertility decline, the time and place at which a singularity of fertility decline was present in Europe can be determined.

First, because France is the first country that went into fertility decline, a singularity must exist in France. The specific district of France in which the singularity of fertility decline existed is to be identified.

The key for a singularity is the differences in the times at which fertility decline began in different countries. From the isomorphism of a progressive wave, the differences between dates of 10 percent decline is found to be nearly equal to the differences between the dates marking the beginning of fertility decline.

Table 2.5: Date of 10 percent decline in marital fertility.

France	Germany	Denmark	Belgium	Netherlands
ca. 1800	1890	1900	1882	1897

Source: J. Knodel & Etienne van de Walle [44], pp. 221–222.

Then, the differences in the time of a 10 percent decline in some countries were obtained, as shown in Table 2.5. These differences correspond to those during the time at which the minimum fertility decline began, namely, in Germany, the front arrived from the the singularity of France in 90 years.

2.2.2.1 Differences in Time from Isomorphism

By determining the date of the beginning of fertility decline in Germany and other places, the time of the occurrence of a singularity in France can be estimated. Therefore, the dates of the beginning of fertility decline in European countries were examined (Table 2.6).

Table 2.6: Initial fertility decline in Germany, Denmark, Belgium, and the Netherlands.

Germany		Denmark		Belgium		Netherlands	
Cohort	CTFR	cohort	CTFR	cohort	CTFR	cohort	CTFR
1852-1860	5.17	1840-1844	4.39	**1836-1845**	4.50	1836-1845	5.06
1857-1865	5.02	1845-1849	4.41	1841-1850	4.43	1841-1850	5.13
1862-1870	4.80	1850-1854	4.44	1846-1855	4.30	1846-1855	5.16
		1855-1859	4.38			**1851-1860**	4.98
		1860-1864	4.16			1856-1865	4.64

Source: P. Festy *Fertility in Western countries from 1870 to 1970* [25], pp. 48–49.

In Germany, Denmark, Belgium, and the Netherlands, fertility decline began between 1852 and 1860: 1855–1859, 1836–1845, and 1851–1860, respectively. The year in which a singularity occurred can be calculated via simple subtractions. For example, for Germany, 1852–90 = 1762.

In this manner, the time when a singularity began was calculated for each item of data.

$$1852 - 90 = 1762, \quad 1860 - 90 = 1770 \tag{2.7}$$

$$1855 - 100 = 1755, \quad 1859 - 100 = 1759 \tag{2.8}$$

$$1836 - 82 = 1754 \quad 1854 - 82 = 1772 \tag{2.9}$$

$$1851 - 97 = 1754 \quad 1860 - 97 = 1763 \tag{2.10}$$

I suppose that a singularity appeared in the period between 1754 and 1772. It is plausible that a singularity appeared among those born during the second half of thc 1750s. Thus, the actual period occurred in the 1770s in France.

2.2.2.2 Where was the Singularity?

If the velocity of the progressive wave is assumed to be universally equal, at approximately 10 km/y, the location of the singularity can be determined. It took almost 90 years for the progressive wave to arrive in Germany, indicating that the singularity was about 900 km away from there. Similarly, it took 82 years for the progressive wave to arrive in Belgium, indicating that the singularity was about 820 km away.

In this work, Frankfurt am Main was considered as a representative point for Germany (Frankfurt was near the population gravity center of Germany), and Brussels for Belgium. A circle with a radius of 900 km was drawn, with its center at Frankfurt[7]. Another circle with a radius of 820 km was drawn with Brussels as its center. One of the intersections of the circles is the point at which a singularity of fertility decline appeared. Specifically, the estimated point is around the *Aquitaine* district (Figure 2.7).

Figure 2.7: The point 900 km from Frankfurt am Main and 800 km from Brussels.

The early diffusionists once doubted that the fertility decline in England-Wales and Austria began simultaneously. This is because the levels of economic development of the two countries were quite different. However, from

[7]If Hamburg is set as the center of the circle, its radius is assumed to be 1100 km. This is because the decline in fertility in Hamburg began around 1880 (from *Coale and Treadway* [13]), indicating that it took about 110 years to arrive. The intersection remains on *Aquitaine*.

the perspective of the reaction-diffusion theory, the distances of the two from Lot-et-Garonne are not very different; rather, considering the Strait of Dover and the Alps, the physical distances appear similar.

2.2.2.3 Lot-et-Garonne

After estimating *Aquitaine* as a singularity, I incidentally noticed a critical fact. A previous study reported that Lot-et-Garonne in *Aquitaine* had the lowest fertility in Europe. Coale and Treadway [13] pointed out that the French *département* of Lot-et-Garonne marked the bottom of marital fertility in Europe.

> A sense of the meaning of these indexes can be gained by considering some examples. Three provinces at the bottom of the range (the lowest 10 percent in the distribution) of overall fertility (I_f) in Europe in 1900 were **Lot-et-Garonne** (a *département* of France), Geneva, and County Tipperary in Ireland. The overall rate of childbearing in the first two areas was only about one-sixth of the Hutterites (I_f of 0.168 and 0.164), while in Tipperary I_f was a little higher at 0.220.
>
> Ansley J. Coale and Roy Treadway [14], p. 34 (boldfaced by the author)

> The French *département* of Lot-et-Garonne had especially low fertility from a very early date, and the age distribution of women in 1900 from 15 to 50 was nearly uniform. I_m[8] in Lot-et-Garonne (0.700) was the highest in Western Europe; with the age distribution of Russia it would have been 0.699. Without the special age distribution generated by its history of especially low fertility, the I_m of Lot-et-Garonne would still have been highest (or nearly so) in Western Europe.
>
> Ansley J. Coale and Roy Treadway [14], p. 159.

There is a relationship between I_f and I_m, I_g, where $I_f \simeq I_m \cdot I_g$. Thus, the marital fertility of Lot-et-Garonne was the lowest in western Europe[9]. In 1831, I_g was estimated to be .351 for Lot-et-Garonne. These data indicate that Lot-et-Garonne had lowest marital fertility in western Europe.

Therefore, it is speculated here that the *département* of Lot-et-Garonne in *Aquitaine* was a singularity. The fact that the point of the lowest fertility rate existed in *Aquitaine* suggests that the coefficients μ, α are universal constants, and that reaction-diffusion precisely traces the fertility decline process in Europe.

[8] An index of the proportion of potentially fertile women who are currently married [13], p. 33.

[9] Geneva's I_g was .458 in 1860. This figure is far greater than that of Lot-et-Garonne. The low fertility of Geneva was caused by the low level of I_m—the rate of marriage. Switzerland remained in the medieval fertility control stage.

Figure 2.8: Geographical location of Lot-et-Garonne.

Source: https://fr.wikipedia.org/wiki/Lot-et-Garonne August, 28, 2018.

2.3 Diffusion of Fertility Decline is Background Independent

We have determined that the constant velocity of the progressive (traveling) wave can be assumed, and that this constant velocity can indicate where a singularity begins. This fact suggests the universality of human behaviors both in the East and in the West. The velocity is approximately the same in Europe and Japan; people of different cultures and socioeconomic conditions have the same coefficients μ, α. As mentioned previously, human fertility behaviors are entirely dependent on neighboring fertility behaviors. This is another expression of "background independence." It is human nature.

As shown in Table 2.2, Germany and Hungary went into fertility decline despite a very high infant mortality rate. There are a wide range of divergences in infant mortality rates at the times when fertility decline began. These facts serve as evidence that fertility decline is independent of mortality, which is a background factor.

Individualism—social nominalism supposes that society consists of aggregated individuals who float like particles, with a "background" constructed by socioeconomic variables. With this background, individuals rationally and objectively play games against each other. Based on this idea, the main factors of social change are inclined to be attributed to the background. From this point of view, alterations in behaviors are adaptations to variations in background, i.e., socioeconomic changes.

Individualism assumes that the individual is a higher organism that adapts to the alteration of socioeconomic background. However, early diffusionists showed that this adaptation did not occur. The discovery of this fact is enormously significant, although they did not recognize it themselves. Humans do not react to background alterations. I call this "background independence."

As Hägerstrand pointed out, human behaviors are dependent on neighboring behaviors. To accept the diffusion hypothesis is equivalent to rejecting individualism. In social science, to argue the diffusion process is not only to apply a mathematical model to the phenomena but to deny the traditional microbehavior model. It is understandable that Hägerstrand was not sufficiently conscious of this problem; after all, he was working in the 1950s.

In the field of social science, the diffusion hypotheses has been forgotten since the 1980s. It was expelled by the rise of household economics alongside the exaggeration of rational individualism, because the diffusion hypothesis of human behavioral adaptation fundamentally contradicts the concept of rational choice. It is important to remember that this philosophical problem is inherent in the diffusion hypothesis of social sciences. This problem will be explained in detail in later chapters.

2.3.1 Why did the Singularity Occur at it Lot-et-Garonne?

As per the "Easterlin Hypothesis," economic growth definitely increases fertility [21, 22, 23]. If a singularity occurred near London, economic growth caused by industrialization eliminated it before it developed. Therefore, economic growth interferes with fertility decline. A singularity of fertility decline had to be remote from industrial revolution.

The confusion created by the French Revolution could have robbed a singularity of its diffusion mechanism. A singularity had to come into existence at a point far from "industrialization" and "revolutions." This coincidence was astounding and providential.

2.3.1.1 Probability of Disintegration of the Balance

Questions on why no singularities existed in the centuries before 1770, and why there were no singularities in Sweden or Italy, are natural and intelligent. I attempt to answer these questions here.

I believe that the singularity is equal to the decisive disintegration of the balance of stochastic fluctuations of the number of children in the direction of decline. Although the appearance of the singularity is a genuine coincidence, its probability is extremely small. Thus, it takes a long time for the decisive disintegration of the balance of stochastic fluctuations.

Here, this probability was experimentally calculated. Medieval villages had a population of 400-450 on average, consisting of approximately 50 families. Suppose the singularity is the size of one village; its occurrence would require that completed marital fertility continuously declines for 10–20 years without any socio-economic disturbance.

The probability that all 30 or 35 sets of neighboring parents (they are neighboring in the time axis) simultaneously reduce their completed fertility is roughly calculated to be $[10^{-3}, 3.64 \times 10^{-5}]$ per year. This is the reason why the singularity had not occurred before 1770.

Because the minimum size of the singularity cannot be estimated, this probability could be smaller or larger. I believe that the occurrence of a singularity is a fortunate coincidence that may or may not take place in human history. This is the reason why the singularity had not occurred before.

As for why the singularity occurred in France, and not Sweden or Italy or others, I believe that this is also a genuine coincidence. However, the probability of disintegration of the balance of stochastic fluctuations was higher in Lot-et-Garonne than in other places. Its marriage rate is known to be the highest in all of Europe. Further, the political, economic, and social conditions were quite stable. The higher uniformity of the major sets of parents in Lot-et-Garonne prompted disintegration. The high uniformity is inclined to break spontaneously.

2.4 Visualized Fertility Decline Process in France

On observing the visualized fertility decline process in France (Figures 2.9—2.14), the wave of fertility decline[10] can be easily imagined to have progressed north from southwest France. It can be confirmed that the fertility of the Ardennes dępartment began to decline from 1841 to 1846. These data consistently correspond to the date when fertility decline began in Belgium, recorded by P. Festy (Table 2.6).

2.4.1 Delay of Fertility Decline in Brittany

It can be confirmed that the fertility rates in Massif Central and Brittany remained relatively high. It is easy to explain that Massif Central acted as a traffic barrier that interrupted the progressive wave. However, why was Brittany not so easily affected by the progressive wave of fertility decline?

Only the reaction-diffusion system can provide a scientific answer to this fascinating puzzle. Socioeconomic factors have been evaluated in vain. Cultural differences from the French, such as the Breton language derived from the Brittonic language, have been assumed. It has also been assumed that the population of Brittany included more religious Catholics or economically weaker individuals than the population of other French dępartments. Each hypothesis is plausible at first glance. However, they can all be eliminated empirically.

[10]This is called the FKPP wave in some contexts.

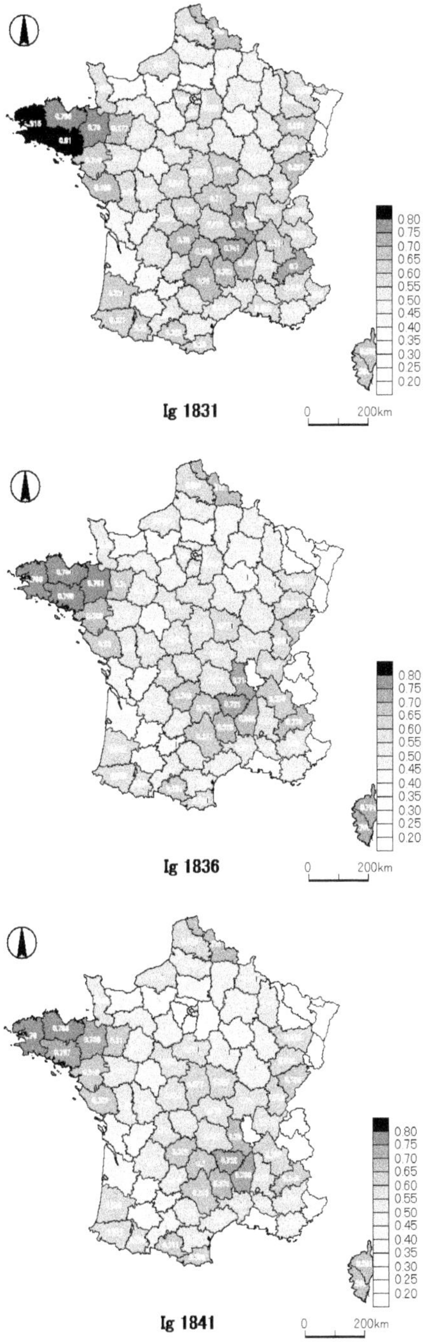

Figure 2.9: Dynamics of I_g in France.

I drawed maps based on the following data source availing MANDARA.

*Source: *The Decline of Fertility in Europe—the Revised Proceedings of a Conference on the Princeton European Fertility Project*—[15], pp. 94–106.

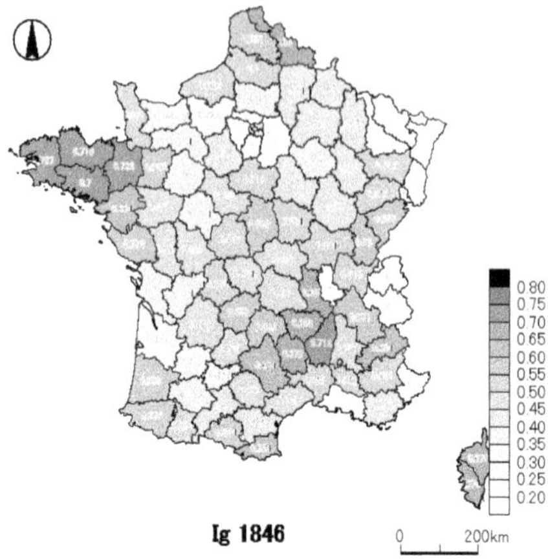

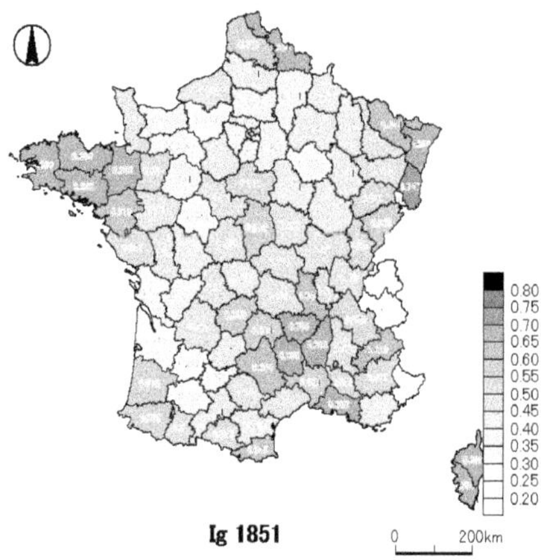

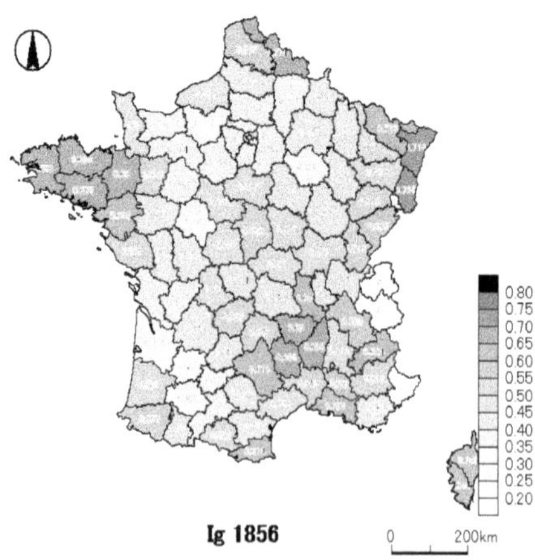

Figure 2.10: Dynamics of I_g in France.

*Source: *Ibid.*

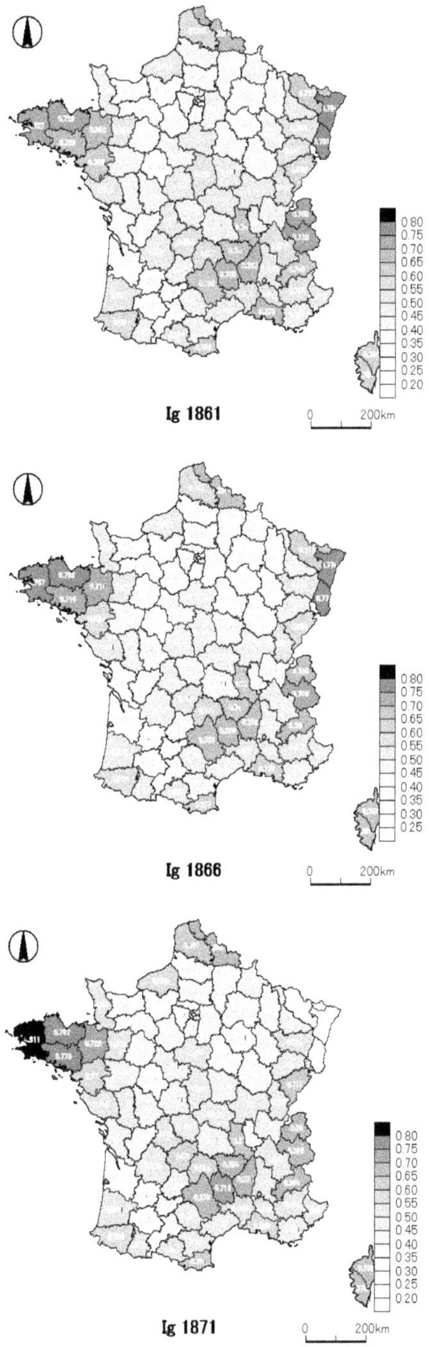

Figure 2.11: Dynamics of I_g in France from 1861 to 1871.

*Source: *Ibid.*

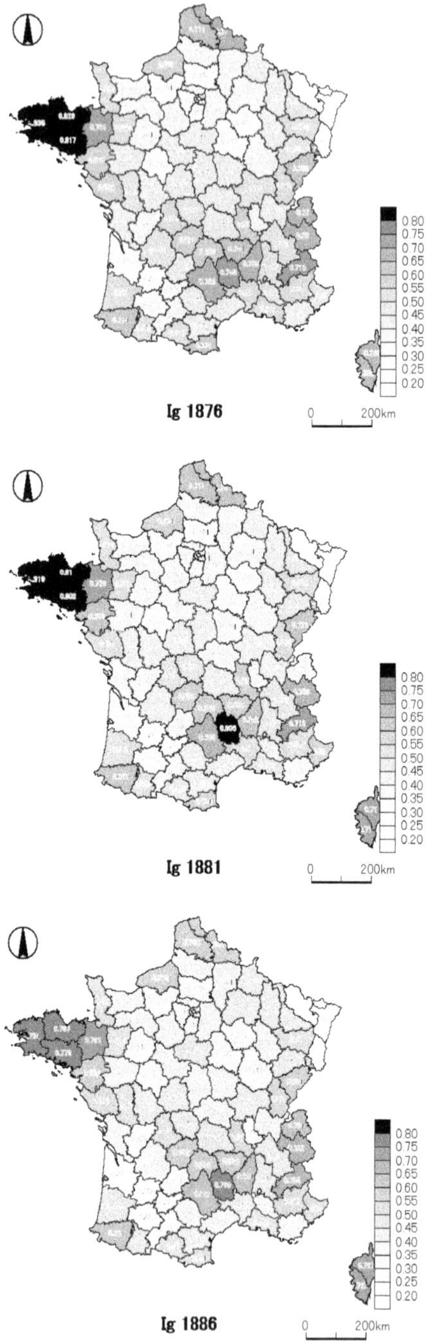

Figure 2.12: Dynamics of I_g in France from 1876 to 1886.

*Source: *Ibid.*

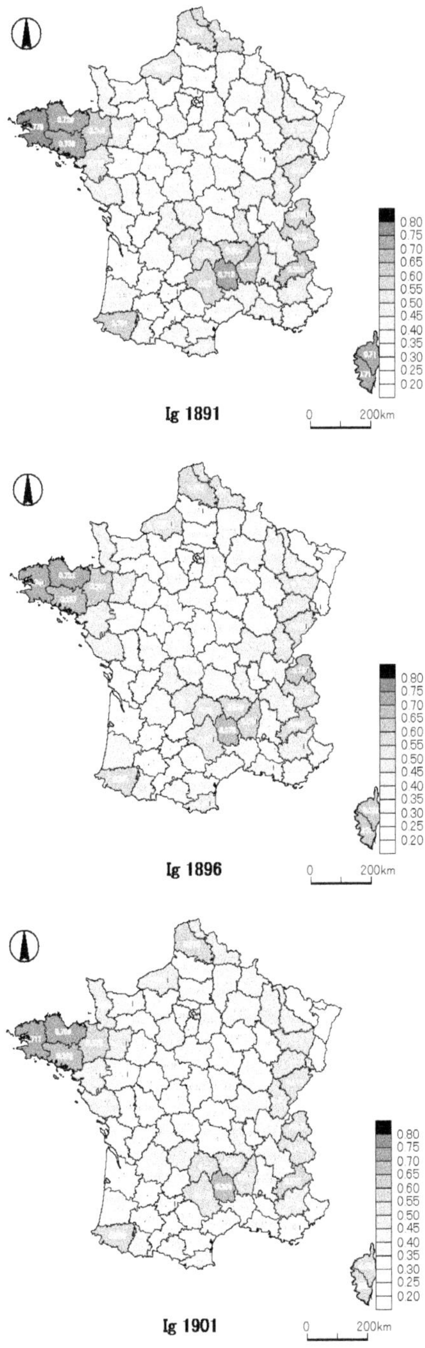

Figure 2.13: Dynamics of I_g in France from 1891 to 1901.

*Source: *Ibid.*

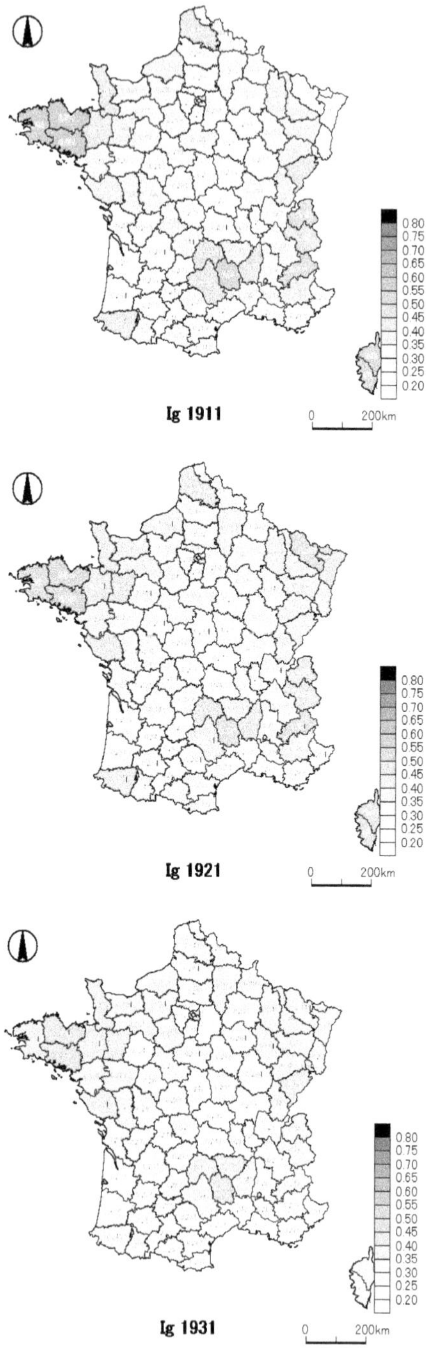

Figure 2.14: Dynamics of I_g in France from 1911 to 1931.

*Source: *Ibid.*

2.4.1.1 Effect of Religion

Several researchers have insisted that religion has some (often obstructive) effects on fertility decline. This hypothesis is entirely dependent on conscious human thought rather than on logical reasoning based on actual data.

It can be seen that in some cases, religion appeared to have some effect on fertility decline, whereas in others, it had none. For instance, Benz [4] presented the following conclusions:

> At least six conclusions seem in order.
>
> **(1)** Early in the fertility transition, the level of family limitation varied strongly by locality. The place of residence was more important than political or religious affiliation in facilitating or impeding the use of contraception.
>
> **(2)** Religion and political affiliation were already linked to each other by 1869. Neither was necessarily linked to family limitations.
>
> **(3)** Political Catholics in Ettenheim county took up contraception at least as early and actively as liberal Catholics.
>
> **(4)** Political Catholics in Ettenheim county took up contraception as early and actively as non-Catholics, on average.
>
> **(5)** Political Catholics in Ettenheim county took up contraception earlier than the Jews, and continued to lead them in this respect as late as 1869.
>
> **(6)** On average, political Catholics in Ettenheim county lagged behind Lutherans in contraception in 1869, but some political Catholics had outdone the Protestant group.
>
> Ernst Benz, Family Limitations among Political Catholics in Baden in 1869 [4], p. 130.
> in *Religion and the Decline of Fertility in the Western World* [68].

As indicated by Benz, religion was not a critical factor in fertility decline. There are no indications that the Germans in Ettenheim were outstandingly free from religious affiliation as compared to the French in Brittany. It is generally assumed that Britons were more religious than other French people. However, their Catholic religious faith has been assumed based on the observation that they used contraception less than the French. Namely, the religious faith as a cause is a presumption based on results, which is not well supported by data.

To begin with, if religion had a significant effect, fertility decline should not have occurred in France first, because Catholic dogma would have a negative effect on fertility control. However, fertility decline began in France. Further, the idea that religious people tend to resist fertility control is logically baseless. The Old Testament said "Thou shalt not commit adultery." This statement does not contradict the concept of birth control. However, in *Genesis*, God blessed Noah and his sons, and said to them, "Be fruitful, and multiply, and replenish the earth", which is somewhat contradictory. In essence, religion involves inconsistent and illogical dogma, which changes depending on the situation. Thus, the effect of religion is an assumption.

As the crow flies, Ettenheim is approximately 710 km away from Lot-et-Garonne. The singularity began around 1770 at Lot-et-Garonne. As mentioned earlier, the velocity of the progressive (traveling) wave of fertility decline is estimated to be approximately 10 km/y. The year 1869, in which E. Benz set his research, appears to be too late for the beginning of fertility decline. This is because, if effective distance is assumed, this is over 20 years after the wave would have arrived there.

However, it must be taken into account that the wave had to make a detour to avoid Massif Central, and thus, it took 79 years to arrive at Ettenheim. Here, the detour is estimated to be approximately 80 km. The year 1848 was a suitable time to observe the fertility decline in 1869. At a conscious level, we cannot recognize the arrival of the wave; the evident fertility decline can only be observed after 10 years. Therefore, Benz's research is consistent with the reaction-diffusion theory. In particular, his first conclusion is equivalent to the reaction-diffusion hypothesis.

2.4.1.2 Linguistic Difference Assumption

As mentioned in *Fertility Decline and Background Independence* [38], the difference between the fertility decline of Walloon and Flemish language groups in Belgium has been reduced to geographical differences. Specifically, it was caused by the time taken for the progressive wave of fertility decline to arrive.

The relationship between language and fertility control cannot be clearly defined. It is unreasonable to state that Flemish language users have a tendency to reject contraception. Breton language speakers might have no special tendency to accept contraception compared to French speakers when they are away from their residences.

If there were equal populations of Wallons and Flemish residents in Belgium, there might have been no statistical difference between the two in terms of fertility decline. In a sense, the difference in languages, and more generally, the difference in cultures, are assumptions. The people constituting the neighborhood and the manner in which they change in time are the primary factors. The true theory of fertility dynamics must be founded on this, and the theory must be applicable in every time space. Theories that cannot satisfy this requirement are simply explanations.

2.4.1.3 Inadequacy of Economic Factors

Compared with other regions, Brittany was certainly late in terms of 19th century industrial development. Some might consider these economic conditions to be significant factors in the delay of fertility decline.

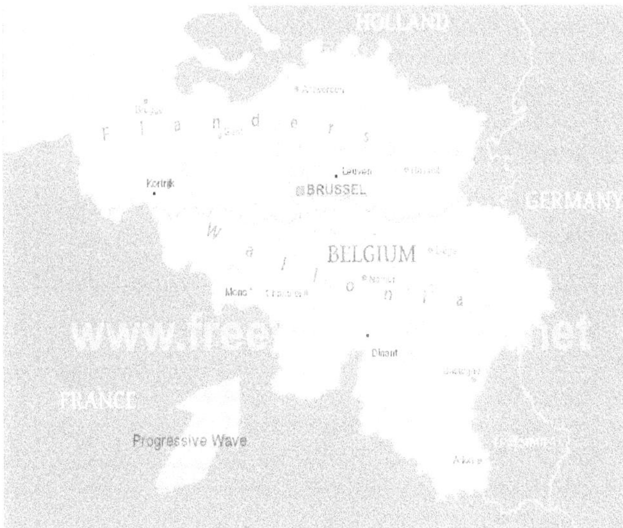

North is a Flemish community/Dutch language area, South is Walloon community/French language area, and East is a German-speaking community.

Source: http://en.wikipedia.org/wiki/Belgium. reprinting from *fertility decline and background independence* [38], p. 65.

Figure 2.15: Communities of Belgium.

Unfortunately, the level of economic development has no consistent relationship with fertility decline. As mentioned previously, despite remaining an agricultural economy, Austria began its fertility decline at the same time as England and Wales. Further, there is evidence that nascent capitalism preferred child labor[11]. Thus, there were no rational reasons for parents in the period of initial industrial development to prefer a smaller number of children. To better understand child labor at the time of the French industrial revolution, I suggest reading *Nobody's Girl: En famille* by Hector Malot [54]. French industries preferred a child labor force. Therefore, it is unfair to state that parents in Brittany alone preferred a large number of children for the agricultural labor force.

To begin with, Lot-et-Garonne, the origin of the singularity of fertility decline, was the only calm countryside in 1770. Economic development has no intrinsic relation to fertility decline. I have repeatedly pointed out this "background independence" of fertility decline. It is a preconception that economic development is the basis of all things. However, a quasi-correlation may be observed when they progress in parallel. Therefore, the relationship between fertility decline and economic development is also a semblance.

[11]See *Family Formation in an Age of Nascent Capitalism* [50].

The fact that Brittany's fertility decline lagged as compared to other regions is a phenomenon particular to the diffusion process. The world is full of phenomena beyond what is imagined in our consciousness.

2.4.1.4 System of Inheritance

Le Play insisted that the alteration of the system of inheritance by *Civil Code* caused a decline in the number of children in France.

> He supposed peasants to dislike the subdivision of their holdings sufficiently strongly to try to keep their families small.
>
> E.A. Wrigley [83], p. 190.

As Wrigley pointed out, in the 19th century, peasants made up the majority of the residents of Brittany. If *Civil Code* were the cause of fertility decline, Brittany's fertility rate might have declined earlier, at least at the same time as other farming regions. In reality, Brittany's fertility decline was delayed compared to other farming regions.

Old Bretons had a system of inheritance by heiresses, as they preserved the Celtic tradition for a long time. They did not easily lose this system.

> ... Thus, though the authors of the *Ancienne Coutume de Bretagne* were certainly influenced by the Roman texts that inspired all French jurists, texts that characterized all women as "irrational and feeble by nature," they were also influenced by the ancient Celtic traditions of rural Brittany.[1] In accordance with these traditions, women could expect a fair share of their parents' estates and the legal right to exploit their own property. To a certain extent, women could even protect their own interests over the objections of their parents and spouses.
>
> Nancy Locklin, *Women's Work and Identity in Eighteen-Century Brittany* [53], p. 83.
>
> [1] is an original footnote.
>
> **Law codes and basic right**
>
> The basics of the Breton law code were laid out in the *coutume* of 1330, written more than a century before the duchy of Brittany was absorbed into France. The social and familial hierarchies described there, however, are hardly different from other codes in practice throughout Europe at this time.
>
> Nancy Locklin, *Women's Work and Identity in Eighteen-Century Brittany* [53], pp. 83–84.

According to Locklin, Bretons in 18th century Brittany continued to subdivide their estates and legal rights to property between their male and female children. If Le Play's hypothesis were true, fertility decline should have occurred earlier among Bretons in Brittany. In reality, Brittany maintained a higher fertility rate until the last stage of the decline in French fertility. Therefore, I cannot help but conclude that his hypothesis is false[12].

If the system of inheritance of the *Civil Code* had been the cause of fertility decline in France, the number of children of Bretons should have been fundamentally smaller than those in other French areas, whose system of inheritance restricted legal heirdom to the oldest son. In 18th century Brittany, Francization was proceeding in all fields, and therefore, the tendency to subdivide was diminishing. Considering these circumstances, the decline in fertility in Brittany cannot be attributed to the system of inheritance there.

2.4.2 Differing Laplacian Δc

Admittedly, I initially explored in vain the idea that the already known causes may be affected by natural obstacles, or that population might be very sparse. Finally, I arrived at a mathematical solution.

From the base to the tip of the Brittany peninsula, the Laplacian Δc differs from that of the non-peninsular region. In Figure 2.16, it can be seen that in the non-peninsular areas, each segment receives the same neighboring effect as the front of the progressive wave. By contrast, at the base of the peninsula (on the left side of the figure), segments are affected by the notably high fertility of the peninsula. Here, the Laplacian represents the net effect of the neighboring flow of a function. Namely, at the base of the peninsula, there was more flux of higher fertility. The Laplacian is higher in the peninsula and

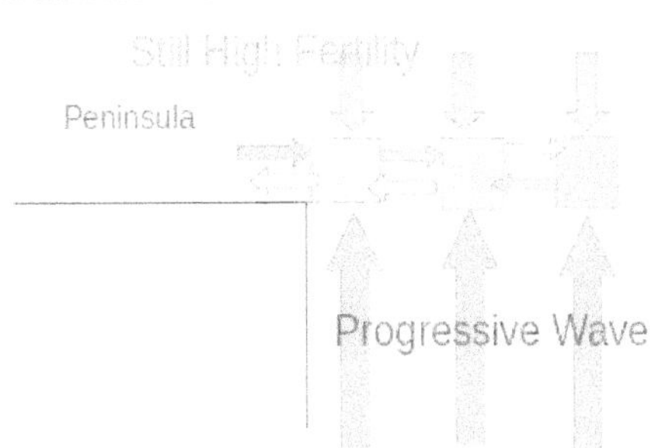

Figure 2.16: Laplacian of reaction-diffusion in the peninsula.

[12] Although his hypothesis is false, I hold great respect for Le Play. He is an eminent pathfinder in social research. His *Lórganisation de la famille* is indispensable in the fields of sociology and social science.

its base than in the non-peninsular land. This is the farthest that could be deduced through approximations alone; to advance more concretely, a model was built and numerical calculations were performed.

2.4.2.1 Simulation of the Peninsula

Figure 2.17 depicts an abstract map of France constructed with $5 \times 5\ km$ grids.

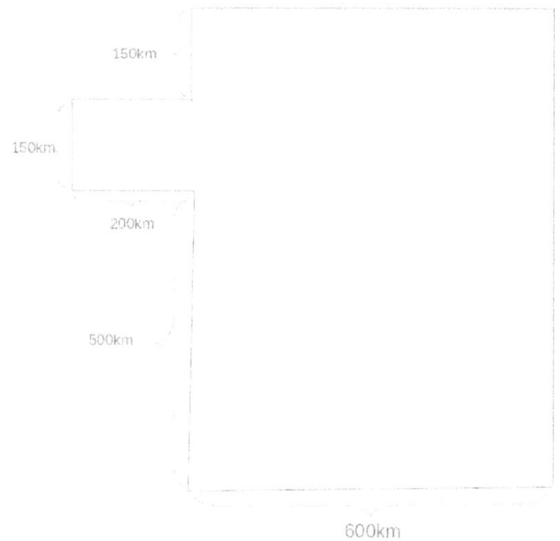

Figure 2.17: Reaction-diffusion model of French fertility decline in the peninsula.

Coefficients

For the numerical simulation, I employed the difference equation (2.11, DESDC:Difference Equation of Spatial Distribution of Children) previously proposed in my book (published in Japanese) *Decline of Marital Fertility and Diffusion Hypothesis* [35].

$$\begin{aligned} & c(x_i, y_j, t_{k+1}) \\ = \; & \mu c(x_{i-1}, y_i, t_k) + \mu c(x_{i+1}, y_i, t_k) + \\ & \mu c(x_i, y_{j-1}, t_k) + \mu c(x_i, y_{j+1}, t_k) + \\ & (1-4\mu)\left[c(x_i, y_j, t_k) + \alpha\sqrt{(\beta - c(x_i, y_j, t_k))^2}\,(\gamma - c(x_i, y_j, t_k))\right]. \end{aligned} \tag{2.11}$$

Here, $(\beta - c(x_i, y_j, t_k))$ is modified to $\sqrt{(\beta - c(x_i, y_j, t_k))^2}$ to maintain the stability of the system.

The problem is to obtain valid coefficients μ, α. The diffusion coefficient μ and the reaction coefficient α must be considered as constants in continuous time space. However, in discrete time space, they must be adjusted based on their scale-time size. This adjustment problem is abstruse.

For a relatively large space grid, α becomes smaller. Conversely, for smaller time intervals, α also decreases. When I first estimated μ, α, I assumed a 1×1 km grid space and 1/12 years. This time, I initially assume a 5×5 km grid space and 1 theoretical unit of time[13].

The numerical simulation eloquently exhibits that the fertility decline is delayed in a peninsula.

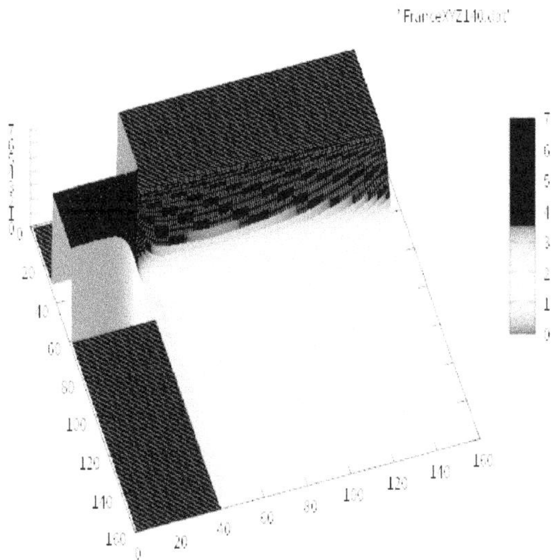

Figure 2.18: Simulation of the reaction-diffusion of French fertility decline for the peninsula, following the passage of 135 theoretical units of time from a singularity.

From Figure 2.14 on page 40, we can suppose that Massif Central retarded the progressive wave of fertility decline such that northeastern France maintained relatively high fertility.

The geographical fact of Brittany being peninsular merely brought about the delay of fertility decline in the region. Namely, if Brittany was not a peninsula, its fertility rate would decline on par with other French regions. Concrete motivations were not required to explain the fertility decline in this case. This is a marvelous finding.

Reaction-diffusion explains the difference of fertility decline in detail

The simulation in this work can explain the fertility decline in the Brittany peninsula for each province in detail. Within the Brittany peninsula, there were differences in the timing of fertility decline for each province (for the location of each province, see Figure 2.21). As shown in Table 2.7 and Figure

[13] I irrationally assumed it to be a year.

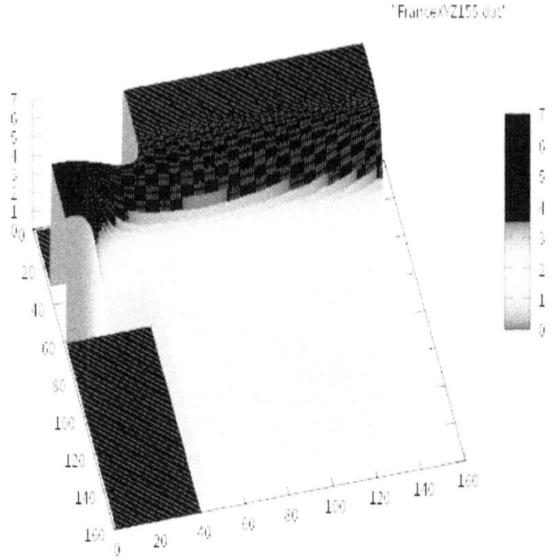

Figure 2.19: Simulation of reaction-diffusion of French fertility decline for the peninsula, following the passage of 155 theoretical units of time from a singularity.

2.22, Côtes-Du-Nord[14] and Finistére had maintained relatively higher fertility, particularly as compared to Irre-et-Vilaine.

The progressive wave first invaded the base of the peninsula, Irre-et-Vilaine. It then diffused within the peninsula in the direction from southeast to northwest. This angle was due to the arrival of the progressive wave in the peninsula located at a horizontal position on the European continent. Thus, the progressive wave invaded Morbihan faster than Côtes-Du-Nord. The reason for the lack of a significant difference between Côtes-Du-Nord and Finistere is the same. If the progressive wave traveled horizontally through the peninsula, Finistere might have been the evident last.

Owing to this consideration, I was convinced that the singularity of fertility occurred in Southwest France. Had it been around Paris, this somewhat shallow penetrating angle may not have occurred.

It is a fantastic fact for the reaction-diffusion theory that Brittany is horizontally located in the European continent.

2.4.2.2 Other Demonstrations

I further obtained other demonstrations of the progressive wave. As per the reaction-diffusion process, the fertility decline wave arrived at France's frontiers last; this was as predicted by the theory. In Figure 2.20, it can be seen that at the upper edge and upper-right corner, higher fertility rates are maintained. The reality was exactly what the simulation presented. In Figure 2.14

[14]In 1990, it was renamed Côtes-d'Armor.

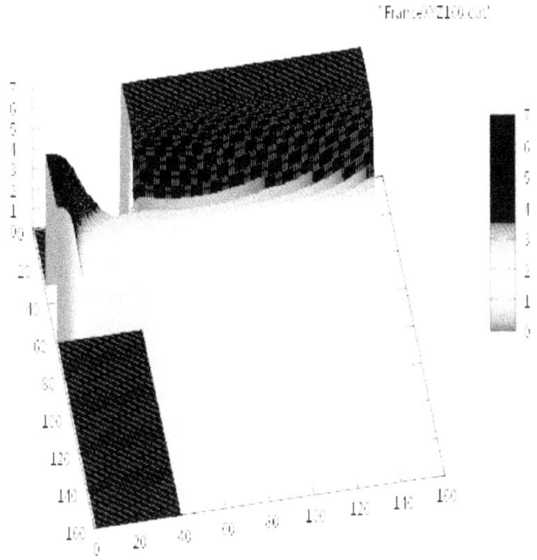

Figure 2.20: Simulation of reaction-diffusion of French fertility decline for the peninsula, following the passage of 160 theoretical units of time from a singularity.

Figure 2.21: Map of Brittany.

Source: https://upload.wikimedia.org/wikipedia/commons/thumb/4/46/Bretagne-Loc.png/300px-Bretagne-Loc.png *Author altered the names of provinces into English.

at p. 40, it can be seen that in 1921, Pas-de-Calais and Alsace-Lorraine were provinces that still had higher fertility rates.

Another more significant fact is that the progressive wave had already arrived at all spaces of the area following the passage of 120 theoretical units of time. However, at the time, the decline of fertility was still minute, and thus, the residents were unable to consciously observe the beginning of the fertility decline. When the wave arrived, one family in the area possibly began contraception for their uncertain fifth or sixth child owing to the neighboring effect, but nobody in the area was conscious of it. Their destiny began in silence; the concealed necessity was never conceived by us.

Table 2.7: Fertility decline of the provinces of Brittany I_g.

	1831	1841	1851	1861	1871	1881	1891	1901	1911	1921
Côtes-Du-Nord	0.765	0.753	0.693	0.729	0.797	0.810	0.728	0.708	0.561	0.542
Finistere	0.815	0.760	0.686	0.727	0.811	0.819	0.773	0.711	0.551	0.500
Irre-et-Vilaine	0.790	0.733	0.693	0.682	0.726	0.729	0.649	0.575	0.420	0.414
Morbihan	0.810	0.737	0.682	0.689	0.773	0.802	0.739	0.652	0.551	0.522

*Source: *The Decline of Fertility in Europe—the Revised Proceedings of a Conference on the Princeton European Fertility Project*—[15], pp. 94–107.

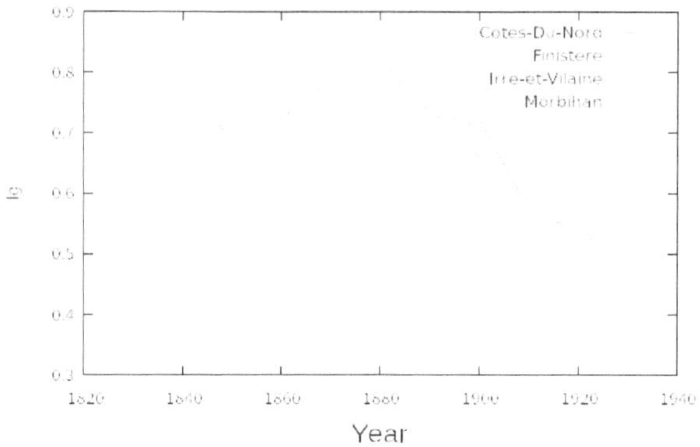

Figure 2.22: Fertility decline of the provinces of Brittany I_g.

Source: *The Decline of Fertility in Europe—the Revised Proceedings of a Conference on the Princeton European Fertility Project–ibid.*

2.4.3 Speculations for Coefficients and Time–Space Scale

With these strange simulation results, the coefficient μ, α superficially contradicts the theoretical equation.

$$v = \sqrt{2\mu\alpha} \tag{2.12}$$

Based on R.A. Fisher's formulation, I speculated that the velocity of the progressive wave of fertility decline soon changed to the aforementioned velocity. At first, I assumed $10km \times 10km$ grids and one year interval. I thought that it is natural setting of the time–space. Therefore, the smaller μ, α might imply a reduction in the velocity.

This numerical simulation was initiated with the coefficients that were previously estimated for two-dimensional space, $\mu = 0.2475$, $\alpha = 1.4142$. However, with these coefficients, the wave velocity is too slow. Therefore, I alterd the time–space setting to $5km \times 5km$ grids, and $\mu = 0.20625$, $\alpha = 1.01$ had to be employed. I suspected that these estimations were still larger than the probable velocity.

By employing μ near 1/4, for any α, I had to review the interval of time that I had postulated to be one year.

These difficulties reminded me of a significant problem: the scale of time space is a primary factor in discreet numerical simulation. I supposed that the grid size of 10 km × 10 km is too large to simulate the reaction-diffusion system of fertility decline. For the discreet grid simulation, the velocity must be adjusted according to the time (k year)–space (h km) scale as per the following equation:

$$v = \sqrt{2\frac{\mu\alpha h^4}{k^2}} \tag{2.13}$$

The time–space scale that I first adopted is likely to be too large to simulate fertility behaviors. If $h = 10\ km$, $k = \ a\ year$, the progressive wave may penetrate France within ten years as the values of equation (2.13). Therefore, k must be larger.

The scale of 10 km × 10 km is too large for the neighboring space, and this setting causes several contradictions such as just above. However, these contradictions suggest that, in reality, the Laplacian Δc is assumed to be defined within a fairly small time–space scale. There may be an adequate timespace scale for the DESDC (2.11).

2.4.3.1 Rounding Errors Cause a Delay in the Wave

Numerical computations cannot handle infinitely small values. At the front of the wave, rounding eliminates the smallest decline, although it occurs in reality. These rounding errors should be handled when the progressive wave (FKPP wave) is computed.

2.4.3.2 Lattice Space Errors

The lattice space is the basis for imagining the neighboring effect. This space is necessary for substantial computation of the reaction diffusion. However, it is an approximation of real-world occurrences, which results in numerical errors.

2.4.3.3 Section Conclusions

I have recognized the possibility that geographical features may cause some specific and plenteous diffusion patterns. The reaction-diffusion system explains the time–space pattern based on the interaction of the terms between the diffusion and reaction. Further, the conditions of space (geographical conditions) may cause a specific pattern of fertility decline.

Regardless, this solution to the Brittany problem is a decisive piece of evidence that fertility decline is a reaction-diffusion process because there are no posibilities of other explanation for this phenonenon.

2.5 Decline of Number of Children in Japan

When estimating the progressive wave velocity, the diffusion process in Japan was partially discussed. However, in Japan, the reaction-diffusion process can be visually observed.

2.5.1 Graphical Representation of the Decline of the Number of Children in the Kanto Area

The monotonic decline of fertility is the evidence of "background independence." The reaction-diffusion process of the fertility decline in the Kanto area from 1896 to 1930 can be observed through an animation, available on "http://Sociology.main.teikyo-u.ac.jp/mandara/."

I alternatively exhibit this in Figures 2.23 to 2.29. The decline in the number of children was reaction-diffusing from the core of the center of Tokyo.

Figure 2.23: Completed marital fertility of each division, 1896–1900 cohort.

In each division, lighter or darker colors indicate higher or lower fertility, respectively. It can be observed that the dark blotch of Chiyoda-ku and Chuo-ku gradually became darker and spreads to neighboring areas. In 1970, the center of Tokyo recorded a total fertility rate (TFR) under 2.0.

Readers can also observe this process of the completed fertility decline on http://Sociology.main.teikyo-u.ac.jp/mandara/ as a GIF animation[15]. Whenever I see this animation, I am impressed at the significance of how the reaction-diffusion system fits the observed phenomena.

[15]Geographic information system "MANDARA" was used to make this animation. "MANDARA" was also used to draw several figures in this chapter. "MANDARA" was built by Prof.Kenji Tani (http://www5c.biglobe.ne.jp/Ÿmandara/). I would like to express my great appreciation for him here.

Figure 2.24: Completed marital fertility of each division, 1901–1905 cohort.

Figure 2.25: Completed marital fertility of each division, 1906–1910 cohort.

I believe that the process of completed fertility decline was independent of the socioeconomic movement of the time. For example, from 1910 to 1920, intensive propaganda, such as "Meshi Houkou,"[16] was spread throughout Japan in the spirit of Japanese militarism. Nevertheless, the number of children substantially declined; in reality, this propaganda had no effect on families.

Generally, socioeconomic factors are largely observed to be fluid and reversible. The fertility decline process that Japan has experienced in the 20th

[16]A historical allusion or slogan. It means to perish one's ego and serve the public or nation. The original source is a Chinese classic. It is ironical that Japanese militarism waged war on China adopting a Chinese historical allusion.

Figure 2.26: Completed marital fertility of each division, 1911–1915 cohort.

Figure 2.27: Completed marital fertility of each division, 1916–1920 cohort.

century is unique in terms of its irreversibility and inertia. Thus, it is very difficult to explain this fertility decline process based on socioeconomic trends. It is compelling that the reaction-diffusion system can longitudinally and coherently describe the process of fertility decline.

2.6 Diffusion as a Fire Leap

Unfortunately, in the present day, diffusion cannot be observed in 19th century France and 20th century Japan. The development of traffic did not confine diffusion to crawling speed; rather, it made it leap to great distances.

Figure 2.28: Completed marital fertility of each division, 1921–1925 cohort.

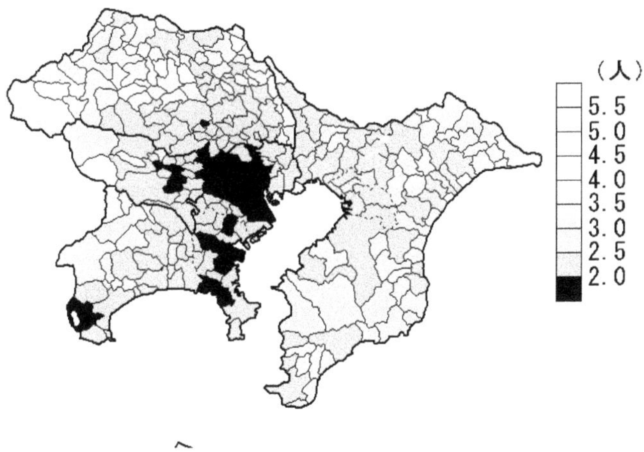

Figure 2.29: Completed marital fertility of each division, 1926–1930 cohort.

Berlin began its fertility decline around the year 1880; however, it was 1,320 km away from Lot-et-Garonne as the crow flies. This initiation began slightly faster than expected by observing the crawling speed of diffusion. Moscow is about 2,900 km away from Lot-et-Garonne, but its fertility decline began immediately after World War II. If it were not for diffusion as a fire leap, the fertility decline in Moscow may have begun around the year 2000.

In other words, fertility decline on a global scale was accelerated by the development of traffic. Brazil serves as an interesting example of this.

2.7 Fertility Decline in Brazil

Brazil's countrywide TFR began to decline around 1965. Even Japan, which is a relatively small territory, took about 50 years to complete its marital fertility decline. Similarly, it took almost 100 years for France to complete its fertility decline process for its crawling speed. These lengths of time must be analyzed considering the width of the area and the development of traffic.

Considering that the flow of TFR also has an inertia similar to the total population, the decline in fertility in Brazil began in the first half of the 1950s.

In 2017, Brazil's TFR was reduced to 1.75. It took less than 50 years for Brazil to decline its fertility. Considering the size of Brazil, this speed of decline is remarkable. In contrast to this fire leap aspect, the old crawling diffusion can be observed in the local space. It is impossible to theorize the fire leap diffusion now, and thus, I observed these local diffusions from the perspective of fire leap diffusion.

2.7.1 *Diffusion from Several Regions (Points)*

It is estimated that there are seven to nine starting points of diffusion in Brazil. Figure 2.30 presents the diffusion process about 25–30 years after the start time.

Manaus, Brasilia, Sãn Paulo, Rio de Janeiro, Curitiba, Florianopolis, and Porto Alegre were certainly starting points. Uberaba, Cuiaba, Juina, and Patos were also added to this list here. These cities were chosen because starting points present the lowest fertility rate during the decline process.

The diffusion of fertility decline in Brazil primarily started in the southeast area of Brazil.

2.7.1.1 Map from 1991 and Ratio of the Total Area of Diffusion

Based on my research results, it was assumed that the velocity of the progressive wave of fertility decline was approximately 10 km/year. If the reaction-diffusion began in 1950 at Sãn Paulo, Rio de Janeiro, and so on, the areas in which the lower fertility had already diffused had grown to a circle of radius 400 km (diameter 800 km). The total area in which the progressive wave of fertility decline had arrived was approximately $410^2 \cdot 3.14 \times 11 = 5,806,174\ km^2$. The total area of Brazil is $8,515,767\ km^2$, and thus, about 68% of the area had already experienced a fertility decline in 1991. This is an approximate calculation, but the logic is steadfast.

In Figure 2.30 [28], the surface area not marked in red on the map on the left side is the area in which the fertility had already lowered. The surface area marked in red appears to be approximately 32% in 1991. The starting date of fertility decline may have been far earlier than 1950. The aforementioned analysis is rather approximate, but the following one is more accurate.

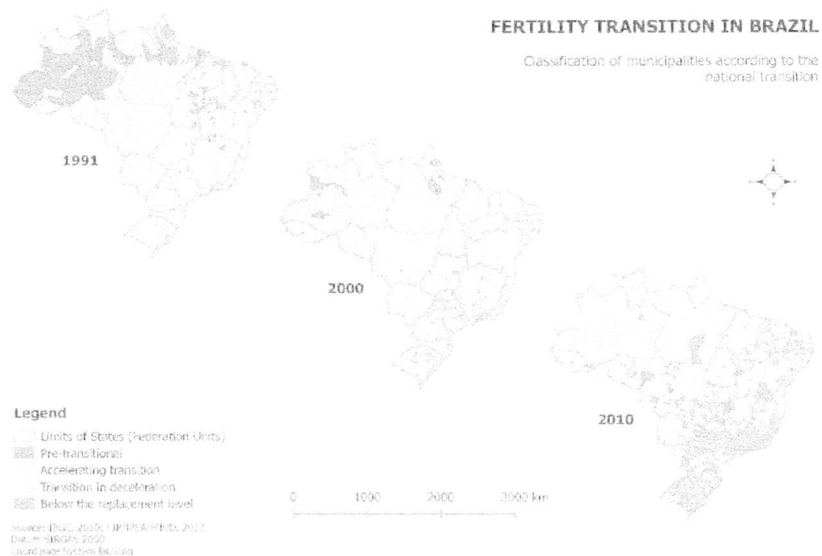

Figure 2.30: Fertility transition in Brazil.

Guilherme Quaresma Gonçalve and Reinaldo Onofre dos Santos and Laura Lídia Rodríguez Wong and José Alberto Magno de Carvalho,
Time and space of fertility transition in Brazil: An exploratory analysis of Brazilian municipalities in 1991, 2000, and 2010. p. 12, Research Gate [28], 2017.
https://www.researchgate.net/publication/320888241.

2.7.1.2 Periods between 1991–2000 and 2000–2010

In Figure 2.30, (deep) blue represents the area in which fertility is below the replacement level. In 1991, these areas were small, shown as spots on the map. On measuring their size in pixels, most blue points were one pixel wide, with the exception of a few that were two or three pixels wide. On the contraction scale, one pixel is about **13.8 km (3000 km/217 pixels)**.

Note these magnitudes. The velocity of the progressive wave of fertility decline is approximately 10 km/year, and the same level of fertility has a front with the same velocity because of its properties. Specifically, the velocity of the progressive wave indicates that the fertility decline down to the replacement level began in 1991.

Within nine years, the front of the wave of fertility below the replacement level had proceeded about 90 km (180 km in all directions), which is equal to 6.5 pixels (13.0 pixels). For the scale of the map, the small spots grew to larger spots. On measuring several cases on the map of the year 2000, the larger spots remained 12–13 pixels wide.

From 2000 to 2010, an isolated spot was observed in Bahia state, which is estimated to be "Vitória da Conquista." This spot remained only 6.1 pixels (84.2 km) long (vertical) in 2000. By 2010, it grew to 21 pixels long (vertical) and 17 pixels wide (horizontal). The value $6.1 + 13.0 = 19.1$ (264 km) is close to 21 (290 km), and $4.0 + 13.0 = 17.0$ (235 km) coincides with the value in 2010. The approximate estimation and calculation by the reaction-diffusion system contributes to a fairly precise prediction.

Another isolated point was observed in Brazil from 2000 to 2010. In the middle of the Piaui "Piauí" state, one blue pixel[17] (< 13.8 *km*) grew to 12.5 pixels (172.5 km) in 2010. In terms of isotropic diffusion, this might be less than 200 km. The diffusion velocity is not disproved in this case.

Specifically, local crawling diffusion remains valid in local space.

2.7.2 Traffic

In reality, Brazil's fertility decline was closely related to the transportation network. Figure 2.31 presents the railway network of Brazil. I have superimposed railroads on Figure 2.30. Figure 2.32 shows the result of this procedure.

Inception points of fertility decline (kernels of fertility decline) were observed on the railroads. In particular, they lie exactly on the railroad from Sãn Paulo to Brasilia. Fire leap diffusion is the result of modern traffic development. It is important to distinguish the factors of modern development that are vital to fertility decline.

Of course, roads for motor vehicles have the same effect. From Figure 2.30, comparing the maps from the years 2000 and 2010, it can be seen that the fertility decline rapidly accelerated in the middle of the Pará state. This is the result of the major arterial road to Marabá that opened after 2003.

[17] It may be "Oeiras" or "Picos".

Figure 2.31: Railway network of Brazil.

https://commons.wikimedia.org/wiki/File:Railway˙network˙Brazil.svg.

Fire leap diffusion is common to countries that began their fertility decline in the latter half of the 20th century (see Figure 2.33 below).

Western and northwestern countries that entered fertility decline in the middle or latter half of the 19th century took almost over 50 years to go below the population replacement level. Japan, whose decline began at the beginning of the 20th century, took 50 years to reach that level. Contrastingly, the countries that entered fertility decline in the latter half of the twentieth century, such as Brazil, China, Republic of Korea, Iran, and Thailand, all took about thirty years.

It is difficult to assume that all these countries were under the same socioeconomic conditions. It is not possible that Thailand, China, and Korea were in the same socioeconomic state. However, the improvement of traffic systems has been relatively common. This improvement of railroads and road networks brought about fire leap diffusion, and the reaction-diffusion process has been accelerated by this traffic network. It is not valid to call this process "the modern lot"; rather, it should be perceived as a simple and common mechanical process.

Regardless of conscious human thoughts, the process develops mechanically. This is because humans are animals dependent on their neighbors. It is clear that the turning point of the global population is closer than previously thought.

Figure 2.32: Fertility transition and railroads in Brazil.

I depicted the railroad map in Figure 2.30 presented in the article by Guilherme Quaresma Gonçalve and others [28].

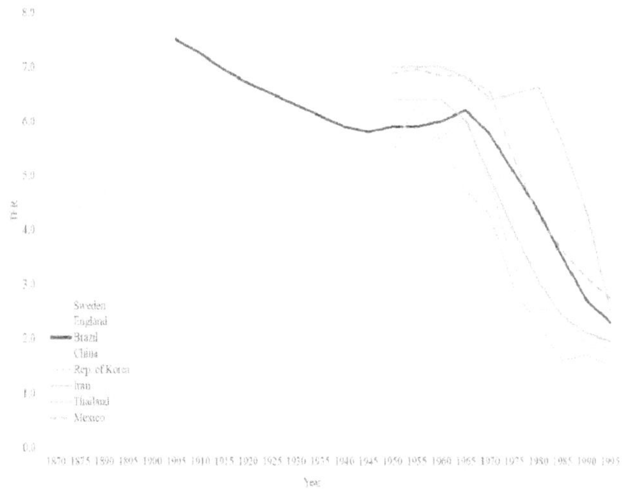

Figure 2.33: Fertility decline in various countries.

Guilherme Quaresma Gonçalve and Reinaldo Onofre dos Santos and Laura Lídia Rodríguez Wong and José Alberto Magno de Carvalho, Time and space of fertility transition in Brazil: An exploratory analysis of Brazilian municipalities in 1991, 2000, and 2010. p. 4, Research Gate, 2017.
https://www.researchgate.net/publication/320888241.

2.7.3 Conclusions Based on the Analysis of Brazil

By assuming several fire leap points and the reaction-diffusion theory, the fertility decline of Brazil (and possibly that of other countries) can be consistently described.

Economic development is a quasi-correlation. Naturally, I am confident in my principle of "background independence of fertility decline." However, there is another empirical reason that economic development does not influence fertility decline (in Brazil). As mentioned earlier, the TFR of Brazil had begun to decline in 1965. Certainly, "Brazil's Economic Miracle" accompanied the fertility decline, but following the high economic growth, ahead of the debt problem, the fertility of Brazil continued to decline. Thus, fertility declined regardless of whether the economy did or did not develop. This is an evident contradiction. If the fertility decline could be explained by both economic development and economic stagnation, it is not explained at all. Clearly, economic development has no consistent relationship with fertility decline.

The improvement of traffic clearly prompted a decline in fertility owing to fire leaps. Some may say that this is also an economic factor. However, in order to accurately analyze fertility decline, the unnuanced use of words should be avoided. The accelerating effect of development of a traffic network on the fertility decline of Brazil must be explained. The conventional crawling diffusion model still applies as an intrinsic locomotive.

2.8 Numerical Consideration of the Geographical Obstacles in the Model

It is difficult to treat geographical obstacles (such as Massif Central) in the reaction-diffusion system in an integrated manner. It is evident that the reaction-diffusion process was impeded by these geographical obstacles. However, it is unmanageable to formally include these effects in the equation. In this section, I would like to consider this problem.

Geographical obstacles impede interactions between neighboring spaces. Therefore, the diffusion term should be manipulated. However, the diffusion coefficient μ must be kept constant. Conversely, the reaction term must be maintained as it is because the reaction is general for all. As a logical consequence, the Laplacian $\frac{\partial^2 C}{\partial x^2} + \frac{\partial^2 C}{\partial y^2}$ must also be changed. In real discrete computation, $c(\acute{x}, \acute{y})$ is manipulated outside $c(x_o, y_o)$ of the geographical obstacles in order not to decline $c(x, y)$ as other points of (x, y).

Namely,

$$\frac{\partial^2 C}{\partial {x_o}^2} + \frac{\partial^2 C}{\partial {y_o}^2} \ll \frac{\partial^2 C}{\partial \acute{x}^2} + \frac{\partial^2 C}{\partial \acute{y}^2} \tag{2.14}$$

In a real fertility decline process, the degree of impediment depends on the geographical situation. For simplicity, $\frac{\partial^2 C}{\partial {x_o}^2} + \frac{\partial^2 C}{\partial {y_o}^2}$ are set as constants. Under the aforementioned assumptions, I added a function to simulate the effect of geographical obstacles to the simulation program.

2.8.1 *Simulation Results*

The effect of the obstacle, Massif Central, was simulated herein. Despite the exceedingly operational procedures adopted here, the effect of geographical obstacles was actualized to the fertility decline of France. In this research process, some characteristics of geographical obstacles could be recognized.

2.8.1.1 Obstacle Size must Exceed a Quantity

During the initial process of simulation, I set the size of the obstacle as 100 $km \times 100$ km. The obstacle of this size does not remain in space for a sufficient time. The Massif Central and the Brittany peninsula maintained the same high fertility from the 19th century to the beginning of the 20th century. When the size was set to 250 $km \times 250$ km, the Massif Central fertility rate remained high along with that of Brittany. When it was set to 275 $km \times 250$ km, it remained clear with the decline in fertility of the Brittany peninsula.

Small obstacles cannot significantly delay diffusion. The gross area of Massif Central is approximately 182,000 $km^2 (\doteq 426^2\ km^2)$. This large area can serve as an obstacle for diffusion. Rivers are easily penetrated by the progressive wave of diffusion. The power of the diffusion is sufficiently potent; specif-

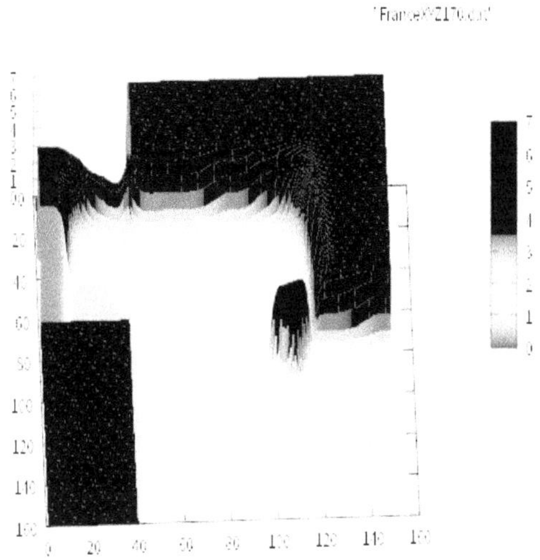

Figure 2.34: Reaction-diffusion simulation of French fertility decline for the peninsula, following the passage of 170 theoretical units of time from a singularity, with massive geographical obstacles added.

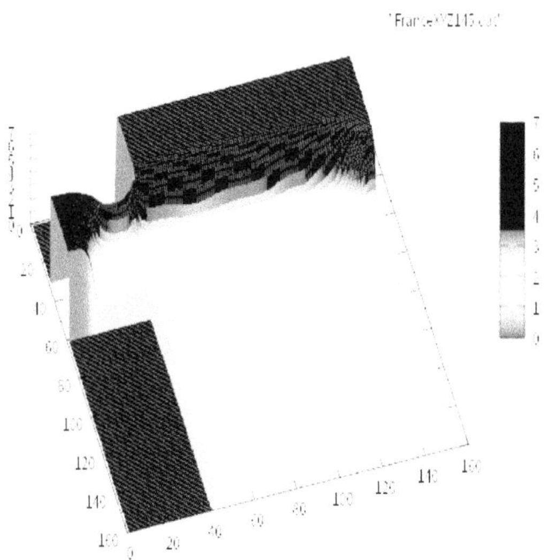

Figure 2.35: Reaction-diffusion simulation of French fertility decline for the peninsula, following the passage of 145 theoretical units of time from a singularity, with small geographical obstacles added.

ically, in the reaction-diffusion system, trivial geographical obstacles can be ignored in most cases.

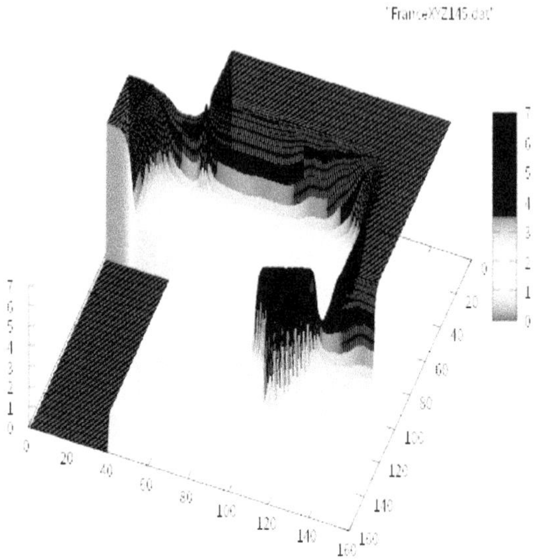

Figure 2.36: Reaction-diffusion simulation of French fertility decline for the peninsula, following the passage of 145 theoretical units of time from a singularity, with valid geographical obstacles added.

2.8.1.2 Laplacian is Extremely Small

Indeed, the fact that $\frac{\partial^2 C}{\partial {x_o}^2} + \frac{\partial^2 C}{\partial {y_o}^2}$ is extremely small but not zero is indispensable, provided the space cannot be eliminated.

During the initial stage of this simulation, $2\Delta C_o = \Delta \acute{C}$ was set; specifically, the Laplacian at obstacles was one-half of that in other normal plain areas. Under this setting, the obstacle could not be interpreted. When the value was set to one-tenth, it visibly presented the effect of the obstacles. This suggests that interactions at geographical obstacles are highly infrequent.

The numerical simulation suggests that the Laplacian between one obstacle and the next is identical to that between a non-obstacle and an obstacle. If not, the obstacles might be penetrated instantly by the wave of diffusion.

To gain further insight into this topic, readers may look into the source program of the simulation. The developmental possibilities to handle these problems are open.

2.9 Chapter Conclusions

The most important result obtained from these investigations on geometrical diffusion models is that the actual spatial definition of fertility decline naturally comes down to the dynamics of fertility. In other words, the mapping defined to the next time in the dimensional space excellently defines the causal dynamics. This is a profound and beautiful phase of the geometric diffusion theory.

Background-dependent theories, such as household economics, which assume conscious and unconscious interest of parents, cannot geographically explain or compute the dynamics of fertility decline in France, Japan, and Brazil like the reaction-diffusion theory can. What is necessary for theoretical explanation and computation is not human motivation, but the formalization of our interactions.

In addition to this eloquent ability of the reaction-diffusion system, it is the sole theory that could predict that the fertility of China would not rise following the repeal of the one-child policy.

The fact that parents were not conscious of their actions during the process of fertility decline should not be ignored. In Japan, the decline in fertility did not become an area of concern until the 1990s. However, it had already begun at the center of Tokyo before the 1930s. At the beginning of the one-child policy in China, policy makers and citizens were unable to conceive its true effects. Bretons in the 19th century did not know why their fertility decline was delayed. Despite these issues, we imagine false causes following mainstream social scientists. They know not what they do.

II

MARRIAGE FUNCTION IN HIGH DIMENSIONAL SPACE

Chapter 3

History of the Marriage Function

> PROP. XXXVI. There is no cause from whose nature some effect does not follow.

Spinoza, *Ethics (Ethica Ordine Geometrico Demonstrata)* [16], p. 34.

The reaction-diffusion model of fertility decline is limited to two geographical dimensions, and the time dimension can be added to it. Generally, this three-dimensional space is not considered to be high dimensional. The probabilities of marriage occurrences depend upon people's ages. Therefore, the abstract "age" dimension must be introduced to describe marriage occurrences. This inevitably creates a high-dimensional space.

3.1 Discovery of the Marriage Function

In the 1970s, by using statistical observation, demographers began to recognize that marriages emerged in a specific pattern. In 1971, A.J. Coale presented an article titled "Age Pattern of Marriage" [12] which demonstrated the uniformity of emergence of female marriages based on the double exponential distribution in several countries. For instance, he proposed the following equation:

$$r_s(x) = 0.174 \exp(-4.411 \exp(-0.309x)) \tag{3.1}$$

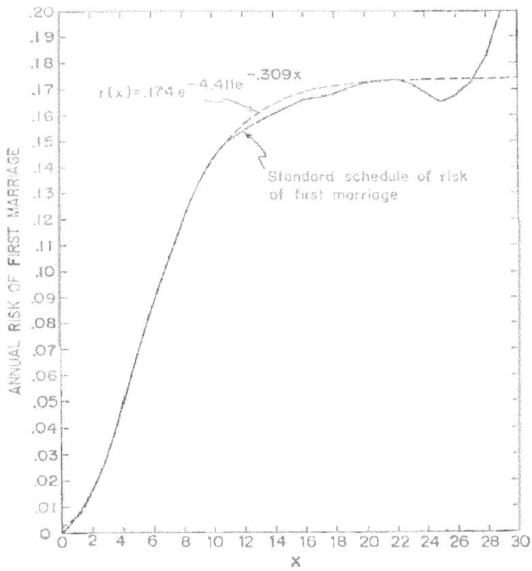

Figure 3.1: Standard risk of first marriage derived from Swedish data, 1965–69, and fitted by double exponential.

Source: A.J. Coale, Age Patterns of Marriage [12], p. 204.

3.1.1 *Uniformity of Marriage Function*

Coale found that this pattern is uniform longitudinally.

> An examination of the rising curve of proportion of women ever-married by single years of age in a number of European populations in the late nineteenth and early twentieth centuries (when, within each population, age at marriage and proportions ultimately marrying were generally stable) suggested that in these populations the curves were essentially the same in structure, differing only in the age at which marriage began, the rate at which marriage increased, and the ultimate proportion ever married – differing, in other words, in origin, horizontal scale, and vertical scale, but with the same functional form.
>
> Source: A.J. Coale, Age Patterns of Marriage [12], p. 193.

3.1.1.1 Adjustment of Scale and Origin

As shown in Figure 3.2 (Figure 3 in his article), the proportions of ever-married individuals of selected countries do differ. The time scale is adjusted by dividing it by k, and the difference between the origins is adjusted by a_0. Using these coefficients, Equation (3.2) is transformed as follows:

$$r_s(x) = 0.174 \exp\left(\frac{-4.411}{k} \exp\left(\frac{-0.309}{k}(x - a_0)\right)\right). \tag{3.2}$$

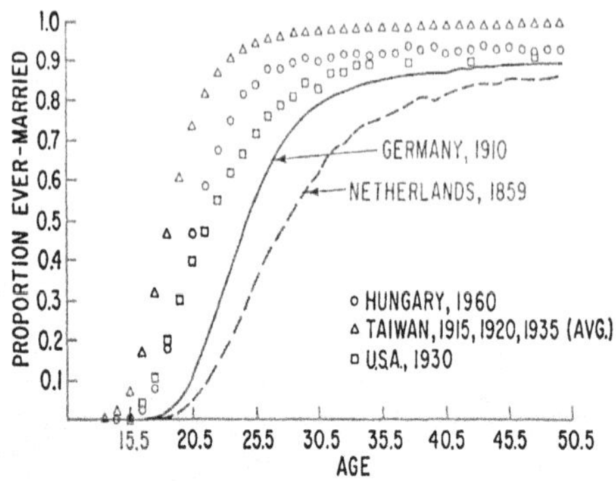

Figure 3.2: Proportions ever-married, selected populations.

Source: A.J. Coale, Age Patterns of Marriage [12], p. 196.

By following these procedures, and by incorporating Coale's analysis, cultural and socioeconomic differences can be excluded to build the marriage function. In this book, this property of the marriage function is referred to as "background independence. Marriages and births, which are the primary behaviors of fertility as a whole, are essentially features of background independence.

Thus, Coale pioneered the search for the marriage function. Those who believe in subjective motivations and that behaviors are controlled by them would not trust the hypothesis of the uniform function of marriage emergence. This "same functional form" is a statistical fact at all. Coale was unable to elucidate the cause of this uniformity of the marriage function; however, he suggested the cross-cultural uniformity of the marriage function.

> It is clear from Figure 4 that the common age pattern of proportions ever-married that characterized western European populations in the nineteenth and early twentieth centuries is also characteristic of non-European populations with very different marriage customs.
>
> Source: A.J. Coale, Age Patterns of Marriage [12], p. 195.
>
> Figure 4 is Figure 3.3 in this book.

In 1972, Coale presented a second article on the marriage function, titled "The Distribution by Age of the Frequency of First Marriage in a Female Cohort" together with McNeil [11]. Although several investigators have frequently referred to and analyzed this article, it was a false mathematical formation. I believe that the basic importance of his observations on marriage patterns is far more significant.

Humans tend to think that our behavior is rational, and cultures and customs are more or less deviant from them in modern times. This common modern belief is entirely fictitious. It is not possible to build any mathematical marriage functions based on it. Rather, it has acted as an intellectual trend or epidemic.

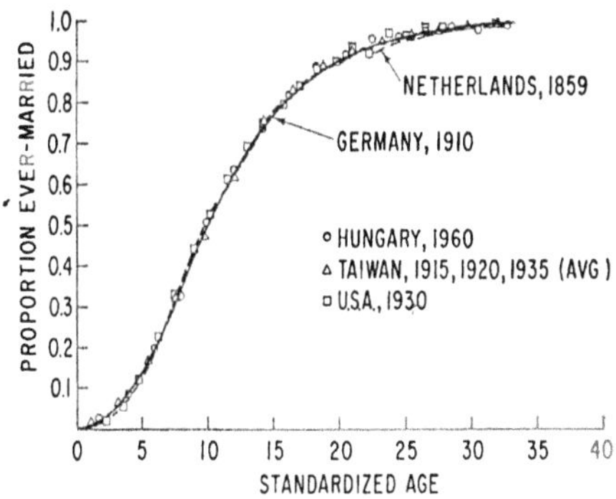

Figure 3.3: Proportions ever-married, adjusted scale and origin for selected populations.

Source: A.J. Coale, Age Patterns of Marriage [12], p. 196.

As a result of this violent torrent of the epidemic of rational choice and neo-conservatism, since the 1980s, all statistical evidence of the marriage function has been erased due to household economics. Hypotheses about our behaviors must be based upon observed data. About forty years have been lost to the illusion of rationality. It is impossible to deduce a mathematical marriage function based on the hypothesis of individual rationality. We must begin with a hypothesis that promises to lead to mathematical development. The scientific starting point for investigating the marriage function should be A.J. Coale, not G.S. Becker.

3.1.2 Flaws of Coale–McNeil Distribution

In the article titled "The Distribution by Age of the Frequency of First Marriage in a Female Cohort" [11], Coale and McNeil presented the widely-acclaimed "Coale–McNeil Distribution." It is a subspecies of the double exponential distribution with particular coefficients.

$$\overline{g}(x) = \frac{\lambda}{\Gamma(\alpha/\lambda)} \exp(-\alpha(x-\mu) - \exp(-\lambda(x-\mu))), \tag{3.3}$$

where Γ indicates the well-known gamma function, $\psi = \Gamma'/\Gamma$ the digamma function, $\mu = a + (1/\lambda)\psi(\alpha/\lambda)$, and a is the mean of $g(t)$.

Source: A.J. Coale & McNeil, Distribution by Age of the Frequency of First Marriage in a Female Cohort [11], p. 744.

This Equation (3.3) presupposes what should be explained by itself. For instance, a, which is the mean of $\overline{g}(x)$, should be calculated by the marriage function itself. Otherwise, it is decided arbitrarily or on the basis of observations. Such a marriage function is purposeless because it cannot deliver any observable predictions. A good marriage function can calculate the unknown mean age of first marriages.

3.1.2.1 Convolution Model

Coale and McNeil insisted that the Equation (3.3) is a convolution of a quasi-normal distribution and three exponential distributions. The quasi-normal distribution corresponds to the physiological maturity for entry into the marriage market. The latter three distributions correspond to acquainted, engaged, and married, respectively.

The Coale–McNeil distribution does appear to bear a relation to the actual marriage process. Several demographers have adopted it because it has comprehensive real objects in the marriage process. Coale & McNeil themselves stated the following:

> The surprising agreement of theoretical and observed values gives some reason for supposing that there is, in fact, behavioral basis for considering $\overline{g}(x)$ the convolution of a normal curve and a few exponentially distributed delays.
>
> Source: A.J. Coale & McNeil, Distribution by Age of the Frequency of First Marriage in a Female Cohort [11], p. 747.

However, when considering common-law marriages that have prevailed over a long period of time, the marriage process assumed within the Coale–McNeil distribution is found to be inaccurate. Legal marriage procedures are not the same as the occurrence of marriage itself. In fact, it has already been reported that the Coale–McNeil marriage schedule did not fit well the Japanese cohort of marriages that followed World War II by Kojima and others [45].

The Coale–McNeil model reflects the traditional values of marriage prevailing in the 1970s. Their subjective model likely applicable to them and demographers having the same social status and values, but it was not universal to every female. This subjectivity is so locally limited that a universal model cannot be built upon it.

In the present day, the marriage process assumed by them has mostly lost those real objects. The plausibility of the Coale–McNeil distribution has collapsed. The model based on temporal subjectivity will not last long. The model that traces human behavioral subjectivity cannot succeed because the dynamics of the marriage function are not caused by our subjective choices.

3.1.2.2 Vital Gap—Never Asymptotic to 0

At present, this function and its epigones are utilized for population projection. However, this function has fatal flaws, as it is a marriage function. As shown in Figure 3.2, the cumulative probability density of marriages does not converge to a real-world scenario. It was negligent of some sociologists to state that every female customarily got married. Even during that period, nearly five per cent of each cohort did not get married by the age of 45 years. In every population, some percentage of females remained unmarried throughout their lives. An accurate marriage function must account for this.

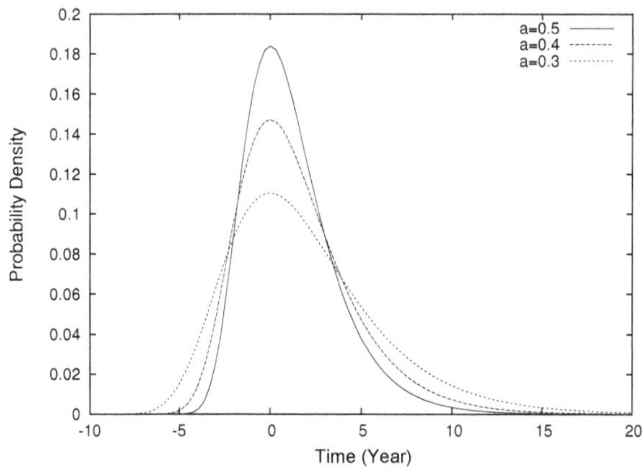

Figure 3.4: Double exponential distributions.

In contrast, the cumulative probability density of double exponential distributions easily converge to one. Namely, the probability of marriage rapidly becomes asymptotic to zero. In general, exponential distributions rapidly converge to zero. The upper probability of a standardized normal distribution at +3 standard deviation is only $> .0013$. This value is far too small compared to the actual values.

See Figure 3.4. When we set $a = 0.3$, the cumulative probability of Equation (3.4) is > 0.99 at 25 time unit (year) passed. Even when $a = 0.25$, the cumulative probability of Equation (3.4) is > 0.97 at 25 time unit (year) passed. This probability of marriage among females is very high; human females have not recorded such a high cumulative probability of marriage.

For population projection, Coale and his epigones had to first estimate the number of females remaining unmarried throughout their lives for each cohort and subtract it from one. However, this number was estimated in an arbitrary manner. A meaningful marriage function can definitively calculate the number of females remaining unmarried during their life time.

For this reason, the family of exponential distributions is entirely incongruous with the marriage function. This asymptotic test is decisive for marriage functions. An effective marriage function must survive this asymptotic test; however, only a few survive.

3.1.2.3 Extreme Event

Despite these fatal flaws, the fact that double exponential distributions provide an adequate approximation to the marriage function indicates significant knowledge about the marriage function.

The expression for the Coale–McNeil distribution is a variation of double exponential distributions. Double exponential distributions (Equation (3.4)) are usually applied to the life of manufactured products or the time between failures. First-marriage occurrences within a cohort resemble the failures of products manufactured during the same period zero.

$$f(t) = a\exp(-a(t-\mu) - \exp(-a(t-\mu))), \tag{3.4}$$

Generally, the weakest part of a product causes its failure, i.e., a single failure can result in the failure of the entire product. **This leads us to infer that first marriages occur as one critical event and not as the sufficiency of several conditions.** This suggestive implication has long escaped our notice and appears outlandish to our subjective consciousness. If double exponential distributions are effective approximations of the function of first marriage, the inference that they only occur as one stochastic event should be accepted.

Convolution models such as the Coale–McNeil distribution assume the sufficiency of four conditions. It has been demonstrated that convolution models essentially contradict double exponential distributions, with the exception of the convolution of infinite random variables. The Coale–McNeil distribution is a convolution of four stochastic variables. Consequently, it differs fundamentally from the double exponential distributions.

As mentioned previously, double exponential distributions (including that of Coale–McNeil) has a critical defect: their integration rapidly converges to one.

3.1.2.4 Why Does the Marriage Function Differ Across Cultures?

Coale forced uniformity among cross-culturally different marriage distributions by adjusting the coefficient for each case. However, the cross-cultural differences have remained evident, as Coale himself pointed out in Figure 3.5.

Neither the Coale–McNeil distribution nor the double exponential distribution can explain this diversity. A meaningful marriage function must reasonably explain the diversity rather than simply obtaining fitting coefficients.

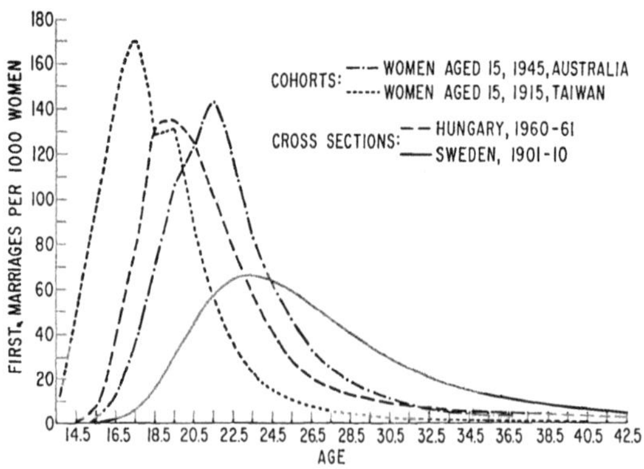

Figure 3.5: First-marriage frequency, selected population.

Source: A.J. Coale, Age Patterns of Marriage [12], p. 198.

3.1.2.5 Why are They not Dynamic?

Double exponential distributions are undoubtedly no more than exclusive approximations of the first-marriage function. This is because alterations of its coefficients cannot be interpreted in the context of real societies.

The Coale–McNeil function (Equation (3.3)) and the double exponential function (Equation (3.4)) are completely static. However, the marriage distribution changes longitudinally, as shown in Figure 3.6.

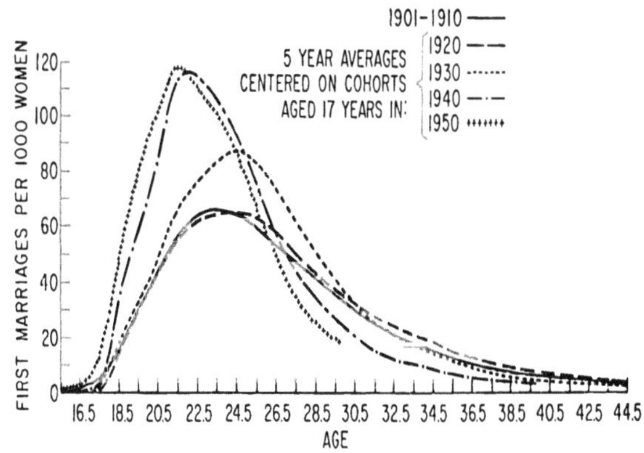

Figure 3.6: First-marriage frequency, selected cohort.

Source: A.J. Coale, Age Patterns of Marriage [12], p. 198.

The possibility of function alterations is the same as that of altering their coefficients. It is intuitive that we query the causes of alterations of the coefficients into socioeconomic conditions. However, this is also the end of the theory.

The reason for these static properties of the functions is that the functions represent the mapping from the age x to the probability of marriage. The age dimension is unvarying and uniform, and the age increases monotonically. The mapping from age to marriage probability should be fundamentally uniform.

To build the dynamics of marriage probability, the mapping must be created from the varying factor to the marriage probability. Here, the varying factor is none other than the previous marriage probability itself[1]. This is the essence of "background independence."

3.2 Hernes Function

In 1972, Gudmund Hernes presented "The Process of Entry into First Marriage" [33]. This marked the birth of the "Hernes function." In reality, the

[1]Of course, when it is extended to geographical space, the factors are "neighboring previous marriage probabilities.

Hernes function can survive the asymptotic test, although imperfectly. It is also semi-background independent. It is rather appropriate that the semi-background independence of the Hernes function enabled it to pass the asymptotic test.

3.2.1 Definition of the Hernes Function

Hernes constructed a model of entry into marriage using a differential equation

$$\frac{dP_t}{dt} = Ab^t(1 - P_t)P_t, \tag{3.5}$$

where P_t is the ratio of the first-marriage population within a cohort at time t. Here, we need to consider the basic relation between the ratio of the first-marriage population and the probability density of first marriage. Namely, the derivative of the ratio of the first-marriage population is the probability density of first marriage. Let $F(t)$ be the probability density of the first marriage at time t. Then, the following equation can be obtained:

$$\frac{dP_t}{dt} = F(t) = Ab^t(1 - P_t)P_t \tag{3.6}$$

Consequently, the probability density of first marriage varies depending on P_t and $(1 - P_t)$. The Hernes function assumes that marriage occurrences depend exclusively on neighboring marriage occurrences, except for the effect of b^t.

The analytical solution of Hernes function is

$$P_t = \frac{1}{1 + \frac{(1 - P_0)\exp(A/\ln b)}{P_0 \exp(Ab^t/\ln b)}}, \tag{3.7}$$

3.2.2 Advantages of the Hernes Function

First, the Hernes function is a parsimonious model. It contains two coefficients similar to the double exponential distribution. This is significantly smaller than the number of Coale–McNeil's coefficients.

Otani [60] reported that the fit of the Hernes function is generally better for Japanese data than any other marriage functions. Hernes himself presented the following figure to demonstrate the fine fit to the census data of the United States.

The Hernes function can calculate the mean age of first marriage by itself. This property makes this function unarguably superior to the Coale–McNeil distribution. Similarly, it is superior to the double exponential distribution, which assumes μ as the mean.

The Hernes function is also decisively superior because it passes the asymptotic test. It converges to zero more glacially than the other two functions.

For the above reasons, the Hernes function is mathematically exceptional compared to other marriage functions. However, it has not been very popular

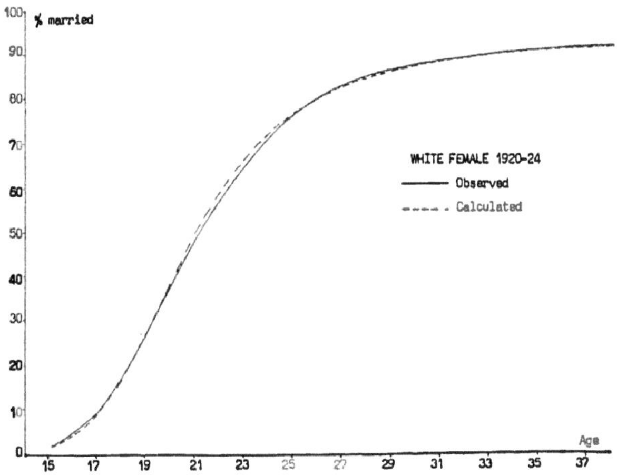

Figure 3.7: Cumulative first marriages for Caucasian-American females born during 1920–24.

Source: G. Hernes, The Process of Entry into First Marriage [33], p. 178.

in the demographic sphere for the reasons discussed in the following subsection.

3.2.3 Flaws of the Hernes Function

The Hernes function has a significant difficulty in interpreting reality. Hernes defined the constants and coefficients as follows: A is the average initial marriageability and $b < 1$ is the constant of deterioration.

Consider Equation (3.5). The term $(1 - P_t)P_t$ is a well-known logistic term that can be easily applied to a real human object or subject. Hernes himself described P_t as "social pressure." However, I do not adopt this interpretation. There is a significant difference between the subjective interpretation of P_t and my perspective. I will explain this difference in depth in a later part of this book.

Hernes estimated $A = 1.046$ for the cohort of Caucasian women born between 1920 and 1924 based on the 1960 United States population census. Why was the average initial marriageability greater that 1? He also estimated the constant deterioration $b = 0.855$. The subjects would already be deteriorated at the initial stage, and therefore, constants are unnatural.

I assume that Hernes was obliged to introduce coefficients A and b^t to fit the equation to the observed data.

An equation explaining a phenomenon must have at least one coefficient. Setting aside the constant A, which is used for the fit with the observed data, the marriage function must have a monotonic decreasing element. Consequently, Hernes adopted the exponential function b^t. However, these definitions and interpretations of the parameters appear unreasonable.

3.2.3.1 Asymptotic Comparison

The exponential function b^t accelerates the convergence to zero. Therefore, the Hernes function does not generally fit within the right tail of the distribution. Let us examine the right tails of the Hernes distribution for a 1960 Japanese cohort.

From Figure 3.8, it can be seen that the right tail probabilities are systematically low for the observation. Hernes adopted R.D. Prescott's procedure to estimate the parameters. The same procedure was adopted for the Japanese data using the Hernes function in the present work. The advantage of Prescott's procedure is that it generally achieves fine fitness for both tails. The tradeoff for this advantage is that the distribution peak deviates from the observed data. Even with the advantage of the estimation method, the Hernes function performs with a lower probability for the observed right tail. This separation is significant because the number of real (common law) marriages occurring at older ages is somewhat greater than that found in statistical data. Elderly marriage tend to not be legally registered.

When the Hernes function is adjusted to fit the distribution peak, the separation of the Hernes function from the observation increases. Although the Hernes function has better characteristics in terms of the asymptotic test compared to the Coale–McNeil and double exponential distributions, this is insufficient for practical scenarios. A meaningful marriage function must converge to zero more glacially than the Hernes function.

A few years ago, I did not have a strong understanding of why b^t is wrong. Like many demographers, I regarded it as uninterpretable, as it can be justi-

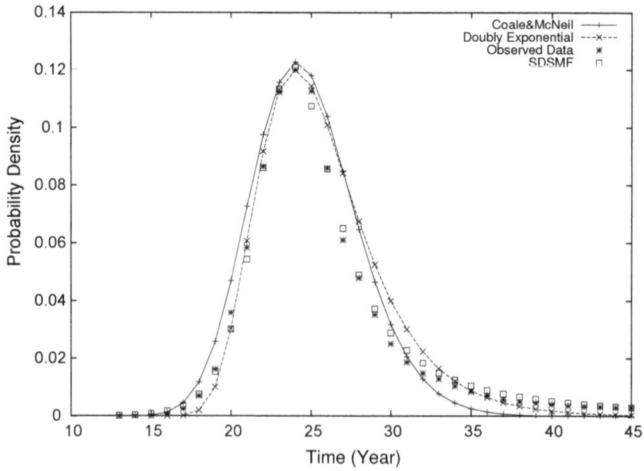

Figure 3.8: Coale–McNeil and doubly exponential distributions for the observed 1960 Japanese cohort, with the space-dependent stochastic marriage function (SDSMF) adjusted at the inflection point.

fied as such. Further, the fit of the Hernes function is inferior to that of my formulation. For these relatively unimportant reasons, I refused the Hernes function. However, I have finally recognized far more essential grounds for its rejection. Now, I can firmly predicate the principle that the variables of the equation should be within the system, in the neighborhood itself. The mapping must be constructed by the variables (tools) within its dimensional space. This principle is extremely valuable, and I have had some opportunities to prove this during my investigations.

3.2.4 Conclusions on the Hernes Function

Hernes proposed the marriage function as a differential equation. As a mathematical sociologist, I am proud that Hernes, who is a sociologist, built the differential equation model of the marriage function, including the genuine causal factor.

The genuine causality should be modeled as a mapping from the present state to the next. Although the Hernes function is modeled in this direction, it is insufficient. Rather, it demonstrates the central principle that the mapping should be constructed entirely within its state.

3.3 Other Marriage Functions

As expected, other marriage functions have already been proposed. They are not as popular as the three functions analyzed in this book. Their fitness to the data is generally inferior to that of the Hernes function. Nevertheless, their names and mathematical expressions are introduced here.

3.3.1 Γ Distribution

The Coale–McNeil distribution is a special parameterized Γ distribution. For a normal Γ distribution, the convolution is not of a quasi-normal distribution and three exponential distributions; instead, the real number k characterizes the shape of the distribution. For application to marriage probabilities, the parameter θ is used to adjust the time scale in many cases.

$$F(t) = \frac{1}{\Gamma(k)\theta^k} t^{k-1} \exp(-t/\theta) \tag{3.8}$$

The parameter k is interpreted as the trial frequency, and the Γ distribution is construed as the waiting time for the event occurrence. As mentioned already for the Coale–McNeil distribution, this Γ distribution also converges to 0 rapidly. This property is not compatible with the marriage function. Otani [60] reported that when he adopted Japanese data, the Γ distribution was inferior to the Hernes function in terms of fitness.

3.3.2 *Generalized* $\log\Gamma$ *Distribution*

As its name suggests, the generalized $\log\Gamma$ distribution considers some lifetime distributions using a more generalized mathematical frame.

$$F(t) = \frac{|k|}{\Gamma(1/k^2)\theta}(1/k^2)^{k-2}\exp\left(\frac{1}{k}\frac{t-\mu}{\theta} - \frac{1}{k^2}\exp\left(k\frac{t-\mu}{\theta}\right)\right). \tag{3.9}$$

The above equation[2] was presented by R.L. Prentice [63], who applied logarithm to the distribution defined by E.W. Stacy. It was originally introduced to implement maximum likelihood estimation in life testing; demographers arbitrarily adopted it to the marriage function.

Despite this extension of parameters, the set of Γ distributions is incompatible with reality for the same reason why the Coale–McNeil distribution was incompatible, which has already been mentioned. The distribution must also cut off the percentage of never-married individuals from the total population.

The disadvantages of these generalizations are that they were obtained by adding parameters, and therefore, the parameters must be interpreted in order to fit the function to a practical scenario. There may be several possible interpretations, and an explanation unique to each scenario cannot be obtained.

Further, the manner in which parameter values are chosen cannot be explained, and consequently, the future parameter values and marriage probabilities cannot be predicted.

It is self-evident that the distribution of marriage probabilities changes and differs between localities. The generalized $\log\Gamma$ distribution is not very useful in terms of complexity. The real marriage process cannot be obtained from this equation. The idea that the marriage process is the convolution of several stochastic variables that compose the emergency process of marriage is no longer valid.

3.3.3 *Log-Logistic Model*

Several researchers have referred to the log-logistic model. In contrast to Hernes' assumption of Ab^t as the monotonically declining element of the marriage function, the log-logistic model assumes $\frac{a}{t}$ as the monotonically declining element. Thus, the differential equation of the marriage function is expressed as follows [43] and [47]:

$$\frac{dP_t}{dt} = F(t) = \frac{a}{t}P_t(1-P_t) \tag{3.10}$$

There is little difference between the respective equations of the Hernes function and the log-logistic model. Their analytical solutions are similar.

$$P(t) = \frac{\exp(\beta)t^{\alpha}}{1+\exp(\beta)t^{\alpha}} \tag{3.11}$$

[2]The parameter symbols have been changed from the original notation to integrate them with the notations used in this chapter.

In the 1980s, Diekmann ([47], pp. 177–183) reported that the Hernes function and the log-logistic model presented finer fits for German female and male data, respectively[3]. He made the following report:

> Computation of a simple goodness-of-fit measure confirms the conclusion reached from graphical inspection (see Table 3.1). It can be seen that the Hernes model yields the lowest average absolute deviation for the female sample, while the log-logistic model yields the best fit of the four parametric models for the male data. For the male sample, surprisingly, the two-parameter log-logistic model leads to better predictions than the three-parameter Hernes model. Is this an accident? Is the good fit of the log-logistic model caused by idiosyncratic data?
>
> **Table 3.1:** Goodness of fit of the four survival models (German data).
>
> | | Exponential Model | Sickle Model | log-logistic Model | Hernes Model |
> |---|---|---|---|---|
> | Women | 0.076 | 0.030 | 0.017 | 0.004 |
> | Men | 0.086 | 0.024 | 0.007 | 0.015 |
>
> Note: Goodness-of-fit measure $= 1/33 \sum |G_{life-table} - \hat{G}_{modelprediction}|$.

A. Diekmann, Diffusion and Survival Model of the Process of Entry into Marriage in *Event History of Life Course Research* [47], pp. 180–181.

The tendency of female marriage processes to proceed more rapidly than male ones is the primary cause of this contrast. Namely, the difference between Ab^t and $\frac{a}{t}$ results in this deviation.

The progress of the late marriage tendency in Germany is significant for German and North American males, and thus, the log-logistic model achieves a finer fit for them as compared to the Hernes function. This does not imply that the log-logistic model is presently superior to the Hernes function. In retrospect, for both the Hernes function and the log-logistic model, there is no theoretical means to determine the parameters of the distributions. The first-marriage timeline depends on these parameters.

Otani also reported that the log-logistic model was parsimonious and achieved adequate fits, and that the Hernes function is comparable to it. It is noteworthy that Diekmann discerningly observed that the marriage function is an expression of the diffusion process and compared the log-logistic model with the Hernes function.

A theory that can determine marriage timelines is necessary. None of the existing marriage functions can achieve this.

[3] In his tests, A. Diekmann used the marriage data reported by Hernes. He stated that the Hernes function presents a finer fit than the log-logistic model for North American male and female data.

3.3.4 Sickle Model

A. Diekmann & P. Mitter proposed [20] the sickle function, which is named based on its assumption of nonmonotonic hazard functions like a "sickle:"

$$r(t) = ct\exp(-\frac{t}{\gamma}). \tag{3.12}$$

The survival function is

$$G(t) = \exp(-\lambda c(\lambda - (t+\lambda)\exp(-t/\lambda))) \tag{3.13}$$

The sickle model is not suitable for the marriage process. Equations having any relation with the diffusion process present a fine fit to the marriage process. Diekmann stated the following:

> The sickle model does not do as well as either variant of the diffusion model. In contrast, the sickle model is highly superior to the log-logistic model (Diekmann and Mitter 1984a) in the analysis of divorce data, probably because of its incorporation of "immunity." The sickle model seems to be less adequate in modeling the marriage process.
>
> A. Diekmann [47], Chapter 9, p. 180.

3.3.5 Problem with Existing Marriage Functions

Including the Hernes function, the existing marriage functions have a common problem: the theories cannot explain the manner in which the parameters governing marriage occurrence timelines are determined. They are mere values estimated by statistical procedures.

For instance, Ab^t of the Hernes function and $k, \theta, and\ \mu$ of the $\log\Gamma$ function are estimations based on data. An effective theory should determine these parameters theoretically. They are evidently changing, and the marriage function theory should explain how. The existing functions, particularly the $\log\Gamma$ function, are unable to achieve this because they are simply mathematical imitations of the marriage probability distribution.

In in a later chapter, I will propose a marriage function that will provide a natural and theoretical answer to this problem. I will state beforehand that it will be done background independently. The late marriage tendency is the result of neither economic development nor the progress of feminism. An effective marriage function theory can answer these questions within the theory itself. Only a theory that can explain the dynamics of the system using its system is capable of making predictions.

3.4 Do Marriages Occur as a Result of our Conscious Choices?

It appears indisputable that marriages occur as a result of our conscious choices. However, this view is commonplace. Ideas intuitive to our consciousness are not suitable for the foundation of a science. I will explain this problem in the case of the marriage function.

3.4.1 *Early or Late Marriages: Mere Individual Differences?*

One individual marries early, another marries late, and yet another never marries. These differences are found everywhere. It is commonly assumed that those who married late decided to do so consciously and individually. Admittedly, this is plausible, but the failure of existing marriage functions suggests that this is not a valid premise.

3.4.1.1 Lifetime or Survival Functions are not Marriage Functions

Within the framework of the lifetime function, a fault or disease occurs due to malfunctions inside individuals. Environmental and physical situations may give rise to these malfunctions, but it is not assumed that neighboring persons or parts could cause one another to malfunction. It is not assumed that the variable $F(t)$ of individuals interact with each other. The coefficients of the Coale–McNeil function, $\alpha, \lambda, and\ \mu$, do not interact between individuals.

Namely, the lifetime function postulates individualistic stochastic variables. Components or individuals must inevitably die sooner or later. Therefore, the convolution of the survival function must easily converge to one.

When the marriage function is built, this is a fatal separation from real marriage occurrences, because marriage occurrences collectively depend on neighboring marriage occurrences. We are not doomed to marry without fail; it solely depends on neighboring marriages. We are embedded in the world (society), and therefore our every behavior affects each other, consciously or unconsciously.

If we were non-interacting individuals, the lifetime function might be an accurate marriage function. However, the log Γ function is not empirically sufficient for the marriage function. Lifetime or survival functions are not fundamentally marriage functions. I positively conclude that marriages do not occur based on individual choices.

The assumption of conscious selection entirely contradicts the stochastic models. It is impossible to develop a marriage function similar to the Coale-McNeil function under this assumption. Simple exponential functions, such as $\lambda e^{-\lambda t}$, are introduced by the postulation of random stochastic variables. Our conscious behaviors are simply an accessible assumption; they cannot formulate a single stochastic variable as $\lambda e^{-\lambda t}$.

This elementary view that social phenomena must be based on the conscious choice of individuals has governed our worldviews since the 1980s, strangely accompanied by neo-conservatism. However, this idea that we consciously select our behaviors is absolutely unscientific. The convolution models that trace our conscious marriage processes are not valid, despite being no less than the stochastic model. This is likely because they are based on the concept of individual conscious choice.

In contrast to these failures of the convolution models, the models that have assumed the diffusion process have survived. This fact suggests that the diffusion process is far more important than individual conscious choices. The diffusion process occurs based on neighboring occurrences of marriages. Essentially, neighboring occurrences alone affect marriage functions. The failures of the convolution models suggest that the conscious process of marriage is unimportant, and the concept of individual conscious choice can be omitted from this phenomenon.

3.4.2 Explaining the Dynamics of the Marriage Function

If the differences in marriage occurrences between two distinguishable socioeconomic groups are formulated in a manner similar to Coale's approach, different coefficients must be provided for the two. Then, it is assumed that conscious differences are dependent upon socioeconomic differences. This leaves the bewildering question of how to determine these coefficients.

There are no reasonable grounds to correlate our conscious choices to these coefficients. Is the marriage function only a fiction created by Coale and other researchers? More than a few demographers, sociologists, and economists seem to believe so. However, this is unjustified. The regularity of social phenomena is never formulated as a mere aggregation of individuals' choices. In reality, no genuine individual choices exist. Every individual choice is affected by others. When we consider marriage occurrences as neighboring-dependent stochastic processes in a dimensional space, the regularity and its dynamics naturally emerge from it. In the next chapter, the simplest case of this space will be introduced.

Chapter 4

Marriage Function as an Integral Function

> ... if I assume as a starting point, what ought to be universally admitted, namely, that all men are born ignorant of the causes of things, that all have the desire to seek for what is useful to them, and that they are conscious of such desire. Here from it follows, first, that men think themselves free in as much as they are conscious of their volitions and desires, and never even dream, in their ignorance, of the causes which have disposed them so to wish and desire.

Spinoza, *Ethics (Ethica Ordine Geometrico Demonstrata)* [16], pp. 35–36.

A meaningful marriage function should be formulated within itself. This is another expression of background independence. We must confront the fact that human behavior is based on a reaction to an immediately previous state, and not arbitrary free will, as imagined by our subjective consciousness.

4.1 Integral Function

The Hernes function was proposed as a differential equation. In most cases of reaction-diffusion models, theoreticians apply differential equations to their models. However, for the phenomenon of marriage occurrence, the integral equation is suitable.

As mentioned earlier, the derivative of the cumulative probability of marriage function is a probability of marriage. Further, the cumulative probability of marriage is the integral of the probability of marriage. Let $F(t)$ be the probability of marriage for time t. Specifically, the below relation is obtained:

$$\frac{d\left(\int_0^t F(t)(dt)\right)}{dt} = F(t) \tag{4.1}$$

Our current objective is to formulate $F(t)$ under the principle of background independence. A concrete mapping for $F(t)$ can be obtained within this setting.

4.1.1 The More Marriages Evident within a Space, the Higher the Marriage Occurrences

We are strongly inclined to get married when several marriages occur within our cohort in a given space. Thus, the marriage probability $F(t)$ is proportional to $\int_0^t F(t)(dt)$. In contrast to this tendency, when there are fewer unmarried persons within our cohort, the probability of being selected for marriage is reduced. $F(t)$ is initially expressed as follows:

$$F(t) = \lambda \int_0^t F(t)(dt) \left(1 - \int_0^t F(t)(dt)\right) \left(\text{unknown element}\right). \tag{4.2}$$

This integral equation corresponds to each of Hernes' elements, $AP_t(1-P_t)$. However, we require a substitute for b^t, which decreases monotonically in the space in which marriages occur. This substitute must be found within the system itself.

4.1.1.1 $\int_0^t F(t)(dt)$ is not Social Pressure

Hernes regarded $\int_0^t F(t)(dt)$ as social pressure.

> Single people experience the social pressure to marry in various ways as the percentage of married people increases.
> (partly omitted.)
> For us, the major point is that the more or less subtle sanctions against the not yet married increase as the percentage of their married cohort increases. At least we shall assume this for the age span we will consider in the ensuing empirical analysis.
> The simplest way to incorporate this assumption in a model, is to suggest that the social pressure on the not yet married is proportional to the percentage of their cohort already married, and furthermore, that the rate of change into the marriage state is proportional to this pressure.
>
> Source: G. Hernes, The Process of Entry into First Marriage [33], p. 174.

To obtain an accurate understanding of what a marriage function is, the concept of "social pressure" is neither suitable nor sufficient. It is a mere metaphor, and it prevents us from understanding the reality of marriage occurrence.

Marriage is not the result of a subjective, thoughtful selection under social pressure. Our brains respond to neighboring marriage occurrences from time to time, and in one of these occasions, if we (our brains) happen to desire to marry, the probability of the desire depends on the neighboring marriage occurrences. I insist that even if we consider it unconsciously, we do not carefully consider the utility of marriage, as household economists assumed.

The idea that we carefully consider our marriages is constructed *ex post* by our subjective consciousness. The evidence of this subjective posteriority is clearly demonstrated by B. Libet, as mentioned in the previous chapter. Here, I add a supplemental illustration. It is clear that "social pressure" changes with the passage of time. Hernes himself was witness to these transitions. To allow the reader to understand the extent to which our subjective consciousnesses change, I quote his explanation of social pressure:

> As a larger percent of the cohort marries, single people will have reduced interaction with old friends. They will be invited less often to parties, dinners, or trips, partly because of the awkwardness of diverging interests, partly because of the threats to the established couples. When singles are invited, hosts often play the role of matchmakers. Thus, the psychological experience of the undesirability of celibacy increases as more of the same age group enter into wedlock. With increasing age, the psychological experience of being unmarried as a kind of deviance is heightened by popular culture. For example, as a woman remains in the pool of the not-yet-married, her social definition changes from maiden to spinster or even reject. In the process, she may pass through the panic-age, trying harder than ever to catch a husband. In her social environment, people will say, "what she need is a man," and perhaps she will pitied for lacking marital bliss. Men who do not marry may have their masculinity questioned, and even be suspected of homosexuality.
>
> Source: G. Hernes, The Process of Entry into First Marriage [33], p. 174.

Gender study has been a major topic in sociology today. Journals of sociology such as *American Sociological Review* do not publish such prejudiced (in today's meaning) sentences. However, they were considered commonsensible in those days. Our social norms have altered unquestionably. A meaningful theory of marriage occurrences must conceive, theoretically explain, and predict these alterations. To achieve this, neither the cultural description of marriage nor the interpretation of the motivation for marriage is necessary. Rather, the background-independent explanation of marriage occurrences illustrates the dynamics of marriage norms.

Social pressure has weakened significantly compared to that in earlier times. What has brought about the undermining of social pressure? Only the present-day diversity and decline of marriage can bring about these subversions of our consciousness. Specifically, our subjective consciousness can be omitted from the depiction of marriage occurrences. Although our and my subjective consciousness could be dissatisfied by this omission, this logical consequence is undeniable. Although many sociologists abominate this perspective as extreme scientism, understanding one's motivation cannot be the basis of a science.

$\int_0^t F(t)(dt)$ does not correspond to social pressure. It indicates that females consciously and unconsciously increase their behaviors and volition for marriage as the percentage of neighboring ever-married individuals increases. This assumption entirely contradicts the concept of social pressure.

The idea of social pressure has its root in individualism. The dyadic antagonism of the social versus the individual is fallacious. This is a simple metaphor created by subjective consciousness rather than scientific objectivity.

4.1.1.2 Monotonic Decreasing Element

The most important considerations for marriage functions are their variables; as mentioned earlier, these must be discovered within the formulation itself. Other important considerations include whether the monotonic decreasing element is appropriate for the fit and whether it provides a reasonable explanation of the phenomenon. In this regard, the right tail of the distribution of marriage functions is the decisive discriminant test for judging the accuracy of existing marriage functions.

It can be readily observed that the distribution of first marriages has a large right tail (see Figure 4.2). This implies that people continue to enter first marriages at relatively older ages (35–50, and even over 60 years old). Empirical data suggests a correlation between lower marriage occurrences at younger ages and higher occurrences at older ages. Therefore, the rapid decrement of the exponential function of t is a lethal attribute for the marriage function. Considering this from another perspective, a monotonic decreasing element in a meaningful marriage function has the characteristic of fewer marriages occurring at younger ages and more marriages occurring at older ages. However, Hernes' b^t and $\frac{a}{t}$ of the log-logistic lack this important property. The monotonic decreasing element in a marriage function must be a variable that is dependent on the timeline of marriage occurrences. Specifically, this feature is one aspect of background independence.

4.1.1.3 Monotonic Decrease in the Never-Married Segment

The probability of first marriage occurrences depends on the percentage of individuals who have never married at time t. This ratio shows a monotonic decrease with t. Thus, the monotonic decreasing element in a first marriage function should be the percentage of individuals who have never married.

4.1.2 Formulation of an Integral Equation

Strictly considering the continuity of time as a real number, $\frac{dP}{dt}$ is zero almost everywhere, because marriage does not occur within an arbitrary and very short interval. Therefore, marriage occurrences must be expressed as a Lebesgue integration[1].

The proportion of the ever-married category is the sum of the probability of the occurrence of marriage. Thus, $P(t) = \int_0^t F(t)(dt)$. Therefore, the following integral equation is obtained for the probability of marriage $F(t)$ at time t.[2]

$$F(t) = \lambda \int_0^t F(t)(dt) \left(1 - \int_0^t F(t)(dt)\right) \left(1 - \int_0^t F(t)(dt)\right) \tag{4.4}$$

$F(t)$ is defined within a measurable space. Consequently, F (*first marriage occurrences) belongs entirely to an additive class. Henceforth, this integral equation will be referred to as a space-dependent stochastic marriage function (4.4) **SDSMF**, because $F(t)$ varies depending on the proportion of ever-married individuals within its space.

4.1.2.1 Characteristics of SDSMF—Thicker Right Tail and Higher Kurtosis

The integral Equation (4.4) cannot be solved analytically; however, it can easily be solved numerically (see Figure 4.1). As various reaction-diffusion models cannot be solved analytically, despite its simplicity, SDSMF seems solely solved numerically.[3]

It can easily be observed that SDSMF has a thicker right tail than that of the double exponential distribution. Consequently, its kurtosis is higher than that for the Coale–McNeil or double exponential distributions (see Figure 3.8). The observed frequencies of first marriages indicate this relatively higher kurtosis.

---

[1] "(dt)" indicates that this integration is a Lebesgue integration. The Riemann–Stieltjes integration is expressed as $\int F(t)dt$.

[2] In the initial stage of this study, I was highly influenced by Hernes' b^t. Therefore, I wrote the marriage function as shown below.

$$F(t) = \mu \int_0^t F(t)(dt) \left(1 - \int_0^t F(t)(dt)\right) \prod_0^t \left(1 - \mu \int_0^t F(t)(dt) \left(1 - \int_0^t F(t)(dt)\right)\right) \tag{4.3}$$

This expression is false because its right tail is far thicker than observed.

[3] This conjecture is probably owing to my incapability.

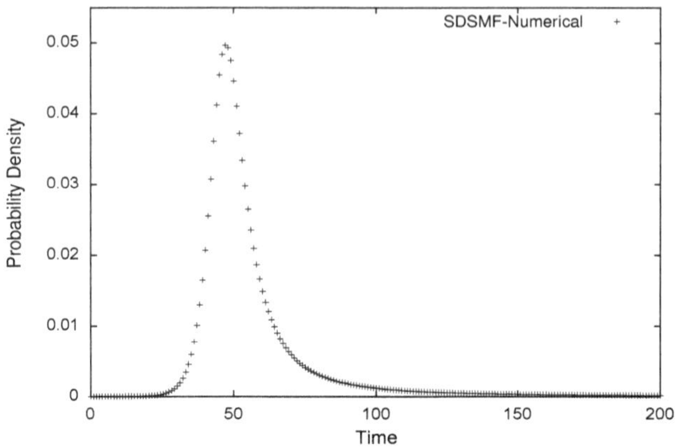

Figure 4.1: Numeric solution of SDSMF.

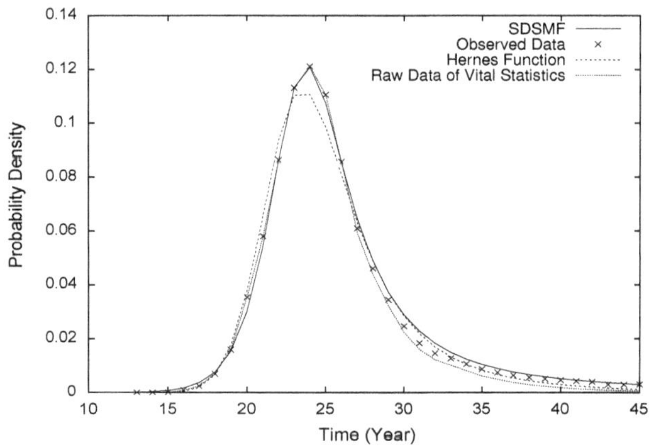

Figure 4.2: SDSMF and Hernes function for the 1960 Japanese cohort and observed data.

4.1.2.2 Background Independence of Marriage Occurrences

It is important that the integral Equation (4.4) is exclusively dependent on the integration of $F(t)$ with respect to t. Thus, the ratio of first marriages develops on its own, independent of other socioeconomic conditions. I refer to this property as the background independence of marriage occurrences. Despite the general vague belief that socioeconomic conditions affect marriage occurrences, individuals themselves decide this stochastically, independent of prevailing socioeconomic conditions. Marriages occur automatically and are dependent on their own quantum in the same manner as any other animal behavior.

Consequently, it can be inferred that the distribution of first marriages is decided by means of itself in the present. The consequences and conclusions of the first marriages of each cohort can be foreseen when the initial occurrences of first marriage in each cohort are examined. However, one exception is the extraordinary and large-scale historical event known as the "baby boom."[4]

4.1.3 Fit of SDSMF to the Data

To fit SDSMF to the observed data, the concrete values of the coefficient λ and the interval of time t should correspond to the actual time (year). The time interval of SDSMF is supposed to be much shorter than a year. Therefore, the appropriate values of the time interval of SDSMF are integrated (summed) to achieve a fit with the probability of first marriage observed by year.

4.1.3.1 Inflection Point Method

To estimate the parameter λ and interval t, a novel inflection point method is developed herein. This method obtains the most likely λ based on the correspondence of SDSMF's inflection point to the observed inflection point.

The ratio of first-time married individuals of a cohort at a specific age (an endpoint) is not sufficient for estimating the parameters. The observed ratio of individuals who have ever been married, aged 45 or 50 years, is far from the true values, because there are many common-law marriages within older age groups that are not publicly declared. I do not consider the endpoint of the marriage process to be suitable for estimating marriage features. However, in contrast to the imprecision of the endpoint, the inflection point is relatively accurate.

4.1.3.2 Estimation Results

In this work, SDSMF was adopted for the first marriages of several Japanese cohorts, applying the inflection point method. A good fit (average absolute error < 0.00159427 for each age) was possible with the observed values[5] of first marriages for the 1960 Japanese cohort. However, the optimum fit was obtained for the 1950 Japanese cohort. The inflection method achieved an average absolute error < 0.000932. It missed less than 0.1%, on average, for annual observations.

The inflection method estimates λ at $\simeq .22$. I have assumed that λ is a constant and the interval t is altered because the dependence on the neighboring ratio of ever-married individuals is universal. Thus, the timeline of neighboring first marriage occurrences alters the macro timeline of first marriage occurrences.

[4]However, even the "baby boom" is not an exception to SDSMF. Its onset occurred as a mere historical accident, and its subsequent durability can be explained by the enhanced SDSMF.

[5]The observed values were calculated from vital statistics that were modified in accordance with the census.

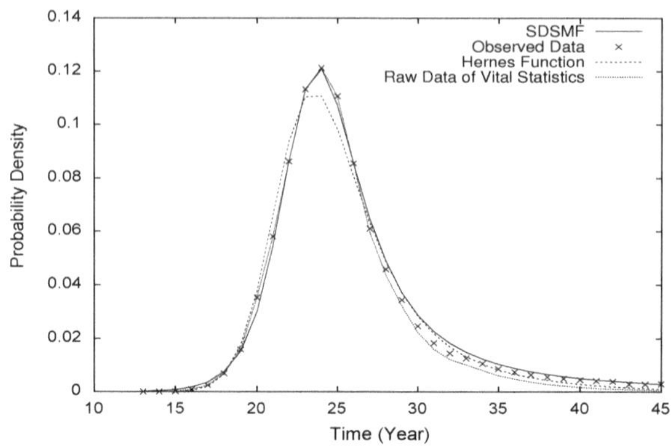

Figure 4.3: SDSMF and Hernes function for the 1960 Japanese cohort and observed data.

4.2 Decisive Evidence of SDSMF

A good fit is not in itself a strong indicator of the validity of the marriage function, as described by Hernes below.

> The general problem with our type of analysis is that the fit between the observed and calculated curve of first marriages in itself is not a strong test of the model. Since the model is the only link between the data and the assumed substantive casual processes (measured by the parameters), it is hard to tell whether the resulting estimates actually reflect the casual forces involved.
>
> Hernes [33], p. 180.

Therefore, he drew attention to the necessity of a demonstrative test:

> A much stronger test of the causal forces represented by the parameters would be to derive alternative measures of them, or to confront other empirical consequences of the model with data.
>
> Hernes [33], p. 181.

Below, I propose a corroborative and novel test for marriage functions that merits the "scientific" label.

4.2.1 A Good Theory can Predict Some Theoretical Values

The critical issue is whether a theory can enable the prediction of real and theoretical values. With SDSMF, the average age at first marriage for a cohort can be predicted as an expected value. Then, because SDSMF is a probability density function, it becomes easy to calculate the mean age of the first-time married population with SDSMF.

The average age at first marriage for a cohort can certainly be calculated from vital statistics. However, this average is always lower than the true value. The reason for this is that people tend not to declare their marriages when they marry at older ages. Thus, the true value of the average age at first marriage cannot be obtained from vital statistics.

$$\text{Mean Age at first marriage} = \int_0^t t\, F(t)(dt) \tag{4.5}$$

The formula for the calculation is as follows:

$$\text{Mean Age at first marriage} = \sum_0^t t\, F(t_i) \tag{4.6}$$

From the estimated values of $F(t)$, the theoretical expected mean age at first marriage was calculated, as shown in Table 4.1. This was then compared with the observed value, the calculated value according to the Hernes function, and the estimations from the Japanese General Social Survey (JGSS)[6] samples. Each of these values is presented in Table 4.2.

Table 4.1: Theoretical mean age at first marriage for the 1960 Japanese cohort.

Age at	SDSMF Prediction	Coale-McNeil	Hernes Function Prediction	Observed from Vital Statistics
45	24.329481	24.728437	23.033714	23.410084

Table 4.2: Average ages at first marriage from samples of the 1960 cohort calculated from social surveys.

Survey	year	$\overline{x}$	n	s
JGSS	2000–2002	24.8219	73	3.42119
NFR, JGSS	1998, 2000-2, 2006	24.7883212	137	3.5405854

When a somewhat higher standard deviation $\sigma = 3.6$ is assumed, the 95% confidence interval is estimated as $[23.996, 25.648]$. The SDSMF prediction falls within this range, although it is still smaller than the estimation for the Japanese General Social Survey (JGSS) samples. I suggest that the SDSMF prediction is the most probable.

The Coale–McNeil distribution computes a valid mean age of first marriage. However, it does not pass the asymptotic test in the previous chapter.

[6]The Japanese General Social Surveys were designed and carried out at the Institute of Regional Studies at Osaka University of Commerce in collaboration with the Institute of Social Science at the University of Tokyo. They were performed under the direction of Ichiro Tanioka, Michio Nitta, Hiroki Sato, and Noriko Iwai with Project Manager, Minae Osawa. The project was financially assisted by the Gakujutsu Frontier Grant from the Japanese Ministry of Education, Culture, Sports, Science and Technology for the academic years 1999–2003, and the datasets are compiled and distributed by the SSJ Data Archive, Information Center for Social Science Research on Japan, Institute of Social Science, the University of Tokyo.

Let us examine this problem as a statistical test. The fact that we can frame an alternative point hypothesis and a null point hypothesis that are not trivial zeroes is a landmark outcome for the social sciences.

$$\begin{cases} H_0: & \mu = 24.329481 & \text{SDSMF's prediction} \\ H_1: & \mu = 23.410084 & \text{Vital statistics} \end{cases}$$

Normally, within the social sciences, we have only been able to test a null hypothesis as $\mu = 0, \beta = 0$. The null hypothesis has been used for falsification alone. This test is equally lenient for the alternative hypothesis, $\mu \neq 0, \beta \neq 0$, which is verified against the null hypothesis. It is almost always true that μ and β are not equal to zero within populations for several reasons.

As long as a concrete null point hypothesis and a point alternative hypothesis having the ring of truth cannot be hypothesized, the null hypothesis can always be rejected by applying a statistical test[7]. This is especially true when calculations can be performed for an unlimited number of samples. We are unable to determine the power of previous statistical tests within the social sciences.

In contrast to this lenient test, I hypothesize H_0 for the verification from a more rigorous standpoint. Even when confined to this statistical test problem, SDSMF is demonstrably more sufficient than other existing marriage theories. No other theory can theoretically predict the mean age at the time of first marriage.

4.2.1.1 Test Results

Samples from the 1960 cohort were selected from JGSS 2000–2002, JGSS 2006, and NFRJ 1998.[8] I then estimated the average age and an associated standard deviation for the 1960 cohort in Japan as follows:
$\overline{x} = 24.7883212, s = 3.5405854, n = 137$. If it is assumed that $\sigma = 3.6$, which is somewhat higher than the expected value, the critical regions of H_0 are $[-\infty, 23.727]$ and $[24.932, \infty]$ for a .05 level of significance. H_0 is accepted.

Then, we can calculate the probability of error of the second type of β, and consequently the power of the test.

$$\frac{23.727 - 23.4100842}{3.6}\sqrt{137} \doteq .98 \Rightarrow P\{z > .98\} \doteq .16354 \tag{4.7}$$

$\beta \doteq .16354$. The power of this test is > 0.83. Accordingly, H_0 is more likely to be true. This implies that SDSMF is a valid function that describes the actual process of first marriage. The fact that SDSMF can precisely predict the average age of first marriage in the 1960 cohort in Japan is an indicator of its theoretical success.

---

[7] This is an incorrect tradition that originated with L.R. Klein. This improper approach to conducting a statistical test must be abandoned in favor of a theory that can hypothesize a nontrivial null point hypothesis.

[8] National Family Research of Japan 1998 was conducted by the Japan Society of Family Sociology.

4.2.2 Comparing SDSMF to the Coale–McNeil and Double Exponential Distributions

4.2.2.1 Decisive Discriminant Test

A critical difference between SDSMF and the Coale-McNeil distribution is that the former does not easily converge to zero. It also takes a considerably longer time than double exponential distributions to integrate SDSMF to 1. This property of SDSMF implies that it includes the never-married population of a cohort within the equation itself. SDSMF can predict the proportion of the never-married population at time t. The percentage of the population that is not married at time t within a cohort can be estimated soon after the marriage process begins (from 16 or 17 years of age).[9]

4.2.2.2 Proportion of Ever-Married Population for the 1960 Cohort

Let us perform a simple statistical test for the proportion of the ever-married population within the 1960 Japanese cohort. Let p be the proportion of the ever-married population aged 45 years. For the 2000–2003 JGSS sample, in which $n = 119$, we obtained the estimation $\hat{p} = 1 - 0.05882 = 0.94118$. Assuming $\sigma^2 \doteq 0.9 \times 0.1 = 0.09$ to be somewhat larger than the population parameter, a 95% confidence interval was calculated for p [0.88728, 0.99508]. The original vital statistics data (without adjustment in accordance with the census), the Coale–McNeil distribution, and the double exponential distribution were excluded.

The number of samples is insufficient for performing a single-proportion z-test for $n = 119$. The following samples were combined: JGSS2006, NFRJ98, NFRJ01, NFRJ03, and SSM1995[10]; Consequently, $n = 445$ samples were acquired for the 1960 Japanese cohort. I obtained the 95% confidence interval as [0.90696, 0.96270]. SDSMF and the observed values were included.

Table 4.3: Calculated proportions of the ever-married populations of the cohort for each function and estimation.

SDSMF	observed	Coale–McNeil	Doubly Exp	JGSS	n
0.93929	0.91332	0.997039	0.99670	0.94118	119
				0.93483	445
*information, Census 2000 for Age 40-44, 45-49				0.91418	0.91795

Let us now consider the single-proportion z-test.

$$\begin{cases} H_0: & p = 0.93929 & \text{SDSMF's prediction} \\ H_1: & p = 0.99670 & \text{Doubly Exp} \end{cases}$$

[9]It is not so easy in realty, when the marriage function is extended to higher dimensional spacc.

[10]Japanese Social Stratification and Mobility Survey, 1995.

Assuming a somewhat large $\sigma^2 \doteq 0.09$, and setting the level of significance at 0.05, the following region of rejection was obtained: $[0, 0.91142]$ and $[0.96716, 1]$. The value $\hat{p} = 0.93483$ does not reject H_0. The power of the test is 0.984. Thus, the double exponential and Coale-McNeil distributions can be excluded from the group of reasonable marriage functions with considerable certainty.

This demonstrates the ability of the background independence property and the integral equation of the SDSMF. It is a natural expression of a marriage process, and it enables us to calculate the proportion of the unmarried population at time t.

4.3 Testing SDSMF in Other Countries

It is necessary to test the validity of SDSMF for other countries. I posit that the integral equation describes a universal process of first-marriage occurrences. Therefore, it is worth examining the background independence of marriage occurrences for several additional countries.

By conducting these test procedures, I was able to determine whether the theory could predict the rate of early marriage occurrences, which is also a decisive feature of the marriage function.

4.3.1 Testing SDSMF using Cohorts within the Swiss Population

4.3.1.1 Similarities to the Japanese Cases

Figure 4.4 shows the result of adapting SDSMF and the Hernes function to the first marriage rates of Swiss cohort from 1924 to 1945.[11] The tendencies of each function that are similar to those observed for Japan can be clearly distinguished.

The SDSMF estimates showed a higher marriage rate for older ages than what was observed. The Hernes function exhibited a good fit for both tails. However, in contrast to that at the tails, the fit was poor at the top.

At the right tail, SDSMF was found to be more reasonable than that in the observations for Japan. This indicates strong grounds for stating that more first marriages occur after the age of 35 years as common-law marriages. These were not even specified in the census questionnaire. The same test that was performed on the Japanese cohort was performed on the 1940 Swiss cohort using the European Social Survey (ESS) data.[12]

[11] I used data from *Two Centuries of Swiss Demographic History—Graphic Album of the 1860–2050 Period* [5].

[12] The ESS is an academically-driven social survey designed to chart and explain the interaction between Europe's changing institutions and the attitudes, beliefs, and behavior patterns of its diverse populations. The ESS was established in 2001 and was led by its founder and coordinator, Roger Jowell, until his death in December 2011 (http://www.europeansocialsurvey.org/).

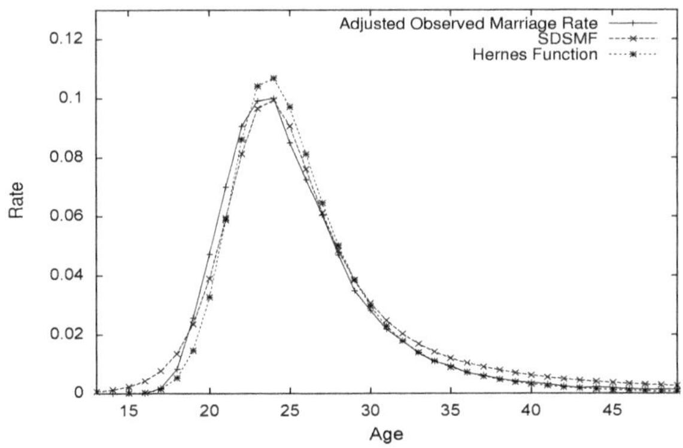

Figure 4.4: Observed first-marriage rate and SDSMF for the 1924 Swiss cohort.

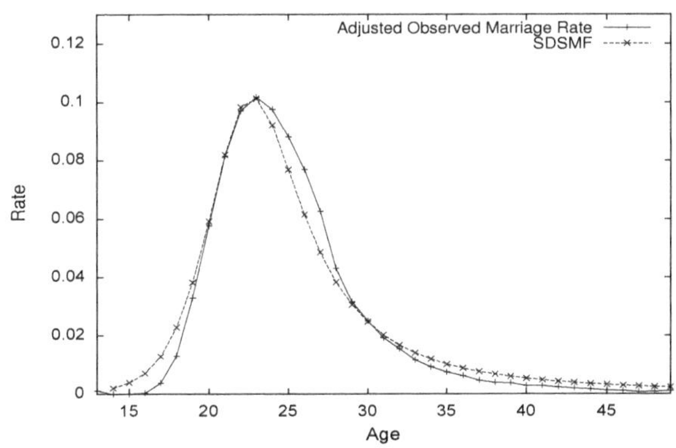

Figure 4.5: Observed first-marriage rate and SDSMF for the 1930 Swiss cohort.

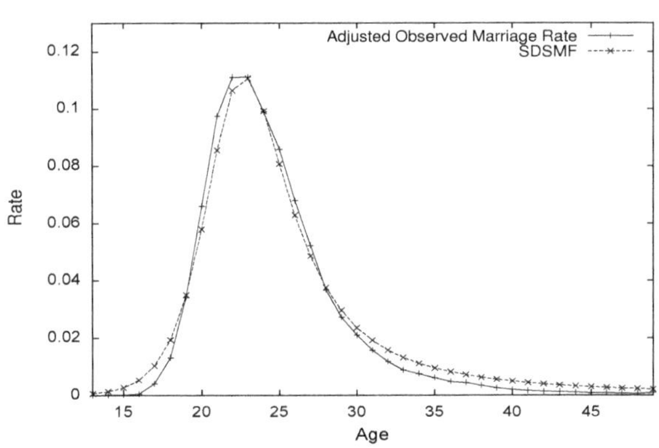

Figure 4.6: Observed first-marriage rate and SDSMF for the 1935 Swiss cohort.

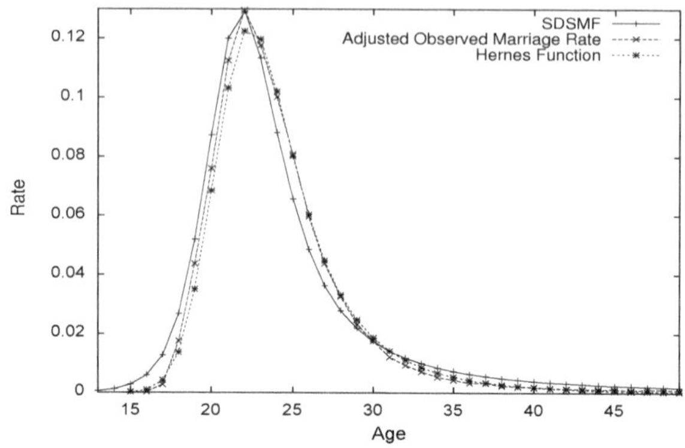

Figure 4.7: Observed first-marriage rate and SDSMF for the 1940 Swiss cohort.

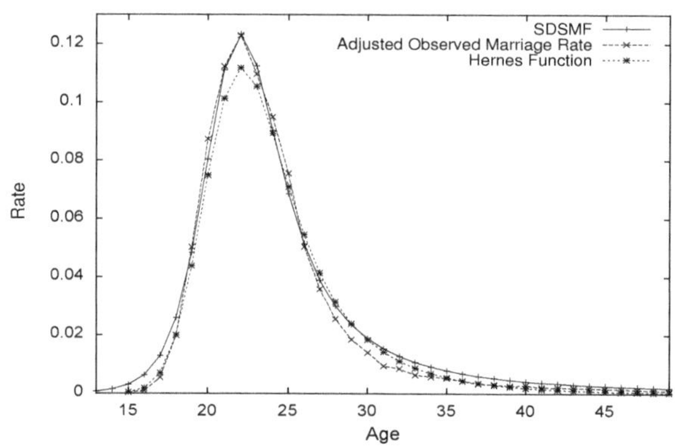

Figure 4.8: Observed first-marriage rate and SDSMF for the 1945 Swiss cohort.

$$\begin{cases} H_0: & p = 0.951243 \\ H_1: & p = 0.918378 \end{cases}$$

The region of rejection is $[0, 0.903031]$, $[1.0, 1.0]$. $\hat{p} = 0.939394$ does not reject H_0. Although the power of the test is poor, SDSMF's ratio is the one that is the most likely.

4.3.2 Marriage at Young Ages: The Second Decisive Discriminant Test

SDSMF systematically predicts that more marriages at young ages would occur compared with the adjusted observed data or estimates using the Hernes function. The question, then, is which of these is correct. Whether or not a theory can predict the real rates of marriage occurrences at young ages is a definitive touchstone for the function of first marriages.

4.3.2.1 How Often do Marriages at Young Ages Occur?

The Swiss federal statistical office does not record the marriages of individuals below the age of 15 years. However, the fact that there are no such records does not mean that marriages of individuals below 15 years do not occur.

Moreover, I believe that the rate of occurrence of recorded marriage of individuals aged 16–18 years does not reflect real marriages, including cohabitant unions or common-law marriages, because they are too few in number. According to the physiological maturity of Caucasoids[13], there should be more than a few cohabitant unions.

Table 4.4: Age at first marriage in the United States derived from General Social Survey (GSS) samples.

Age	Freq	Prob	Prob(vaild)
12	4	.00013	.00026
13	37	.00125	.00240
14	103	.00347	.00668
15	263	.00886	.01706
16	783	.02639	.05079
17	1247	.04202	.08090
⋮	⋮		
Total	14260	/26675	

The ages at first marriage in the United States were extracted from the cumulative GSS[14] samples. Table 4.4 presents the frequencies of occurrence of young marriages for the total cumulative samples.

SDSMF is an integral equation. If the validity of SDSMF is accepted in relation to first-marriage occurrences, they absolutely depend on the initial first-marriage occurrences. This can be logically deduced from the background independence of marriage. Thus, the question of whether there are more marriages than registrations is decisively important.

In actual processes of marriage occurrences, neighboring occurrences of early marriages determine subsequent marriage occurrences within that cohort. On reflecting upon this consequence of the second decisive discriminant test, I have concluded that the occurrence of more marriages than those that are registered is significantly more probable. Nonetheless, a social survey that is meticulously designed to count early young marriages will resolve this problem.

[13] The concept of race is not scientific. Nonetheless I use "Caucasoid" to denote ethnic groups such as Anglo-Saxons and Germans.

[14] http://www3.norc.org/gss+website/.

4.3.3 Test using Algerian Data

L. Henry [32] recorded the marriage distribution of Muslims in Algeria in 1948. Consider the application of SDSMF and the Hernes function to these data (see Figure 4.9).

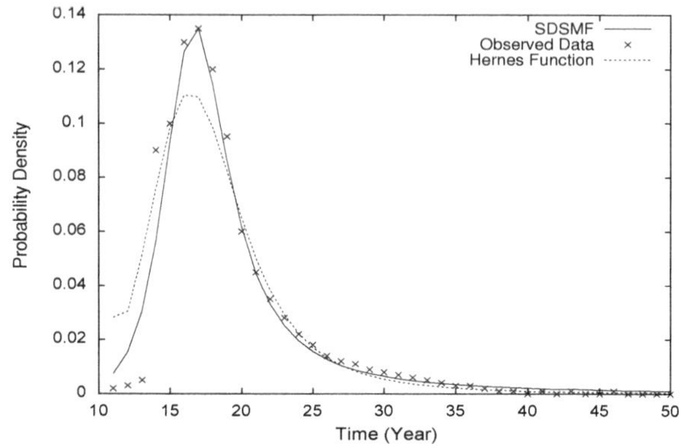

Figure 4.9: Observed data and the application of several marriage functions for Algerian Muslims in 1948.

Both SDSMF and the inflection point method can effectively describe the distribution peak. This presents an intriguing contrast in that the Hernes function did not demonstrate a good fit for the first half and the peak of distribution when it was adjusted to fit the latter half.

The observed data exhibit unreasonable jumps for ages 13 and 14. Although societal norms caused this jump, young marriages must exist. Only their declarations and registrations are suppressed by the married couples or by others.

By applying SDSMF, more plausible values could be estimated and interpolated for the first half of the distribution. This indicates that by using SDSMF, the first marriages of those below the age of 13 (whose declarations are below their actual numbers) can be calculated more accurately. It is suspected that the sum of undeclared marriages of individuals below 13 years of age is almost equal to the difference between SDSMF and the observed value. Namely, SDSMF is sufficiently robust to estimate the true parameters, even when the data contain significant noise.

As A.J. Coale stated, despite cultural differences, SDSMF can effectively describe marriage occurrences. This example demonstrates the universality of neighboring marriage occurrences as the cause of marriage occurrences. I can confidently conclude that previous marriage occurrences themselves cause the desire for marriage. I must emphasize that every sociological theory must be constructed on this causality of events, not on our conscious motivations. This is the only scientific approach to the subject.

4.4 Marriage Function Tests Using Kurtosis

I believe that it is somewhat excessive to repeatedly test marriage functions. However, it would be better to add another statistical analysis to confirm the effectiveness of SDSMF. The comparison between marriage functions by using kurtosis is significant in the sense that SDSMF alone can approximate the kurtosis of real marriage data.

4.4.1 Definition of Kurtosis

Kurtosis is the fourth standardized moment, defined as

$$\frac{\displaystyle\int_a^b (x-\mu)^4 f(x)dx}{\left(\displaystyle\int_a^b (x-\mu)^2 f(x)dx\right)^2} \tag{4.8}$$

To compare the kurtosis of a given distribution with the normal distribution, the next equation is defined as follows:

$$\frac{\displaystyle\int_a^b (x-\mu)^4 f(x)dx}{\left(\displaystyle\int_a^b (x-\mu)^2 f(x)dx\right)^2} - 3 \tag{4.9}$$

The kurtosis of the normal distribution is three. Functions in Microsoft Excel, statistical packages such as SPSS Statistics, and so on can be used to calculate this Equation (4.9)[15]. Distributions with kurtosis greater than three imply that the distribution has a larger tail (or tails) as compared to the normal distribution.

4.4.1.1 For Discreet Distribution

In order to perform numerical calculations, the discreet distributions need to be defined.

$$\frac{\displaystyle\sum_{i=1}^{n}(x_i-\mu)^4 f(x_i)}{\left(\displaystyle\sum_{i=1}^{n}(x_i-\mu)^2 f(x_i)\right)^2} - 3 \tag{4.10}$$

[15]They call it excess kurtosis to distinguish it from kurtosis.

Because SDSMF is a probability density function, kurtosis can be calculated directly. When calculating for an entire population, the following formula is used:

$$\frac{\frac{1}{N}\sum_{i=1}^{N}(X_i-\mu)^4}{\left(\frac{1}{N}\sum_{i=1}^{N}(X_i-\mu)^2\right)^2} \tag{4.11}$$

Note that the population size N is used in the formula without $f(x)$. N is cancelled out.

$$N\frac{\sum_{i=1}^{N}(X_i-\mu)^4}{\left(\sum_{i=1}^{N}(X_i-\mu)^2\right)^2} \tag{4.12}$$

4.4.1.2 Kurtosis of Sample

Statistical packages assume sample data for the calculation. Therefore, they utilize another formula for kurtosis. Here, the degree of freedom must be considered.

$$\frac{n(n+1)}{(n-1)(n-2)(n-3)}\frac{(x_i-\overline{x})^4}{s^4}\frac{3(n-1)^2}{(n-2)(n-3)} \tag{4.13}$$

4.4.1.3 Calculation of Kurtosis of each Marriage Function

I calculated the kurtosis of the observed data and each marriage function. The kurtoses of the Coale-McNeil distribution and the double exponential distribution were much less than the observed data.

Table 4.5: Kurtosis of each marriage function with respect to probability density until age 45 in Japan.

Cohort Year	Coale-McNeil	Double Exponential	Hernes	SDSMF	Observed
1950	3.772	5.867	7.607	8.350	8.753
1955	3.525	4.804	7.423	7.186	8.599
1960	3.330	5.057	6.599	6.769	7.455

The Coale–McNeil and double exponential distributions are too far from the observed kurtosis. From the perspective of their kurtoses, the Coale-McNeil and double exponential distributions also drop out.

4.5 Late-Marriage Trend in Japan from SDSMF

Since the 1990s, the late-marriage trend has become evident throughout Japan. Unfortunately, in the range of SDSMF, the dynamism of the marriage function cannot be effectively described. I will explain why SDSMF cannot be applied in the real case of Japan. Through this process, readers may understand the necessity of extending the marriage function to spatial dimensions.

4.5.1 *Earlier-than-Expected Outset of Late-Marriage Trend*

It is significant that, based on Japanese census data, the outset of Japan's late-marriage trend is recognized to have occurred around 1925.

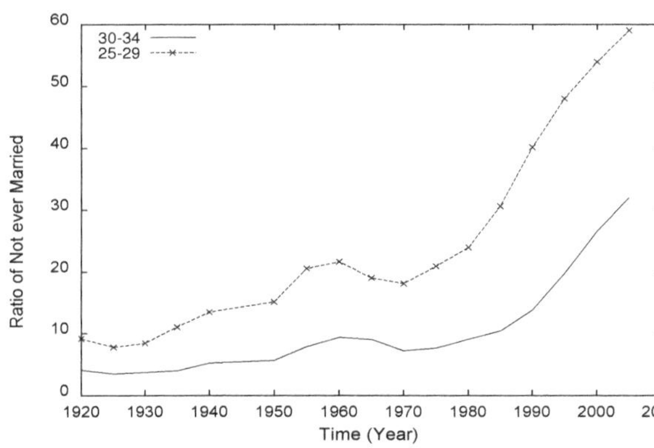

Figure 4.10: Proportion of unmarried females from 1920 to 2005 in Japan based on census data.

Japanese people have generally taken it for granted that the serious low fertility, mainly caused by the late-marriage trend, began around 1970. However, the late-marriage trend had begun far earlier; the human consciousness was simply unable to conceive this slight inclination at that time. The late-marriage trend was established by around 1925; we only obeyed and are obeying these gradients. This is one important reason why I reject the idea that conscious motivations drive our behaviors.

Of course, some scholars called the late-marriage trend around 1970 the "second demographic transition," distinguishing it from the prior trend. It is a vice of some social scientists to simply name a phenomenon and believe that it has been understood. If the "second demographic transition" hypothesis was a theory, it would be able to predict the time period for which the late-marriage trend would last and the degree to which it would proceed. No element of these predictions can be found in the "second demographic transition" hypothesis. The "second demographic transition" is always attributed to free will, which is a typical tautology. Naming or branding is not a theoretical understanding at all.

4.5.2 Late-Marriage Trend from the Perspective of Marriage Functions

Here, I try to analyze the late-marriage trend in Japan based on the coefficients of marriage functions. This analysis soon meets to its boundary.

Within the framework of SDSMF, the late-marriage trend is indicated by the increase in Δt. On the other hand, within the framework of the Hernes function, the late-marriage trend is indicated by the decrease in A (the average initial marriageability) and b (the initial deterioration). In 1940, Japan was undergoing the late-marriage trend; this can be confirmed by the census trends. In contrast to SDSMF, the Hernes function displays poor performance; the estimated A and b even resulted in reversed directions for some cohorts.

Table 4.6: Coefficients of SDSMF and the Hernes function in Japan.

cohort	SDSMF	Hernes Function	
	Δt	A	b
1935	0.255	2.641	0.765
1940	0.242	4.228	0.758
1950	0.208	2.371	0.829
1955	0.251	2.000	0.843
1960	0.269	1.989	0.844

4.5.2.1 Temporal Backward or Stationary Trends

From the late 1950s to the early 1970s, which Japanese people call the period of rapid economic growth, there were temporal backward or stationary trends against the late-marriage trend in Japan. This backwardness dazzled several people and concealed the long, longitudinal late-marriage trend beneath it.

As shown in Figure 4.10, the proportions of unmarried females in the ages of 25–29 and 30–34 did not grow; instead, they declined between 1960 and 1970. This was the effect of the rapid economic growth in Japan. The coefficients of SDSMF dependably reflect its fluctuation.

This phenomenon has been regarded as counterevidence against the concept of background independence. This view is inaccurate because the trends have revived and proceeded even more seriously after the rapid economic growth. The momentum of the late-marriage trend continues to govern marriage occurrences in Japan. Although the period of rapid economic growth was a huge event, in which a continuous 10% annual growth rate was recorded over 15 years, its effect was relatively small and restrictive. The economic effect is no match for the neighboring dependence, which is the only disturbing factor.

4.5.3 Limitations of SDSMF

The SDSMF framework has limitations in terms of its ability to describe the late-marriage trend. Until the 1960 cohort, SDSMF effectively described the late-marriage trend; however, it lost its description power after this period.

Within the framework of SDSMF, the late-marriage trend must occur as shown in the following Figure 4.11. In Japan, the late-marriage trend followed a different path. In order to describe it more profoundly and accurately, the marriage function must be extended to a higher-dimensional space. The limitations of a low-dimensional function (where the marriage function is based on a single dimension of age) are now clear.

The reason for these limitations is quite apparent. The late-marriage trend in Japan has not varied as shown in Figure 4.11. It must be understood that marriages do not occur in a single cohort.

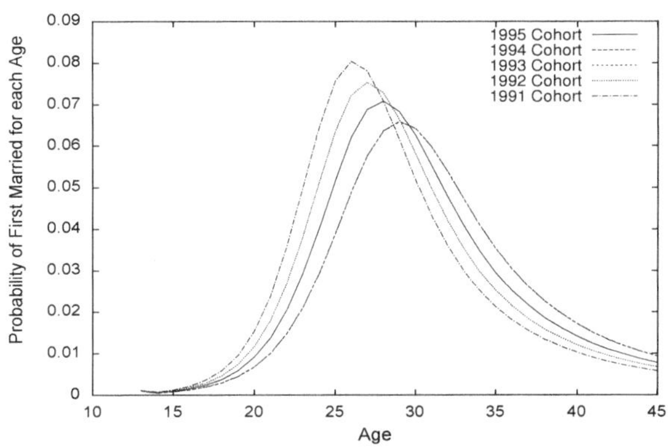

Figure 4.11: Late-marriage trend as described by SDSMF(in single cohort = dependent on an initial $\Delta t \simeq \alpha$).

Chapter 5

Marriage Function in High-Dimensional Space

> Experience teaches us no less clearly than reason, that men believe themselves to be free, simply because they are conscious of their actions, and unconscious of the causes whereby those actions are determined.

Spinoza, *Ethics (Ethica Ordine Geometrico Demonstrata)* [16], p. 96.

Marriage occurrences depend upon previous neighboring marriage occurrences. The marriage function must be extended to the physical and cognitive neighborhood in order to understand its dynamics.

5.1 Diffusion of Late-Marriage Trend

The elementary marriage function $\int_0^t F(t)(dt)(1 - \int_0^t F(t))(dt))^2$ must be extended to the geographical surface. The reason for this extension is that the diffusion of the late-marriage trend has been observed in Japan. The following figures illustrate this phenomenon in the Kanto area of Japan.

The areas marked in white indicate an early-marriage trend, while the areas marked in a darker color indicate a late-marriage trend. Readers may also watch an animation of this dynamic alteration of marriages in Japan by visiting "http://www.sociology.main.teikyo-u.ac.jp/bankonka".

As described in the previous chapter, during the 1960–70s, the late-marriage trend was temporally reversed due to the high economic growth of Japan. In the 1980s, the trend began again and has continued since, despite the bubble economy. Although it is continuing, this process appears asymptotic to the fixed points.

It may be observed that the late-marriage trend does not diffuse as smoothly as the number of children. This is because the marriage function is formulated as an integral equation.

It is significant that for all the older age groups, identical patterns have occurred, as shown in Figures 5.1–5.10. Thus, it can be considered that the late marriage trend diffuses from younger to older generations. Specifically, two geographical dimensions and one age dimension (a total of three dimensions) must be considered to substantially describe marriage occurrences.

Consequently, the marriage function is extended to a three-dimensional space plus time. Geometrically, it is a four-dimensional hypersurface.

5.2 Formulation

Let $F(x, y, a)$ be the marriage probability in the three-dimensional space at x, y (geographical coordinates)and age a. At the outset, let us consider the discrete lattice space. I express the probability as $F(x_i, y_j, a_k)$ instead of $F(x, y, a)$ in the discrete lattice space.

$$\lambda \int_0^{a_k} F(a_k)(da) \left(1 - \int_0^{a_k} F(a)(da)\right)^2 \tag{5.1}$$

Which neighboring probabilities affect $F(x_i, y_j, a_k)$? Marriage occurrences are only affected by neighboring ages. In fact, the proportion of marriages in the parental generation does not affect the marriages of the children's generation. In contrast, individuals are deeply impacted (sometimes unconsciously)

Figure 5.1: Diffusion of the late-marriage trend in Japan, 1960, for the age group 20–24.

Figure 5.2: Diffusion of the late-marriage trend in Japan, 1965, for the age group 20–24.

Figure 5.3: Diffusion of the late-marriage trend in Japan, 1970, for the age group 20–24.

Figure 5.4: Diffusion of the late-marriage trend in Japan, 1975, for the age group 20–24.

Figure 5.5: Diffusion of the late-marriage trend in Japan, 1980, for the age group 20–24.

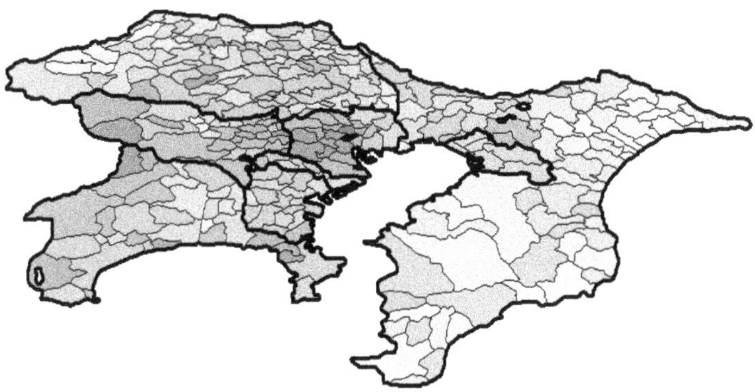

Figure 5.6: Diffusion of the late-marriage trend in Japan, 1985, for the age group 20–24.

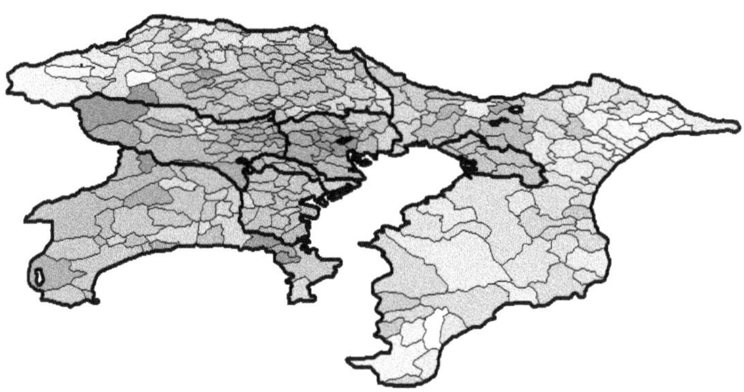

Figure 5.7: Diffusion of the late-marriage trend in Japan, 1990, for the age group 20–24.

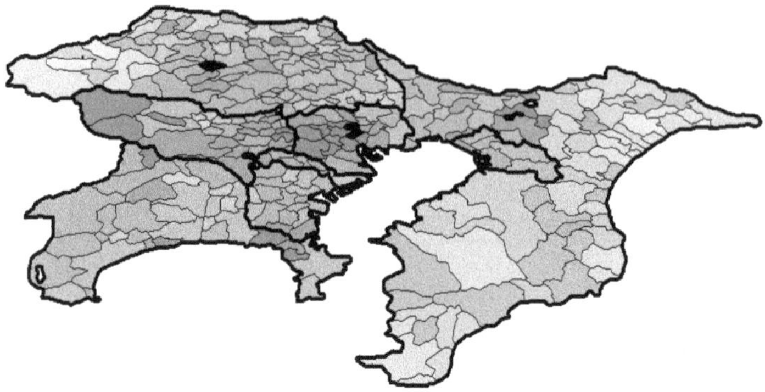

Figure 5.8: Diffusion of the late-marriage trend in Japan, 1995, for the age group 20–24.

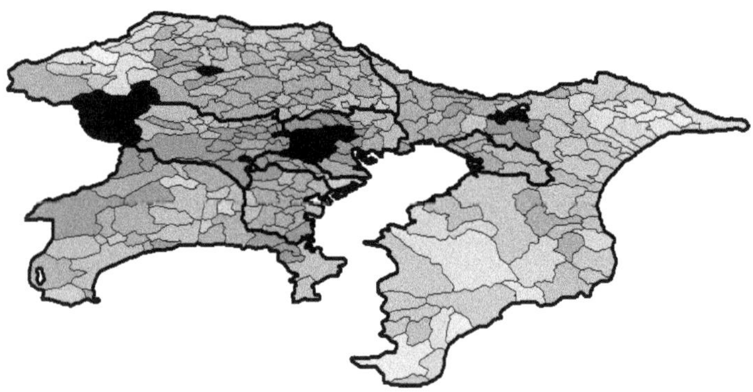

Figure 5.9: Diffusion of the late-marriage trend in Japan, 2000, for the age group 20–24.

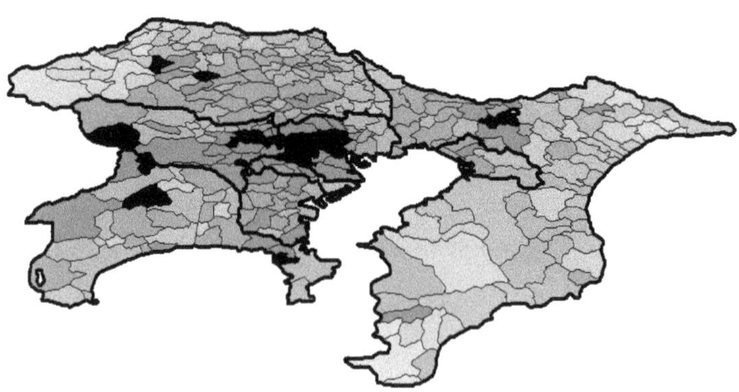

Figure 5.10: Diffusion of the late-marriage trend in Japan, 2005, for the age group 20–24.

by neighboring ages. Therefore, Equation (4.4) must first be re-expressed for the age variable.

The marriage function varies by age; it also varies with time depending on the marriage occurrences of neighboring cohorts. The following difference-integral equation is considered:

$$
\begin{aligned}
F(a_k, t_{\ell+1}) \\
= \lambda \Bigg(\mu \int_0^{a_{k-1}} F(a_{k-1}, t_\ell)(da) + (1-2\mu) \int_0^{a_k} F(a, t_\ell)(da) \\
+ \mu \int_0^{a_{k+1}} F(a_{k+1}, t_\ell)(da) \Bigg) \left(1 - \int_0^{a_k} F(a_k, t_\ell)(da) \right)^2 \qquad (5.2)
\end{aligned}
$$

Changes to the marriage function are not exclusively dependent on the λ coefficients. The marriage function changes in relation to neighboring marriage occurrences. In this high-dimensional space (society), every marriage occurrence is related in a causal chain process. The marriage function alters itself at every point in time, except for the fixed points.

The marriage function also alters itself at every location. Marriage occurrences depend on other geographically neighboring marriage occurrences. Therefore, new terms must be added to the difference-integral Equation (5.2). However, this extension is temporarily withheld in order to focus on an in-depth explanation of the implications of extending the age dimension.

5.2.1 Unification of Cohort and Period

Through this age dimension extension, the marriage function can be naturally unified for the cohort data and period data. The marriage probabilities of a cohort change depending on the influence of marriage occurrences in previous and successive cohorts. This simple fact is crucial. Our conscious and unconscious desires to marry are activated by the effect of temporally ($\doteq$period) neighboring marriage occurrences. This is an undeniable and unyielding fact.

The idea that we are controlled by social norms and customs of marriage is an illusion produced by our minds. We are controlled by neighboring behaviors themselves; however, we remember social norms and customs rather than actual factors. This is because human memory is limited and cannot recall every factor that affects us; therefore, some type of abstraction(norm or custom) is necessary.

Human consciousness is not the cause of our behaviors. Causality must be constructed from previous behaviors to present behaviors to successive behaviors. The causality of social events flows continuously and fluently. When we interject our conscious motivations, this causality is not perfectly assured because we always receive unconscious effect from neighboring behaviors. In this regard, *"Der Verstehen Soziologie"* by M. Weber is completely false.

Thus, this unification also compromises the dynamics of the marriage function. The natural extension of the marriage function from a single cohort to several cohorts comes down to the foundation of the function's dynamics.

Although, in this direction, the extension to geographical dimensions is inevitable, I will demonstrate the performance of this unification and extension.

5.2.2 Baby Boom

The occurrence of the "baby boom" depended upon this mechanism[1]. Let us exclude birth behaviors from the ongoing temporal consideration. The baby boom primarily occurred in the Allied countries post World War II due to the recommencement of marriages, which were suspended by the war. Teitelbaum & Winter described the baby boom as follows:

> In a number of countries, the post-1945 baby boom was one of the most dramatic demographic shifts in recent history. In the United States, fertility rose by over two-thirds, from the record-low fertility levels of the 1930s to levels in the last half of the 1950s, approaching those of the period 1905–1910 (U.S. Bureau of the Census, 1975, Table B-3136). However, it is important to understand that this baby boom in its complete form characterized only a few of the industrialized countries, albeit also some of the largest.
>
> Source: *The Fear of Population Decline*, M.S. Teitelbaum and J.M. Winter [71], p. 68.

It is important for the marriage function theory to explain *terra incognita* the manner in which the diverse baby boom occurred and passed. There remain unsolved problems that elucidate the characteristics of the marriage function.

> However, in contrast to the sustained booms of the above four countries, most of these could be fairly described as "boomlets," that is, rising trends in both the late 1940s and later in the 1950s, deriving partly from realization of deferred births and a surge in marriage, but which never reached the levels of true baby boom countries. As may be seen in Figures 4.2–4.7, most of the countries in this group had already had fertility levels at or near replacement by the 1930s. The end of the Second World War was followed by a brief baby boomlet that appears as a 3- to 4-year "spike" in the graph, followed by fully two decades of relatively unchanging fertility levels in the general range of 2.0 to 3.0.
>
> Source: *The Fear of Population Decline*, M.S. Teitelbaum and J.M. Winter [71], p. 68.

The reason for this heterogeneity of the baby boom has not been clarified thus far because almost all demographers supposed that the cause of its beginning cannot explain its entire process. Teitelbaum & Winter depicted these circumstances.

[1] Every marriage occurrence is stochastically governed by Equation (5.2).

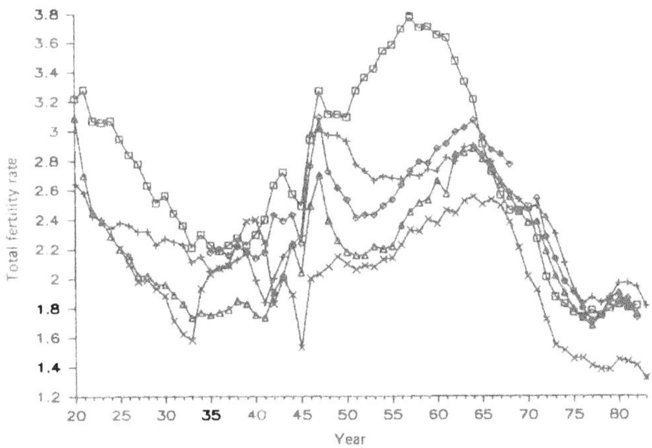

Figure 5.11: Total fertility rates, 1920–1983, for England and Wales (Δ), France (+), Scotland (O), the United States (□), and West Germany (X).

Source: *ibid*, M.S. Teitelbaum and J.M. Winter [71], p. 70.

Various explanations have been offered to account for the striking demographic phenomenon of the baby boom. One school of opinion holds that the fertility experience of the postwar decades was an aberration arising from the "random shocks" of wartime and postwar developments, and not a fundamental departure from prewar trends. The baby boom is thus an oddity, of significance to policymakers no doubt, but of little importance in the longer-term history of population change. Some saw a parallel between the post-1918 baby boom and the one following the Second World War. Since the former gave way to then-record interwar low fertility rates within a short period of time, it was logical to believe that the same pattern would prevail once the repercussions of the 1939–1945 war had passed.

The problem remained, though, that the arrival of postponed births after 1945 constituted only the beginning of the surge in postwar births. What could not be explained solely by reference to the war experience was why the prewar decline in fertility was arrested, and in a number of cases dramatically reversed over a 20-year period.

Much of the theoretical and empirical debate over the meaning of the postwar baby boom surrounds the American case. It is well established that the sharp increases in U.S. fertility derived principally from two major changes in behavior: (1) the shift toward earlier and more nearly universal marriage; and (2) a larger proportion of women having at least two children. While there were also increases in the number of families with three or four children, the number of very large families (i.e., five or more children) actually declined (Ryder, 1982, pp. 286–292).

Proposed explanations have included changes in sociocultural norms surrounding childbearing, social-psychological effects of growing up during the Great Depression, and the effects of economic change between the 1930s and the 1950s (Bean, 1983, pp. 353–365).

Source: *ibid*, M.S. Teitelbaum, and J.M. Winter [71], pp. 70–71.

Nearly all scholars and researchers have held a preconception that "only the beginning of the surge" could not continue for an extended range of time. Therefore, they pursued conscious causes—e.g., "sociocultural norms surrounding childbearing, social-psychological effects of growing up during the Great Depression, and the effects of economic change."

This direction is false for two primary reasons. The first is that the explanations are merely eclectic and provide no unified perspective of marriage occurrences. A meaningful theory of the marriage function would have no aberrations nor unconditional fundamental trends, and all phenomena would be described by the equation with minimum divisions. Consider the attribution of two different phases of the baby boom, i.e., the phases of early and late marriage tendencies to conscious will. This would necessitate the consideration of ad hoc plausible conscious motivations and their interactions with socioeconomic changes. This is not a reasonable explanation. When theories are based on these considerations, demographers elucidate the reoccurrence of demographic events in human history using every plausible explanation added later to explain the changes. This can be seen in the theories of the "first demographic transition" and "second demographic transition."

The second reason is that these explanations are not falsifiable. Once the cause is attributed to human conscious will, we need to possess the conscious will necessary to bring about the result. Thus, the predictability, and hence the falsifiability, is lost.

5.2.2.1 Boom or Boomlet

The difference between the boom and a boomlet should be reasonably explained. It is essentially the quantity of surges in marriage. In Equation (5.2), the durability and size of the boom are determined by $\int_0^{a_i} F(a_i)(da_i)$ at a certain time point.

Beginning in 1945, a significantly high number of marriages of individuals of marriageable age occurred in the United States for two years. This is because the triumphant soldiers could return. In Germany, on the other hand, as few soldiers could return to their "Vaterland," there was no increase in the number of marriages. This is the only difference.

I shall demonstrate this by a numerical simulation. Because Equation (5.2) is a type of diffusion equation, it has a smoothing effect. In a diffusion equation system, the protuberance caused by the outside power (shock) is smoothed with the passage of time, and its magnitude and continuity are determined by the initial magnitude of the protuberance.

I simulated the marriage function system for several initial magnitudes of protuberance. For this simulation, it was considered that female resources are always ample during and after wartime in European countries and the United States. In other words, addition and subtraction need not be considered rigorously.

Simulation plot

I chose Swiss data for the initial simulation. The statistical marriage data of Switzerland could be prepared easily, and the marriage function of the Swiss cohort was already analyzed in a previous chapter (p. 96).

- Begin the marriage probabilities at 1941 as 1945.
- For the age group of 17–25, the marriage probabilities increased by 1.2 times in 1946.
- During the baby boom, male resources were plentiful[2], and therefore $1 - \int F(a)(da)$ was not equal to their male counterparts. Specifically, the assumption of male and female symmetry in the marriage function theory is temporarily abandoned[3]. Because of this exceptional measure, in 1947, the marriage probability for the age group of 18–26 increased by 1.1 times.
- The protuberance occurred for only one year. For the subsequent years, the marriage function entirely followed the difference-integral equation (5.2).

Figure 5.12 shows the result of the simulation of the longitudinal dynamics of the marriage function, which causes the one-year protuberance. This elucidates "the shift toward earlier and more nearly universal marriage" and explains the end of the baby boom as an inevitable conclusion. A one-year surge may continue for years in the world of the difference-integral Equation (5.2).

These slight increases in marriages over several years in U.S. may lead to a chain reaction of birth behaviors[4]. It is natural that the earlier women marry, the more children they will bear. Thus, the phenomenon of the "increase in the number of families with three or four children" can be naturally explained.

[2]If it were not for this ampleness of male resources, the baby boom phenomenon would not have occurred.

[3]I believe that rigorous monogamy is not ideal for human reproduction.

[4]To explain the birth behaviors, I will provide a deliberate, theoretical, and in-depth consideration of the birth function in later chapters.

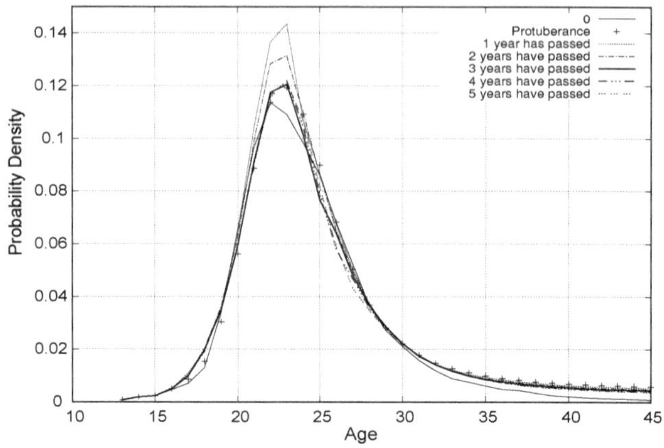

Figure 5.12: First baby boom simulation of Swiss.

5.2.2.2 Baby Boom as a Chain Reaction

My hypothesis is that the baby boom was the result of a chain reaction. This view answers the question of why booms or boomlets occurred in different countries.

Table 5.1: Number of individuals in military service and war casualties.

Country	Number Served	Killed/Missing	Percent Killed
United States	16,353,639	407,316	2.5
Germany	18,200,000	5,318,000	29.2
Japan	8,400,000	2,121,955	25.3

Source: *World Almanac and Book of Facts 2007* [1].

According to the *World Almanac and Book of Facts 2007* [1], during World War II, about 16 million American soldiers were mobilized, and about 15 million of them returned to the United States after several years. This massive influx of marriageable males created a large and continuous protuberance, immensely increasing marriage probabilities. This chain reaction continued for a long time. In contrast to the United States, in Germany and Japan, the majority of the mobilized soldiers did not return to their homeland. Thus, the protuberance was low, and the marriage probabilities increased only slightly, resulting in boomlets.

Consider the case in which the boom is prolonged for one more year. This simulation is presented in Figure 5.13. One more year prolonged chain reaction for more four years long.

In any case, the past analysis of the baby boom from the 1960s proposed that industrial development (alteration) was related to the boom, and its logical consequence was that our conscious reaction to this development might have caused the boom. This is not scientific and logical reasoning but faith in

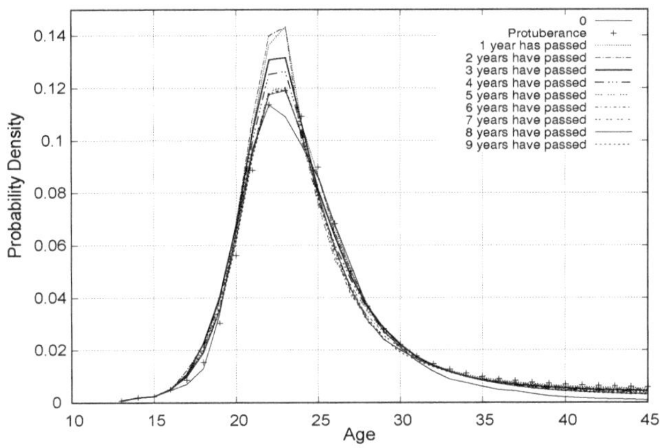

Figure 5.13: Second baby boom simulation of Swiss.

human nature. Since the nineteenth century, it was implicitly assumed that industrialization has been the motivating force of social change. This implicit assumption is a bourgeois ideology (also shared by K. Marx), which should be abandoned. Human nature must be inferred through empirical data, and not by self-consciousness.

E.A. Wrigley, a prominent historical demographer, rightfully admitted the crucial importance of time and place. He even stated the following:

> No demographic model which seeks to show something of the complexity of the relationships between the demographic characteristics and other social and economic features of a society can be successful unless it does justice to the peculiarities of place and time.
>
> Source: *Population and History*, E.A. Wrigley [83], p. 106.

It is time to accept the relative independence of demographic phenomena and the precise effects they bring onto social and economic factors.

As for marriage occurrences, the postulation that we depend on neighboring marriage occurrences can lead to a more integrated, falsifiable, and metrical view.

5.2.2.3 Boomlet or No Boom Following World War I

It is known that following World War I, either small boomlets or no boom occurred in European countries and the United States. The difference-integral Equation (5.2) can explain these facts.

In Germany, a very small baby boom (boomlet) was observed. During the 1920s, the TFR suddenly rose in contrast to its low during wartime. World War I resulted in an enormous loss of life compared to World War II. Even Allied countries such as the British empire and France sacrificed about at least 15 percent of their mobilized soldiers. When the wounded are included, this value jumps to 40 percent. The central powers, especially their leader Germany, sacrificed even more. At that time, the protuberance of Germany was so small that it could last for only a single year. For the United States, wartime mobilization lasted for about a year and a half at maximum. In that situation, the number of prolonged marriages was not high.

Furthermore, during this period, European countries and the United States were in the last stage of fertility decline. In particular, the number of children per family was declining. In such a situation, marriage increase had little effect on fertility.

5.3 Extension to Geographical Dimensions

5.3.1 Purposes of Extension

As discussed in Chapter 1, the natural extension to the geographical dimensions provides the dynamics of the number of children per family. The extension to the age dimension already provided the dynamics of the marriage function. The rightful definitions of the neighboring effects naturally reveal the dynamics of the phenomena. By adding geographical neighboring effects, a more prolific marriage function and its dynamics may be obtained. This is one purpose of the extension.

Another purpose is to explain the observed phenomena. For the marriage occurrences, as shown in Figures 5.1–5.10, a geographical dynamic alteration was definitively observed in Japan. The importance of geographical differences during the industrial revolution was also pointed out by the famous and prominent historical demographer E.A. Wrigley.

> A second consideration which helps to underline the diversity of demographic change at the time of the industrial revolution is in a sense an extension of the first. It is essential to recognize that during the early decades of the industrial revolution in England and in other countries, the economic changes were local or at most regional phenomena. Employment structures and habits of life were revolutionized only in small areas at first. In time, whole countries were caught up in the process, but the leaven took several decades to affect the whole loaf. Local demographic changes were often the results of purely local situations. To deal with national aggregates can then be highly misleading. The chronology of change was local, too. The onset of rapid change varied considerably from area to area in demographic history just as in the history of industrial change, and developments big enough in scale to affect national aggregates occurred much later than the first local changes.
>
> At the local scale of analysis, however, simple relationships between economic and demographic change are no longer easy to find. It may be plausible to attribute the sudden alteration in

> the balance of baptism and burial in, say, Worcestershire or Lancashire to the changes in train in the local economies. It is more difficult to do so in a parish like Hartland, where the change was also marked and began about 1750, as early as in many industrial areas. Hartland was far from the small centers of new industrial growth. Its economic base did not begin to change for many years after the sudden appearance of substantial baptism surpluses.
>
> Source: *Population and History*, E.A. Wrigley [83], pp. 154–156.

As Wrigley suggested, the alterations in marriage probabilities were indirectly brought about by economic development. However, these alterations also occur without economic bases. To bring these possibilities into the theory of the marriage function, the marriage function must be extended to geographical dimensions. This is an extension of the high (three) dimensional space in order to explain the overall environment.

5.3.2 Marriage Function as a Hypersurface

The "age(a)" and "geographical(x, y)" dimensions are necessary to describe marriage occurrences. Namely, the marriage function is a four-dimensional hypersurface $F(a, x, y)$. It is considered to be changing with respect to time t.

Why are other dimensions not required? Almost all demographers, sociologists, and economists suppose that socioeconomic variables (conditions) affect the marriage probability of each individual. However, I recommend the abandonment of these preconceptions, because socioeconomic variables, such as income, education, and social class, are incapable of quantitatively predicting future marriage probabilities.

There is still another reason that these conscious variables are almost always not the actual reason behind individuals remaining unmarried. Higher-income individuals are inclined to marry later than lower-income individuals. The fact is that neighboring high-income individuals are not yet married. Human conscious will is created based on neighbor interactions.

5.3.2.1 Formulation

I assume isotropy of the geographical dimensions. From any direction, an equal effect is obtained. Consequently, the marriage probability at $F(a_k, x_i, y_j, t_{\ell+1})$ is an injective mapping from $F(a_{k-1}, x_i, y_j, t_\ell), F(a_k, x_i, y_j, t_\ell), F(a_{k+1}, x_i, y_j, t_\ell)$ by the following operator—NDSMDE(Neighboring Dependent Stochastic Marriage Difference Equation) in a lattice space.

Isotropy

"Isotropy" indicates that every direction except the age direction is equal and there is no direction unique to the marriage occurrences. Essentially, the effect is considered equally in every direction. Therefore, a constant diffusion coefficient μ can be assumed for every direction.

$$
\begin{aligned}
&F(a_k,x_i,y_j,t_{\ell+1}) \\
&= \lambda\Bigg[\Bigg(\mu\int_0^{a_{k-1}} F(a_{k-1},x_{i-1},y_j,t_\ell)(da)+\mu\int_0^{a_k} F(a,x_{i-1},y_j,t_\ell)(da) \\
&+ \mu\int_0^{a_{k+1}} F(a_{k+1},x_{i-1},y_j,t_\ell)(da) \\
&+ \mu\int_0^{a_{k-1}} F(a_{k-1},x_{i+1},y_j,t_\ell)(da)+\mu\int_0^{a_k} F(a,x_{i+1},y_j,t_\ell)(da) \\
&+ \mu\int_0^{a_{k+1}} F(a_{k+1},x_{i+1},y_j,t_\ell)(da) \\
&+ \mu\int_0^{a_{k-1}} F(a_{k-1},x_i,y_{j-1},t_\ell)(da)+\mu\int_0^{a_k} F(a,x_i,y_{j-1},t_\ell)(da) \\
&+ \mu\int_0^{a_{k+1}} F(a_{k+1},x_i,y_{j-1},t_\ell)(da) \\
&+ \mu\int_0^{a_{k-1}} F(a_{k-1},x_i,y_{j+1},t_\ell)(da)+\mu\int_0^{a_k} F(a,x_i,y_{j+1},t_\ell)(da) \\
&+ \mu\int_0^{a_{k+1}} F(a_{k+1},x_i,y_{j+1},t_\ell)(da) \\
&+ \mu\int_0^{a_{k+1}} F(a_{k-1},x_i,y_j,t_\ell)(da)+\mu\int_0^{a_{k+1}} F(a_{k+1},x_i,y_j,t_\ell)(da)\Bigg) \\
&+ (1-14\mu)\left(\int_0^{a_{k+1}} F(a_k,x_i,y_j,t_\ell)(da)\right)\Bigg]\left(1-\int_0^{a_k} F(a_k,x_i,y_j,t_\ell)(da)\right)^2
\end{aligned}
\tag{5.3}
$$

When this formalization is made[5], we should have a geometric image of the marriage function. The marriage function is a geometric object, and its gradient is significant for its analysis. By availing the theorems of geometry and mapping, the end of the dynamic process can be predicted.

On writing the formulation, I realized afresh the number of factors that affect marriage probabilities, and how the probabilities of marriage occurrences relate to each other. This is the true nature of the social order. Countless sociologists have insisted that social interaction is the main process of society. In the present work, the interaction is formulated more concretely than ever before.

5.3.3 Marriage Function in Tokyo

Readers may doubt the existence of a marriage function as a hypersurface. There is a precise hypersurface; I will present it for Tokyo. In Japan, the late-

[5]In my book [35] (*The Mathematics of Marriage Function*), I proposed an incorrect marriage function that had a reaction term to the Equation (5.3). I was influenced by the reaction-diffusion model of the number of children per family that I had proposed earlier. I supposed that the reaction term was a substitute for the micro process of the gradient of the high-dimensional space.

marriage trend was most prominent in the Tokyo region. Within Tokyo, there exists a regional undulation of marriage probabilities. Although the marriage function $F(a,x,y)$ is actually a hyperspace, the integral function of $F(a,x,y)$, namely, $\int_0^a F(a,x,y)(da)$, is more visually revealing in itself. The integral function $\int_0^a F(a,x,y)(da)$ is a four-dimensional undulation, and a marriage function is a thin four-dimensional membrane on it.

However, humans cannot visually perceive the four-dimensional world. Therefore, to observe the undulation of $\int_0^a F(a,x,y)(da)$, the degree of dimensions must be reduced. For this, the four-dimensional world is cut at $y = 0$ (the local coordinate of the Tokyo area set herein) to first show $\int_0^a F(a,x,y)$ for each age-specific representative.

5.3.3.1 Fit Marriage Function to Observed Data

The following lattice space (Table 5.2), which is composed of the territories of local public organizations in Tokyo, is considered.

Although there are some differences between the lattice space and the geographical map (Figure 5.14), the lattice space reliably reflects the neighboring spatial order of the real geographic space.[6]

The starting time point is set at 1965, and $F(a,x,y)$ and $\int F(a,x,y)(da)$ for Tokyo between 1975 and 2005 are calculated using Equation (5.3). These are then compared with the observed data.

In a sense, the difference-integral Equation (5.3) is simple. The concrete calculation for the real observed data is more complicated in its concrete procedure. The method used to calculate these is explained using an example.

It is somewhat surprising that the difference-integral Equation (5.3) is still valid in this large-scale lattice space (about 4 km × 4 km square) and large time interval. I was frequently deluded by the idea that the fit of the equation to the observed data was common or accidental. I was unable to believe the difference-integral Equation (5.3).

The marriage function dynamically changes itself to lower the probability of marriage at a younger age and simultaneously improves the probability of marriage at an older age. A simple mapping, such as that to multiply a coefficient, cannot map these dynamics. The integral and difference functions for the age dimension and the geographical space dimensions must be requested for these dynamics.

Indeed, for some cases, the difference-integral Equation (5.3) failed to calculate (predict) valid marriage probabilities. However, I speculate that these relatively humble deviations can be attributed to socioeconomic factors or geographical mobility. Somewhat apologetically, I will explain in depth the difference between the observed and theoretically expected/calculated marriage probabilities in Tokyo. For the time being, mobility will be considered as the enemy of the demographic function theory.

[6] Although Sagamihara city belongs to the Kanagawa prefecture, it has been adopted here in order to reflect the neighboring spatial order. Cities of the same name are duplicated for the same reason.

Table 5.2: Lattice space supposed as Tokyo Area.

Akiruno	Futsa	Higashiyamato	Kodaira	Kodaira	Nishitokyo	Nerima	Nerima	Toshima	Bunkyo
Hachiouji	Akishima	Tachikawa	Kokubunji	Koganei	Musashino	Suginami	Nakano	Shinjyuku	Chiyoda
Sagamihara	Hachiouji	Hino	Kunitachi	Fuchu	Mitaka	Setagaya	Setagaya	Sibuya	Minato

Figure 5.14: Geographical Map of Tokyo.

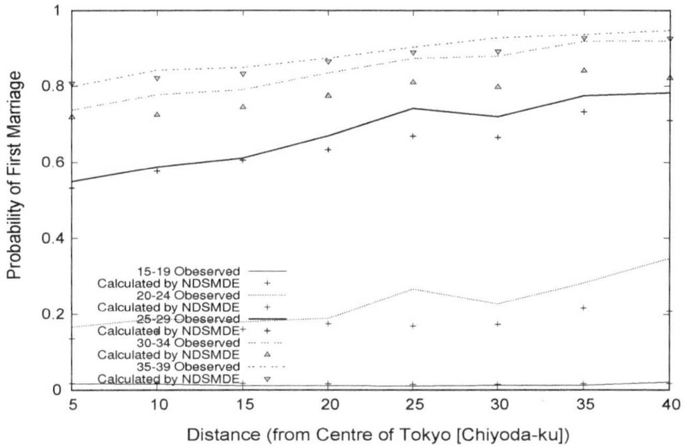

Figure 5.15: $\int F(a,x,y)(da)$ at Shinjyuku, Nakano, Suginami, Musashino, Koganei, Kokubunji, Tachikawa, and Akishima, calculated by NDSMDE and observed in 1975.

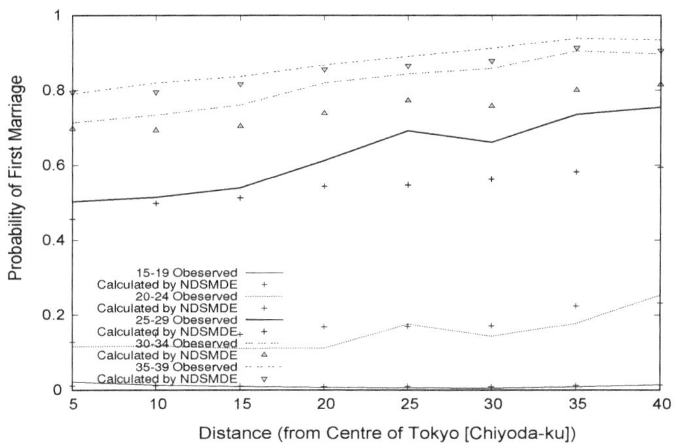

Figure 5.16: $\int F(a,x,y)(da)$ at Shinjyuku, Nakano, Suginami, Musashino, Koganei, Kokubunji, Tachikawa, and Akishima, calculated by NDSMDE and observed in 1980.

5.3.3.2 1970–1975

For instance, from 1970 to 1975, predictions by $\int F(a,x,y)(da)$ underestimated the observed values in the outer suburbs of Tokyo. In contrast, for the central districts of Tokyo, the predictions are nearly perfect fits. During the 1970s, there were large movements of families to the outer suburbs of Tokyo. This large population inflow might account for this divergence.

It is worth noting that from 1970 to 1975, the gradient of the late-marriage trend already existed in the hypersurface of Tokyo. The late-marriage trend has been a collective neighbor-dependent phenomenon that could not be denied.

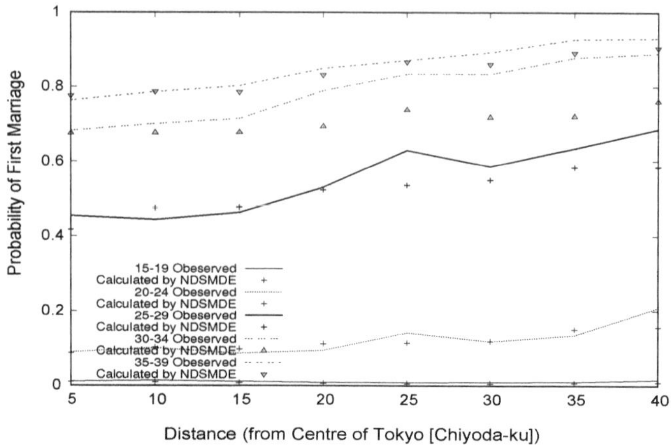

Figure 5.17: $\int F(a,x,y)(da)$ at Shinjyuku, Nakano, Suginami, Musashino, Koganei, Kokubunji, Tachikawa, and Akishima, calculated by NDSMDE and observed in 1985.

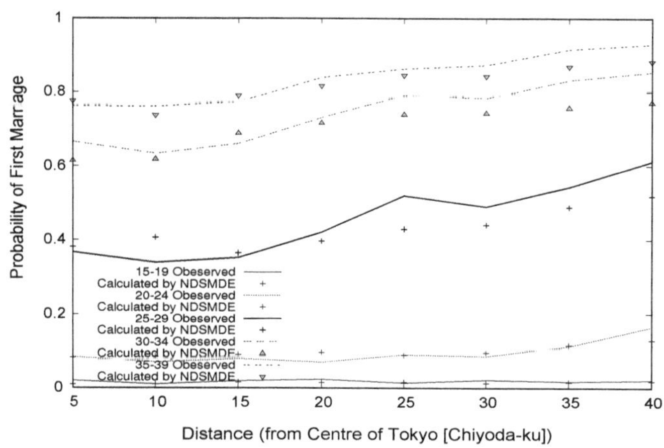

Figure 5.18: $\int F(a,x,y)(da)$ at Shinjyuku, Nakano, Suginami, Musashino, Koganei, Kokubunji, Tachikawa, and Akishima, calculated by NDSMDE and observed in 1990.

Marriage occurrences are described by an integral equation, and therefore lower limits exist. At what point of time does the marriage function reach that lower limit? Additionally, how long does this take?

5.3.3.3 1975–1980

For the period 1975–1980, the marriage function produces a fairly fine fit to the observed values. This period marks the prelude to "Japan's economic bubble." The Japanese economy was developing auspiciously in those days, and the marriage probability of the age group 25–29 was a few percent over the calculated values.

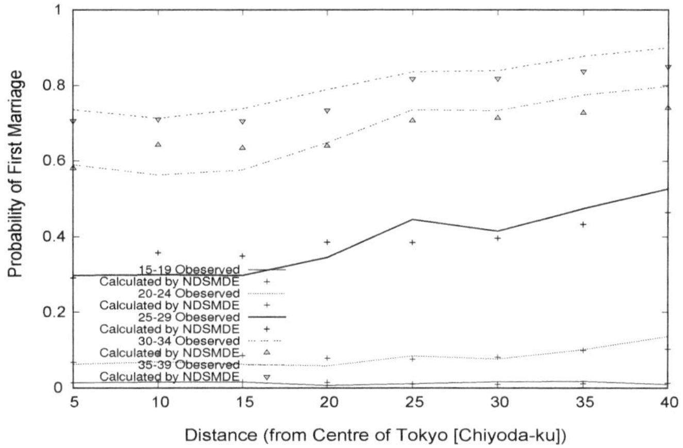

Figure 5.19: $\int F(a,x,y)(da)$ at Shinjyuku, Nakano, Suginami, Musashino, Koganei, Kokubunji, Tachikawa, and Akishima, calculated by NDSMDE and observed in 1995.

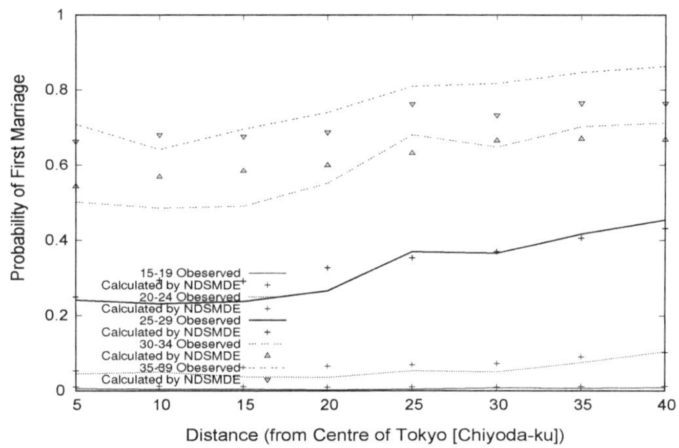

Figure 5.20: $\int F(a,x,y)(da)$ at Shinjyuku, Nakano, Suginami, Musashino, Koganei, Kokubunji, Tachikawa, and Akishima, calculated by NDSMDE and observed in 2000.

These excess values are caused by the effect of economic development on marriage occurrences. It should be noted that these economic effects are merely temporal, and the gradient of the late-marriage trend in the hypersurface is permanent and more potent. Comparing Figure 5.24 to Figure 5.27, the probability of young (age 15–20) marriage had already declined in the latter half of the 1970s.

5.3.3.4 1980–1985

In 1985, Japan was in the middle of an economic bubble. Nearly nobody knew that we were approaching the burst of this bubble and low probabilities of marriage occurrences. The decline in the probability of young marriages had

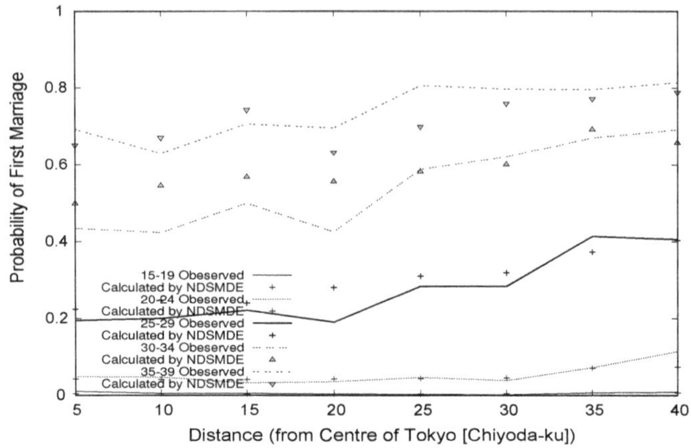

Figure 5.21: $\int F(a,x,y)(da)$ at Shinjyuku, Nakano, Suginami, Musashino, Koganei, Kokubunji, Tachikawa, and Akishima, calculated by NDSMDE and observed in 2005.

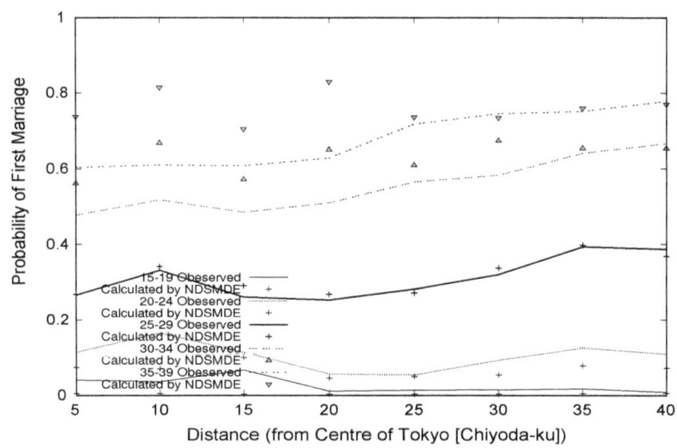

Figure 5.22: $\int F(a,x,y)(da)$ at Shinjyuku, Nakano, Suginami, Musashino, Koganei, Kokubunji, Tachikawa, and Akishima, calculated by NDSMDE and observed in 2010.

accelerated. Thus, $\int F(a,x,y)(da)$ had already declined in older generations. The present-day low fertility rate in Japan was laid out in ancient days, and it finally revealed itself to us appreciably, although almost no one had recognized it as yet.

5.3.3.5 1985–1990

Following the burst of the bubble, marriage probabilities fell short of the expected values predicted by the theory of the marriage function.

As you have already understood, the economic effect on fertility has to be adjusted sooner or later. The probability of early marriage declined further.

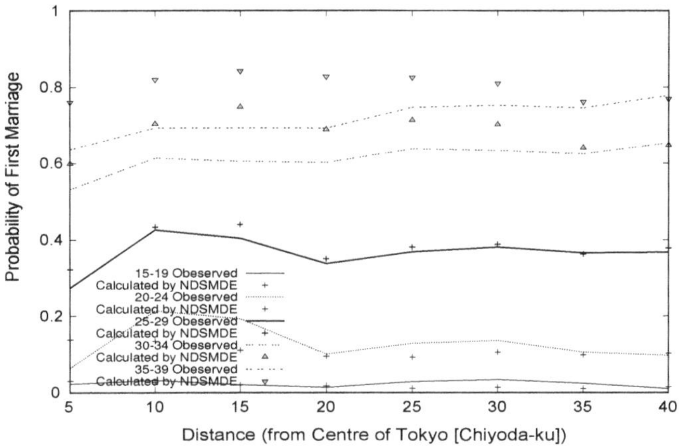

Figure 5.23: $\int F(a,x,y)(da)$ at Shinjyuku, Nakano, Suginami, Musashino, Koganei, Kokubunji, Tachikawa, and Akishima, calculated by NDSMDE and observed in 2015.

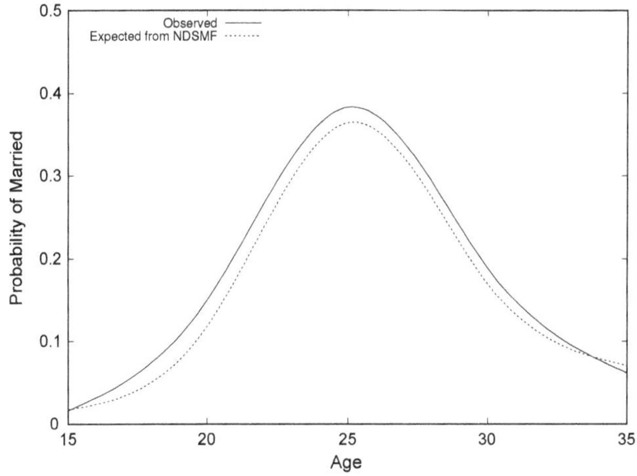

Figure 5.24: $F(a,x,y)$ at Shinjyuku-ku calculated by NDSMDE and observed in 1975.

The gradient of $\int F(a,x,y)(da)$ was more essential to marriage occurrences than the socioeconomic phenomenon of the "bubble."

5.3.3.6 1990–1995

From 1990 to 1995, the late-marriage trend had propagated significantly throughout Japan. Pro-natalist policies were examined and did not come into effect at this time.

It was not until the manifestation of low fertility was perceived that we attributed it to our conscious will. This is because we were unconscious of the true causes determining those actions. These true causes are described by the

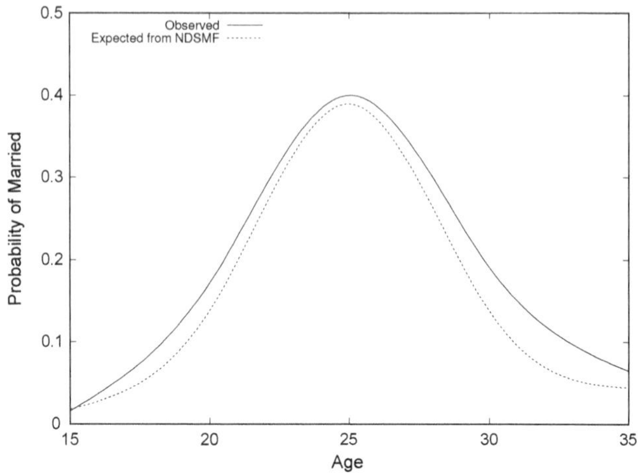

Figure 5.25: $F(a,x,y)$ at Nakano-ku calculated by NDSMDE and observed in 1975.

difference-integral Equation (5.3) and its law of neighboring dependence. It is evident that marriage probabilities presented a fine fit to Equation (5.3).

What governs marriage occurrences is not conscious will but neighboring marriage occurrences as formulated by the difference-integral Equation (5.3). It is a failure on our part that we have only considered conscious will as the cause. As a matter of course, a set of Japanese pro-natalist policies, called the "Angel Plan," exhausted about 6000 million dollars, but it had no effect on fertility rates. What a waste!

What must be noted is that Shinjyuku-ku might have already arrived at the lower limit of the marriage function. Namely, the existence of fixed points can be induced. Therefore, it is concluded that the late-marriage trend has a saturation point; it does not continue indefinitely.

When the existence of non-zero fixed points is assumed, its stability is easily proven.

Lemma 5.1 Existence of Fixed Points

If the points were zero for all $F(a,x,y)$, the integral equation would not develop. Thus, when they remain zero, it is a contradiction to reality. Further, if they were not fixed points, there would be fewer $F(a,x,y)$ in the space. The difference-integral equation inevitably moves towards them, and this would lead to a chain reaction by which all $F(a,x,y)$ would be zero. This contradiction shows that non-zero fixed points of the marriage function exist in the space.

5.3.3.7 1995–2000

It can be seen that between 1995 and 2000, the low fertility of central Tokyo had been diffusing into the suburbs of Tokyo. $\int_{15}^{25\ 29} F(a,x,y)(da) \doteq 0.2$, which is the lower limit of the probability of ever-married individuals in the age group 25–29, was diffusing.

The observed data are below the values predicted by the equation. In the latter half of the 1990s, large corporations and banks became bankrupt, and economic recession depressed marriage probabilities temporally.

5.3.3.8 2000–2005

Between 2000 and 2005, the late-marriage trend proceeded exactly as calculated by the equation. Tokyo entered a stage of low fertility.

5.3.3.9 2005–2010

Around 2010, Tokyo entered the state of the lowest marriage function. Some communities recorded about 60% marriage occurrence for the age group of 35–39, which is the lowest percentage of ever-married individuals in Japan.

5.3.3.10 2010–2015

On comparing Figure 5.26 and Figure 5.35, a dramatic alteration of the marriage function can be observed. It can be considered that a smoothing of the hypersurface occurred, due to which the isotropy of the marriage function could be observed near the fixed points throughout. If Japan wants to recover from the present-day low fertility rate, these very low marriage probabilities must be raised first.

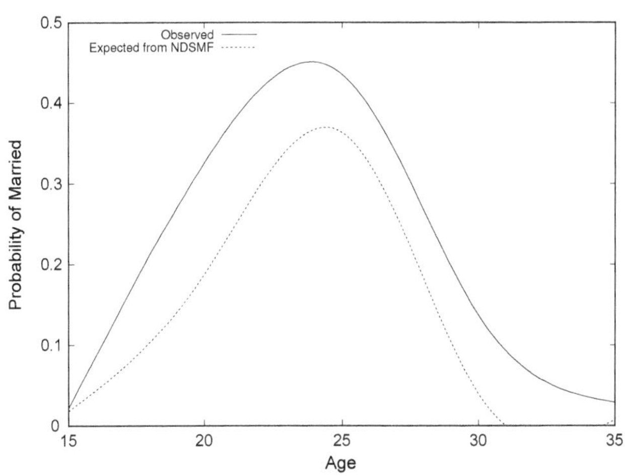

Figure 5.26: $F(a,x,y)$ at Akishima-shi (suburb of Tokyo) calculated by NDSMDE and observed in 1975.

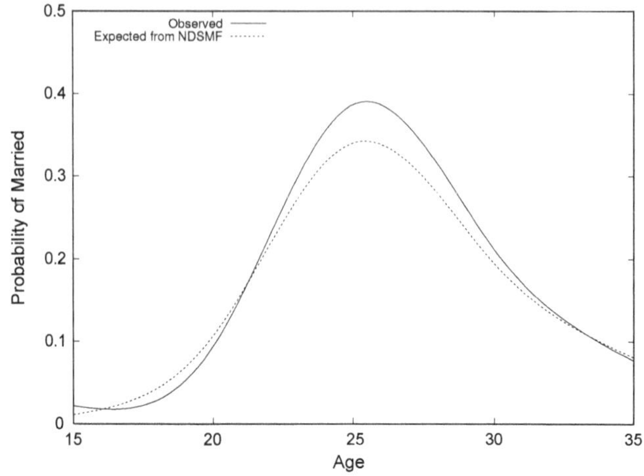

Figure 5.27: $F(a,x,y)$ at Shinjyuku-ku calculated by NDSMDE and observed in 1980.

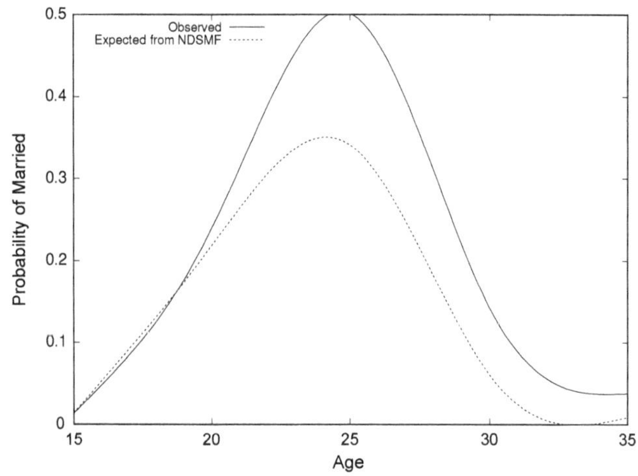

Figure 5.28: $F(a,x,y)$ at Akishima-shi (suburb of Tokyo) calculated by NDSMDE and observed in 1980.

5.4 Drawing the Three-Dimensional Surface via Interpolation

Using the interpolation method, the age-specific $\int F(a,x,y)(da)$ is estimated for each community. Further, the three-dimensional surface $\int F(a,x)(da)$, which is composed of a single geographical dimension x (degraded to omit the y dimension) and the age dimension a, is obtained. Figures 5.36 and 5.37 illustrate the three-dimensional plot.

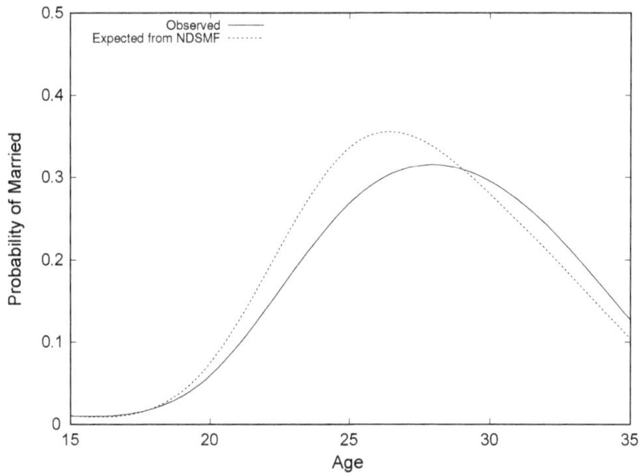

Figure 5.29: $F(a,x,y)$ at Nakano-ku calculated by NDSMDE and observed in 1990.

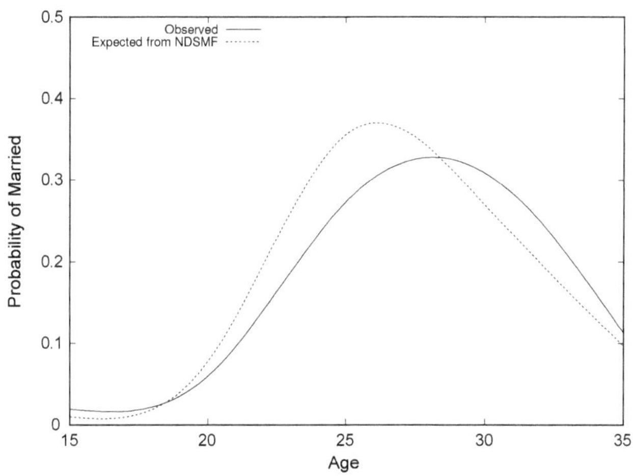

Figure 5.30: $F(a,x,y)$ at Suginami-ku calculated by NDSMDE and observed in 1990.

It is meaningful that the three-dimensional surface of $\int F(a,x)(da)$ can be visualized first. The two-dimensional gradient of the cumulative marriage probabilities to the low marriage probabilities of central Tokyo can be realized.

The actual marriage function is a four-dimensional hypersurface. It must be imagined considering a three-dimensional surface. Marriage occurrences depend on geographic and generational factors, and thus, every marriage is causally connected.

We do not marry because the majority of our neighboring cohort has not married yet, not because we desire not to marry. The volition to marry or not to marry is determined by neighboring marriage occurrences. Our desires to marry are constructed by the chain of marriage occurrences. Therefore, the

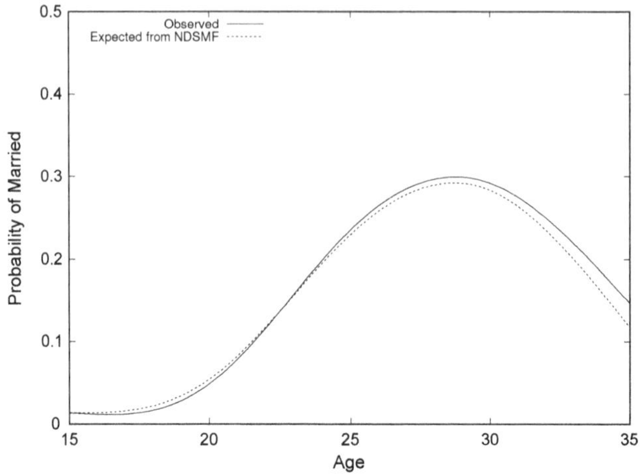

Figure 5.31: $F(a,x,y)$ at Shinjyuku-ku calculated by NDSMDE and observed in 1995.

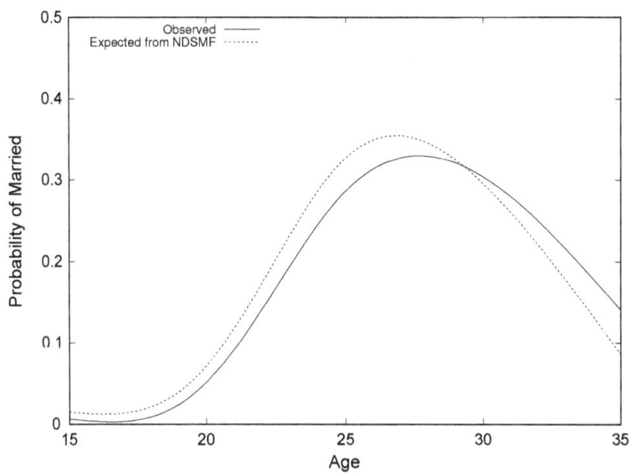

Figure 5.32: $F(a,x,y)$ at Musashino-shi calculated by NDSMDE and observed in 1995.

late-marriage trend has diffused geographically and across generations. The high-dimensional space is the optimal formalization of this phenomenon.

5.5 Time-Space Pattern Caused by Geographical Extension

When first considering the marriage function as a hypersurface, I supposed its flattening and smoothing in terms of the reaction-diffusion equation of the number of children. However, when it is extended to geographical dimensions,

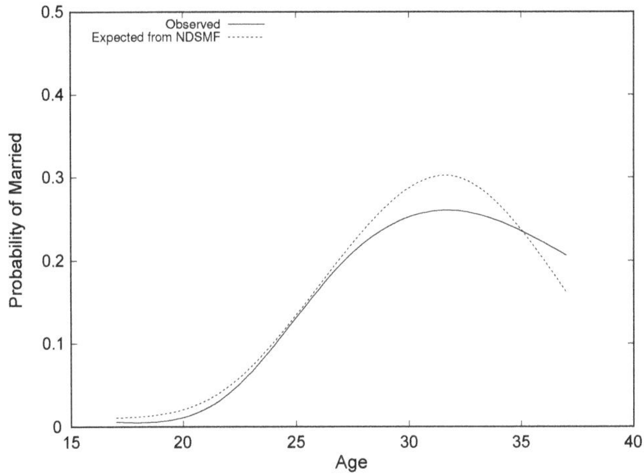

Figure 5.33: $F(a, x, y)$ at Shinjyuku-ku calculated by NDSMDE and observed in 2000.

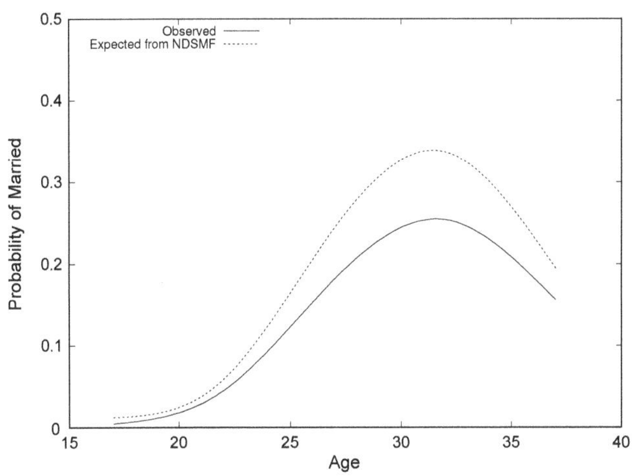

Figure 5.34: $F(a, x, y)$ at Nakano-ku calculated by NDSMDE and observed in 2000.

the difference-integral function has a peculiarity that I could not conceive. This peculiarity becomes particularly apparent when we consider the difference-summation function in the discreet world.

5.5.1 Fundamental Instability

The marriage function as a cohort is stable in itself. On the contrary, the high-dimensional extended marriage function is essentially dynamic and unstable. When uniform marriage probabilities are set for the geographical dimensions for each age, the function changes dynamically. This is because there might have been gradients in the marriage probabilities of ages. The extension of the

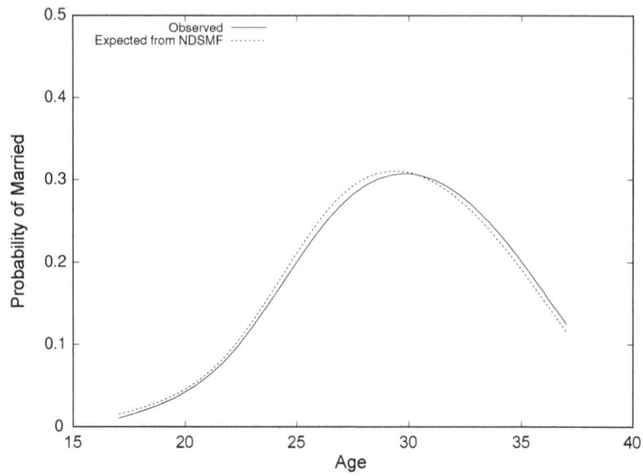

Figure 5.35: $F(a, x, y)$ at Akishima-shi (suburb of Tokyo) calculated by NDSMDE and observed in 2015.

marriage function to a multi-dimensional space fundamentally conceives the instability of the system, similar to some types of reaction-diffusion systems. The interactions between ages serve as intrinsic generators of the strange dynamics of the marriage function. Therefore, the coefficient λ in Equation (4.4) no longer has the same value as that for a cohort in the high-dimensional space.

The interaction between the speed of diffusion and the speed of development of the marriage function may cause a strange effect. The high-dimensional marriage function behaves like **the exothermic reaction-diffusion equation**. Combustion is a typical exothermic reaction-diffusion process. When we consider the phenomenon from a different perspective, marriage occurrences can be regarded as an interaction between married and unmarried groups. The unmarried react to the married as reactant, the increase in the marriage reaction gains intensity, and the decrease in the unmarried (i.e., the decrease in the reactant) reaction calms down. Thus, a time-space pattern can be expected in the hypersurface.

For a long time, I supposed that the false difference-integral function has a reaction term. This false supposition made me ignore the true figure of the marriage function. See Figures 5.36 and 5.37 again; regular geographical undulations of the marriage functions can be imagined. See Figures 5.15 to 5.22. I critically overlooked the undulations that presented themselves in the real figures of the marriage function.

Until this time, I neglected these undulations because I could not explain them. However, there are no plausible socioeconomic causes; the genuine mathematical peculiarity makes it appear identical to the Brittany peninsula case. Our consciousness does not understand the regularity of our behaviors; we must question it.

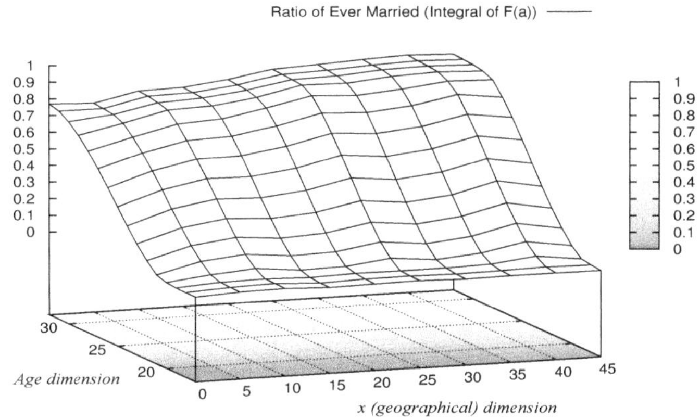

Figure 5.36: Proportion of ever-married individuals in Tiyoda-ku, Shinjyuku-ku, Nkakano-ku, Suginami-ku, Musashino-shi, Koganei-shi, Tachikawa-shi, Akishima-shi, and Hatiouji-shi in 1970, represented as a three-dimensional hypersurface.

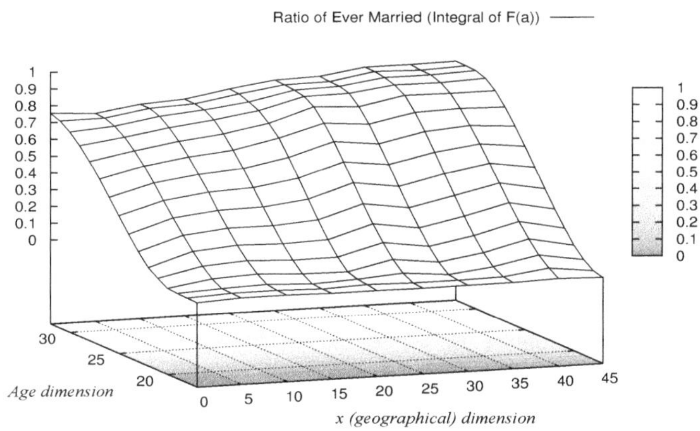

Figure 5.37: Proportion of ever-married individuals in Tiyoda-ku, Shinjyuku-ku, Nkakano-ku, Suginami-ku, Musashino-shi, Koganei-shi, Tachikawa-shi, Akishima-shi, and Hatiouji-shi in 1975, represented as a three-dimensional hypersurface.

5.5.2 *Brief Convergence of Patterned Undulations*

In general, reaction-diffusion systems, such as exothermic reaction-diffusion and undulations (patterns), last with alterations throughout as long as the reaction materials last. In the world of marriage occurrences, the reactant is severely limited, and it is exhausted within a short time. Therefore, the dynamic patterned undulations only reveal themselves for a short time, and their differences are so negligible that we cannot conceive them. Although they cannot be observed statistically now, I will report the strange patterns unique to the reaction diffusion system of the marriage function for my scientific interests.

5.5.2.1 Initial Heterogeneity

The time zero of marriage occurrences (about 13 years of age) may be governed by stochastic randomness. Geographical undulations were created using a normal random number. Other age–space marriage probabilities are geographically flat. Further, I traced the processes within the system.

See Figures 5.38 to 5.42. On considering the theoretical age of 100 (corresponding to 33 years old), it is strange that at about 85 time steps (depending on the initial condition), the undulated surface has been smoothed. In the following time step, unknown terraced fields are revealed, and then another uundulation patterns develops.

The previous terraced field surface was smoothed to the undulated surface. The dynamic alteration does not indicate an end. The steepness of the surfaces then increases for a certain period, and as a total surface, its probabilities continue to decline. Despite such potent diffusion power being active (the diffusion coefficient was set to $\sum \mu_i = 0.99$), the geographical differences of probabilities develop within a short period.

Being educated in the humanities and having studied social sciences for a long time, I would normally consider that socioeconomic conditions cause the geographical differences in the probability of ever-married status. In Figures 5.1–5.10, it can be seen that there are geographical undulations in the proportions of ever-married individuals in Japan's Kanto area. I had assumed that these undulations are brought about by differences in social conditions. However, I consider that they form genuine mathematical patterns depending on the initial perturbation. It is a bad practice to assume that every social phenomenon has socioeconomic and conscious reasons. The genuine complexity of geographical conditions and reaction-diffusion has already been observed in the Brittany peninsula.

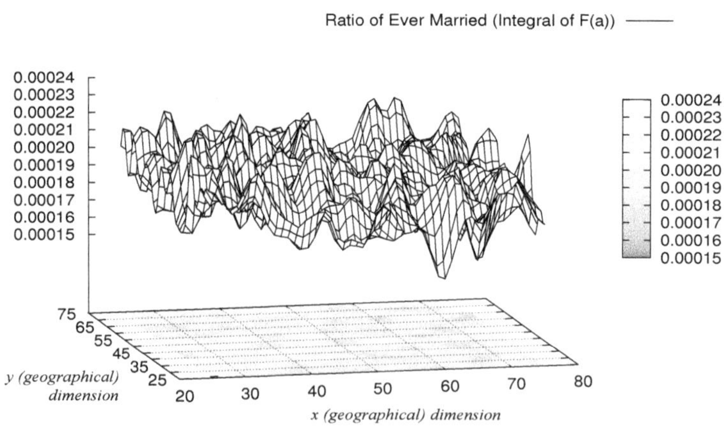

Figure 5.38: A25 at time 1.

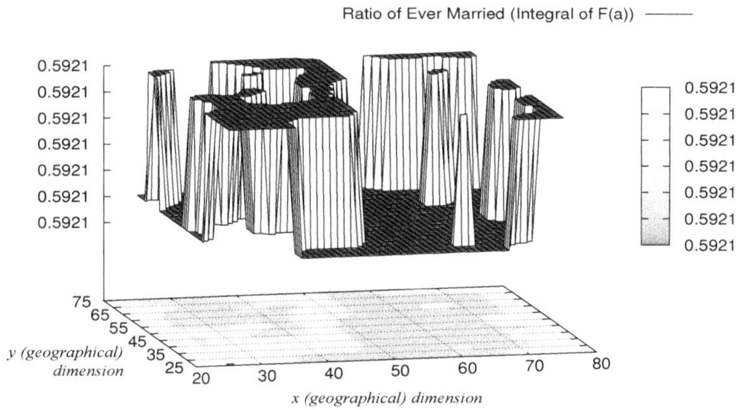

Figure 5.39: A100 at time 85.

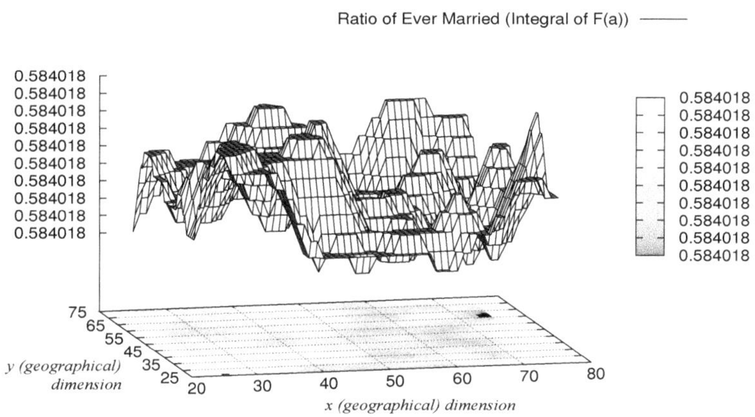

Figure 5.40: A100 at time 86.

5.5.2.2 Pattern Dynamics

When we begin from the flat (geographical) surfaces of the marriage function, if any undulations are added initially, complex patterns develop in the system. I believe that because initial random occurrences of marriages are inevitable, strange pattern development is unavoidable. Therefore, marriage probabilities are more uncertain in the extended age–space dimensional space than in the one-dimensional space.

The period of this strange development is relatively short. For a theoretical age of 25 (corresponding to 18 years old), strange patterns begin at one theoretical time step. They approach fixed points following 25 theoretical time steps.

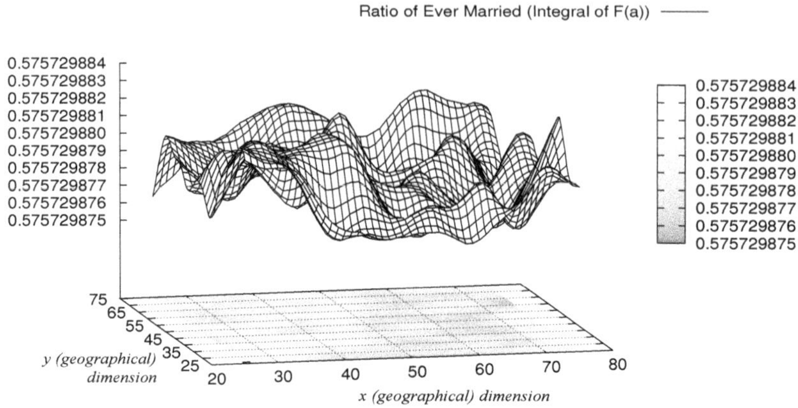

Figure 5.41: A100 at time 87.

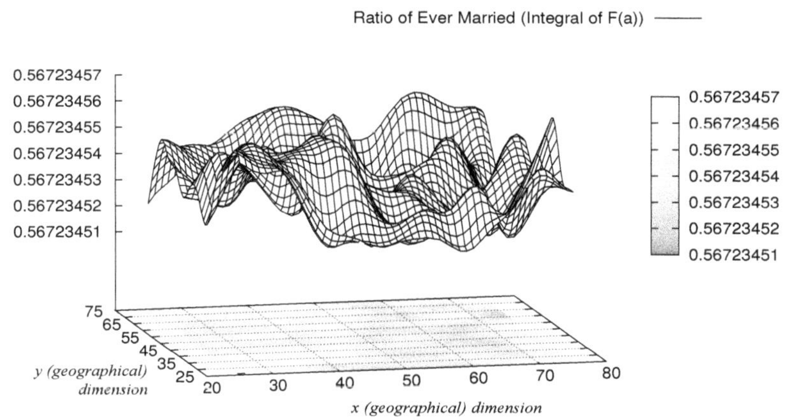

Figure 5.42: A100 at time 88.

For a theoretical age of 50 (corresponding to 23 years old), strange patterns begin at 39 theoretical time steps. At 55 theoretical time steps, they have already approached fixed points.

For a theoretical age of 150 (corresponding to 43 years old), strange patterns begin at 135 theoretical time steps. At 160 theoretical time steps, they have already approached fixed points. Meanwhile, the surface becomes steep, maintaining isomorphism. In this manner, the marriage function behaves like other reaction-diffusion systems. These pattern formations can be considered to depend on μ and λ. However, for the human marriage function, it was evident that μ is quite large, and further analysis was not conducted.

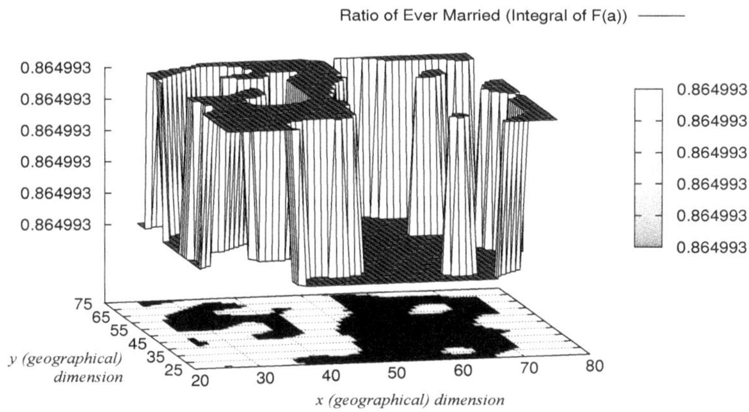

Figure 5.43: A150 at time 135.

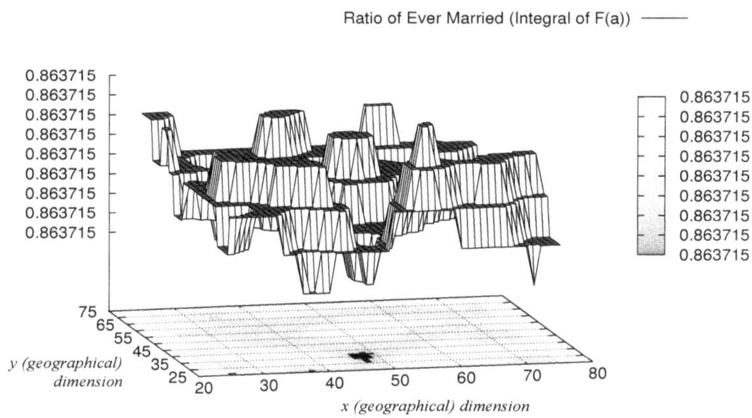

Figure 5.44: A150 at time 136.

5.5.2.3 Qualitative Analysis of the Alteration of the Hypersurface

These temporal sharpening of the hypersurface is caused by a mechanism that is active in the exothermic reaction-diffusion system. The proportion of ever-married individuals is a type of thermal energy that activates the diffusion mechanism reaction. The areas in which the proportion of ever-married individuals is high generate larger $F(a, x, y)$ through their chain interaction. In contrast, the areas in which the proportion of ever-married individuals is relatively low are less activated in their chain interaction. Thus, differences in the proportions of ever-married individuals develop between areas. This is the mechanism of the sharpening of the hypersurface.

As these reaction-diffusion processes take place, noncombustible matter (the ever-married population) accumulates. Therefore, the system is inacti-

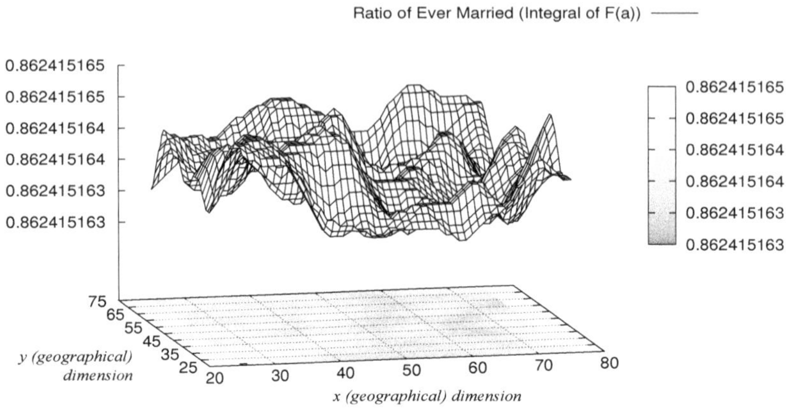

Figure 5.45: A150 at time 137.

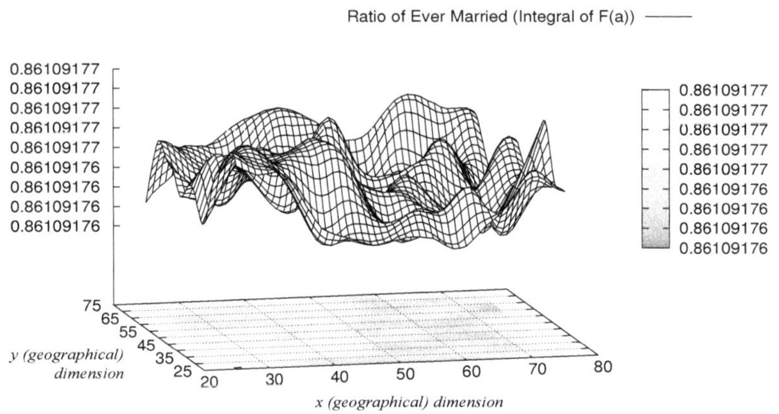

Figure 5.46: A150 at time 138.

vated, and from one time step to the next, $F(a,x,y,t) \rightarrow F(a,x,y,\acute{t})$ mappings switch to contraction mapping.

5.5.3 Fixed Undulated Surfaces Remain

Over time, these movements stabilize, and smoothing begins at the fixed points. However, the surfaces remain undulated despite the active diffusion power. As I mentioned before, it is surprising that the flattening does not proceed in a one-sided manner. The geographical differences in the proportions of ever-married individuals may exist due to this mechanism.

I regard the postulation of the initial random occurrences of marriages of teenagers as reasonable. The undulation patterns remain in the space. These perturbations occur continuously, and small undulations are observed at every surface of the marriage function. Although the marriage function can be com-

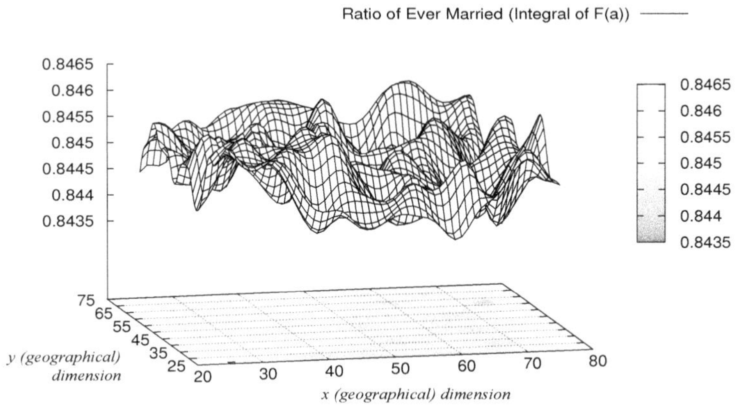

Figure 5.47: A150 at time 150.

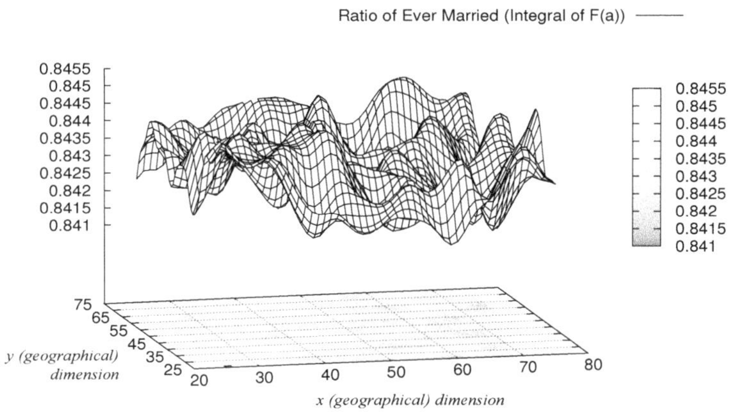

Figure 5.48: A150 at time 160.

puted based on the assumption of global asymptotic stability, uncertainties are supplied by initial perturbations, and the patterns are constantly changing.

Every age has the same pattern of fixed points. This isomorphism of the hypersurface of each age conforms to the observed data. This simple pattern preservation serves as evidence that marriage occurrences obey the difference–integral equation of marriages. It is not possible that our individual conscious selection forms these patterns voluntarily.

It is the marriage function established on the principle of the neighboring dependence that determines human marriage behaviors. These behaviors form the pattern of marriage occurrences. Marriage occurrences must be understood as a scientific object. They are not simple aggregations of the conscious selections of individuals; instead, they are pattern formations in the high-dimensional age–space hypersurface. In fact, this is the basis of our coexistence.

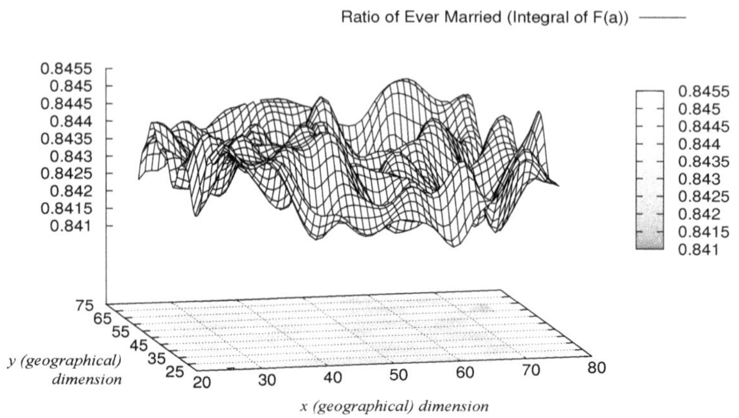

Figure 5.49: A150 at time 250.

Chapter 6

To Alter Marriage Function

> It has not been considered before that time and space are essential elements of the social sciences.

S. IKE, *The Mathematics of Marriage Function* [37].

The late-marriage trend is a primary cause of the low fertility that began in 1920s and had grown evident for everyone in the latter half of the 1970s in Japan. Japan's total fertility rate fell below the population replacement level in 1974. Since the end of the twentieth century, several policies and programs have been implemented to address this low fertility rate, which is called "Shoushika Taisaku[1]," or pronatalist policies. Since the 1990s, the Japanese government has set aside budgets for them every year. In 2017, the total budgets related to these policies amounted to over 4,800 billion yen (4.8 billion dollars). However, the TFR of Japanese has been performing poorly, at around 1.4. It appears that this massive spending has had no effect on Japanese fertility. Essentially, from historical experiences, pronatalist policies seem to have no effects.

It is clear, and must be emphasized, that the primary cause of fertility decline in present-day Japan is the late-marriage tendency. Recently, this fact was finally understood, and therefore, policies aimed at promoting marriage have been adopted. However, from a theoretical standpoint, their potential effectiveness is doubtful. As long as marriage occurrences are governed by the

[1] Japanese expression. These words are fundamentally inaccurate because the low fertility in Japan is primarily caused by late marriage. Nevertheless, this word illustrates that the phenomenon of fewer births per family is significant.

difference-integral equation, conventional policies and programs are wasteful endeavors.

In this chapter, I examine measures that can be taken to alter the marriage probabilities in terms of the marriage function.

6.1 Dynamics of the Hypersurface

The marriage function is a difference-integral function. The solution of this difference-integral function is a geographically four-dimensional hypersurface. Therefore, it is necessary to understand how the hypersurface is altered and the peculiarity of these alterations. I primarily employ numerical simulation for this purpose.

6.1.1 Simulation Settings

As the number of dimensions of a space increases, the quantum of data increases exponentially. It is impossible to visually reproduce a real geometric situation. Therefore, I have drastically simplified the model in order to illustrate the dynamics of the hypersurface. The concrete settings used were as follows:

- x,y as geographical dimensions, a as the age dimension.
- Origin — $(x,y,a)=(0,0,0)$.
- Lattice space extends to x from west to east, y from north to south, and a from young to old.
- Alter the difference-integral function in this space as each $t=a\leq 25$.
- Set the marriage function at the level of the 1950 cohort in Japan for every age and space.
- Set the marriage function at the level of the 1960 cohort in Japan in the center of the space only ($t=a\leq 25$).

The dynamics of the marriage function are considered in this virtual space.

Next, the coefficients were chosen as follows:

- $\lambda \doteq 0.2$
- $\mu > 1/14$

$$\frac{\partial F(x,y,a)}{\partial t}=\lambda\left[\mu\Delta\int_0^t F(x,y,a)(da)\right]\left(1-\int_0^t F(x,y,a)(da)\right)^2 \tag{6.1}$$

Following Poincaré's approach, the hypersurface was cut at the age dimension to analyze its alteration.

It was observed that neither the late-marriage trend nor the early-marriage trend diffuse geographically in the manner of the reaction-diffusion of the number of children. However, both trends diffuse in the age dimension. Alongside this diffusion in the age dimension, both tendencies diffuse geographically to a certain degree.

6.1.2 Slight Marriage Delays at Young Ages Caused Every Delay

The interval of the age dimension at which the late-marriage trend began is not known for certain. I suppose that the teenage group is the most probable.[2] Therefore, I allow the ever-married rate to decline from $a = 1$ to $a = 25$ (it appears appropriate to decline the ever-married rate under 18 years old).

Then, the alterations on the hypersurface of the following figures are observed (6.1, 6.2, 6.3, 6.4, 6.5). As the dynamics of the four-dimensional hypersurface cannot be viewed directly, it was cut into three dimensions at the respective ages $a = 5, 50, 150, 200$, and 225. A hundred time steps correspond to approximately 20 years of age.

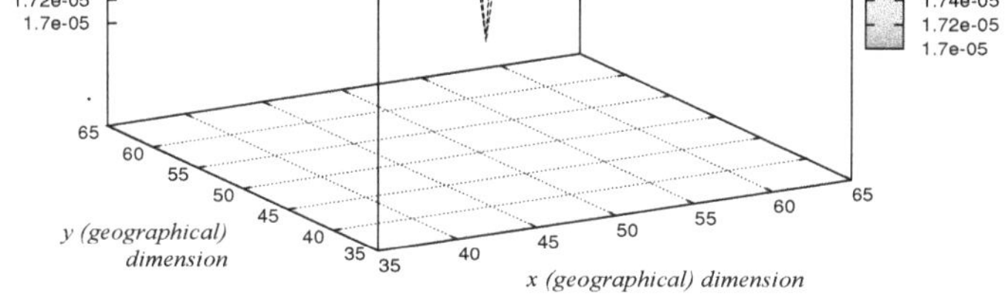

Figure 6.1: Ever-married rate as a three-dimensional surface at (theoretical) age 5.

The late-marriage trend diffused stably from the teenage group in the high-dimensional space. Thereafter, these trends developed significantly at relatively higher age groups.

However, for the x, y dimension (geographical dimension), the low marriage probability did not diffuse as much. I conjecture that the marriage function is not as diffusible as the reaction-diffusion process of the number of children because it contains an integral equation. This weak (geographical) diffusion property of the marriage function is observed in several instances.

[2]For the justifications of my supposition, please read Chapter 10.

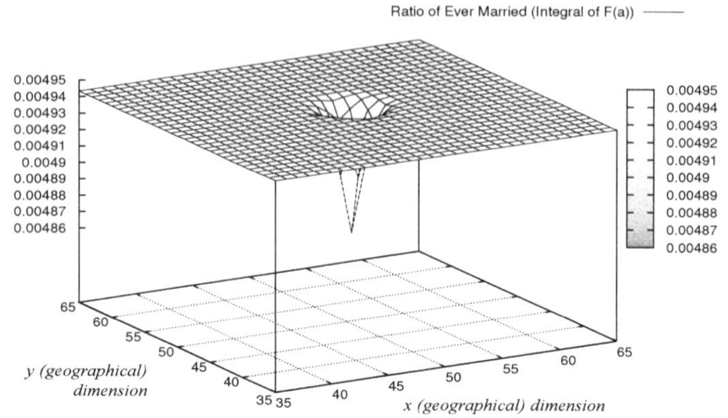

Figure 6.2: Ever-married rate as a three-dimensional surface at (theoretical) age 50.

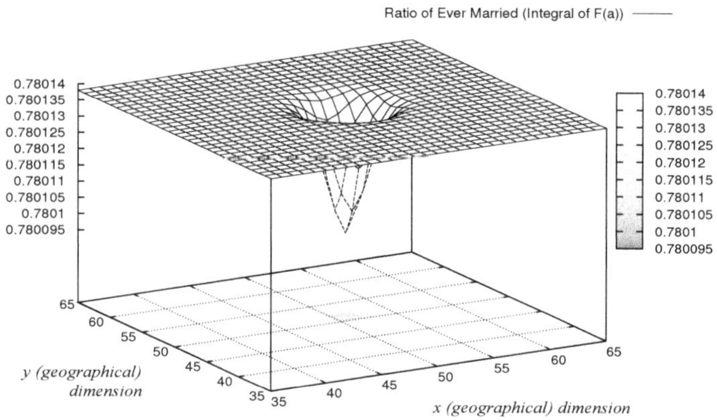

Figure 6.3: Ever-married rate as a three-dimensional surface at (theoretical) age 150.

See Figures 9.9 and 9.5 on pages 260 and 257, respectively. The palest point of the lowest child–woman ratio (CWR) is Nagoya, which is the center of Middle Japan. This low fertility does not diffuse into neighboring municipalities. Although the lowest marriage probability was recorded at the center of Nagoya city, low fertility did not diffused to its peripherals.

6.1.2.1 Beginning of the Late-Marriage Trend in Japan in 1950

When I set the marriage function at the level of the 1950 Japanese cohort for all spaces, the late-marriage trend proceeded throughout the hyperspace. Specifically, Japan had already begun the process of delaying marriage. Japanese census data recorded the late-marriage trend as early as 1920. Since then, the fertility decline of present-day Japan was destined to happen.

However, the late-marriage trend will not continue forever. The numerical simulation in this work suggests that the marriage function may have fixed

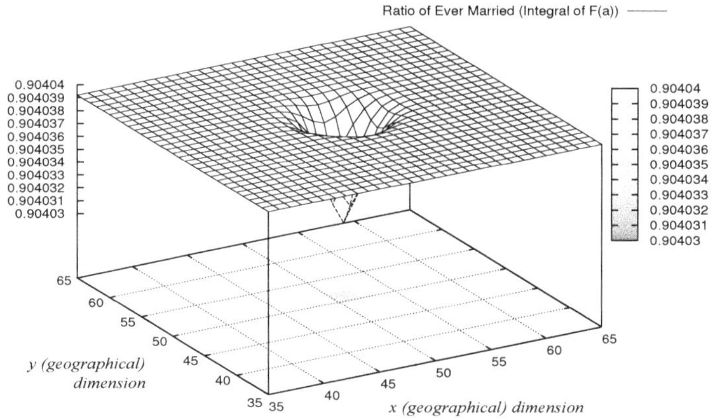

Figure 6.4: Ever-married rate as a three-dimensional surface at (theoretical) age 200.

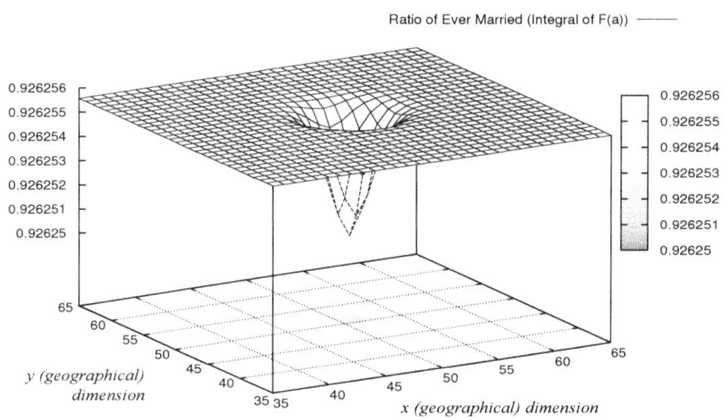

Figure 6.5: Ever-married rate as a three-dimensional surface at (theoretical) age 225.

points on the hypersurface. Therefore, the late-marriage trend in Japan will eventually come to an end. Nevertheless, it is certain that the time points marking this end do not enable maintenance of the population replacement level. Further, it is evident that the concept of a constant population, which demographers once dreamed of, is merely an illusion. The undulations of the population appear to be cyclic.

Where in the age dimension did the late-marriage trend begin? I have already implicitly assumed that it began with the teenage group. Following the decline of the marriage probabilities of teenagers, the general late-marriage trend began to proceed silently.

From Figure 6.6, it can be observed that in 1920, nearly 0.2 teenagers were married. In the 1960s, the marriage rate of teenagers had completely declined. The traveling wave of the late marriage tendency spread to the age dimension over this time. From the 1920s to the 1960s, nobody in Japan was conscious of

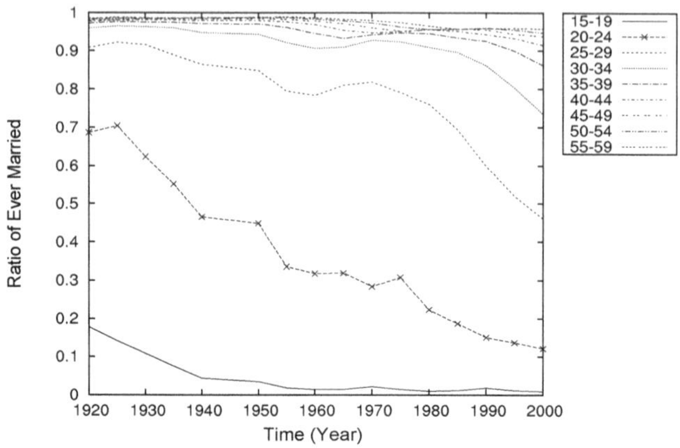

Figure 6.6: Dynamics of ever-married status based on age groups in Japan.

this late marriage tendency. The people of Japan had unconsciously adapted to neighboring marriage emergencies.

It took a long time for the marriage rate of older age groups to begin declining because the marriage function is an integral equation from age = 0. While marriage occurrences in the older age groups had declined, the change was too small for humans to recognize it. As the late-marriage trend began unobtrusively amongst teenagers, the initial fertility decline was very slight. Therefore, people failed to recognize it. Such a trend can continue to develop over a long period of time.

6.2 Constancy of Undulation of Hypersurface

When the hypersurface is flat at one time, it remains flat as time passes. In reality, undulations can be caused by several factors or even by chance. What happens to these undulations?

If the effect of diffusion is potent, the hypersurface may flatten over time. However, A.M. Turing indicated that an odd structure emerged and reserved itself in the reaction-diffusion system. What types of peculiarities does the marriage function have?

Here, the dynamics of the undulations on the hypersurface are simulated using a random variable. In Figure 6.8, it can be seen that these undulations are fairly significant. The undulations have not completely flattened in time.

It is surprising that the flattening of the hypersurface does not simply proceed as time passes. I refer to these constant undulations as "ripples." Ripples are always present on the hypersurface of the marriage function. These ripples grow longitudinally through the age dimension. I call this phenomenon the ripple effect.

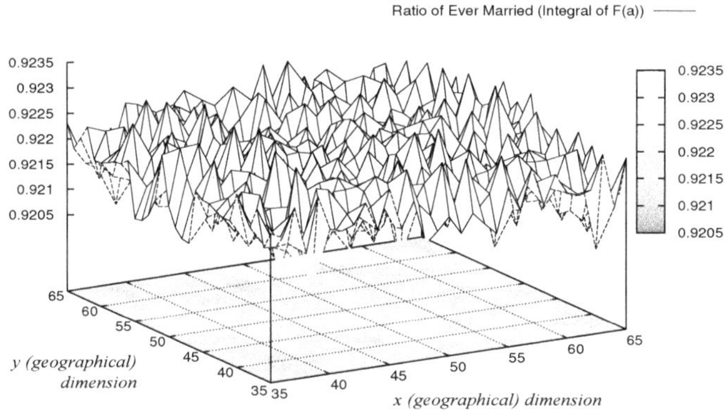

Figure 6.7: Initial undulation of the hypersurface of a marriage function at a theoretical age of 100 (corresponding to 30 years old.)

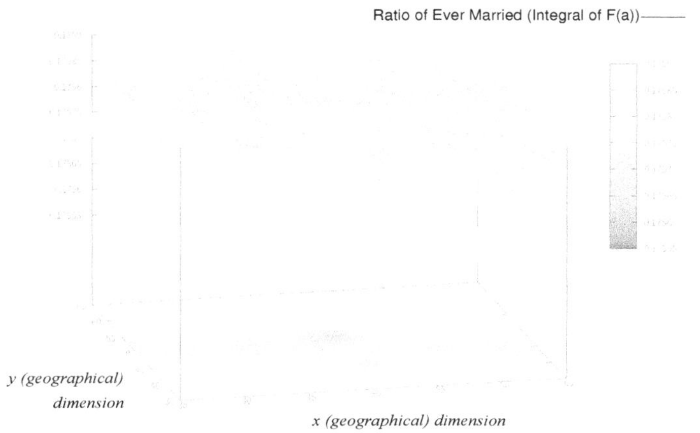

Figure 6.8: Remained undulation of the hypersurface of the marriage function at a theoretical age of 100 (corresponding to 30 years old).

6.3 Altering the Hypersurface

Altering the marriage function as a hypersurface appears to be a difficult task. Specifically, we cannot control marriage occurrences easily. We cannot change our behaviors through our conscious intentions; this is a twentieth century precept.

The power of diffusion is extremely potent, as discussed in Chapter 2. Natural obstacles such as Massif Central can only gain time. In general, the higher the order of dimensions, the more potent the growth of the diffusion power.

If marriage probabilities can be raised for all age–geographical dimensional space, marriage probabilities can be permanently increased. However, is this practically possible? However enthusiastically the government conducts its pronatalist programs, for young to middle-aged individuals and for all geographical space, it is impossible to increase the marriage probabilities of all of them. Several political resources may be exhausted in vain.

What is realistically possible is to raise the local marriage probabilities. When local marriage probabilities are raised, the diffusion effect will soon flatten and smoothen it. Therefore, pronatalist policies are fundamentally ineffective.

Nevertheless, I examine an effective approach to raise the hypersurface, primarily because I want to learn more about the features of the hypersurface and secondarily because I find the pronatalist policy to be wasteful. The basic starting point is that, accounting for the ripple effect, raising older marriage probabilities is less effective. The question is what quantum of the ripple effect can be expected based on the initial investment.

6.3.1 Flattening Effect vs Ripple Effect

When marriage occurrences tend to decline, the ripple effect grows spatially as age increases. During the decline of marriage probabilities, the flattening (smoothing) effect is the ripple effect of decline. In contrast, when marriage probabilities are raised at a point, the smoothing effect countervails the ripple effect.

Fundamentally, because it is an integral function, a local convex-concave surface of the marriage function does not diffuse well in geographical dimensional space. This is the basic reason behind the difficulty of marriage promotion. The simple solution to reverse the late-marriage trend is to raise the marriage probabilities of teenagers for the entire (geographical) space. This is the most effective means to remedy the decrease in marriage occurrences.

6.3.1.1 Ineffectiveness of Raising the Marriage Probability of Older Individuals

When the marriage probabilities of older females are raised as part of the pronatalist policy, such as raising the marriage probabilities for the theoretical age of 100–105 (corresponding to 33 years old), the artificial convex surface disappears in time. During the late-marriage trend, local convex surfaces have no significance.

It is wasteful to invest in the promotion of marriage at older ages. Almost all pronatalist policies that appeal to our conscious motivations are fruitless, because humans are deeply governed by our high-dimensional space.

An effective approach to raising the probabilities of marriage in a sustainable manner is to increase teenage marriages. If individuals of that age group are disengaged from certain constraints, the TFR will recover naturally without pronatalist policies.

Further, artificially increasing the marriage probabilities of older individuals is dangerous. An excess increase in marriages may destroy the system. The spatially extended marriage function has already lost its stability. Exten-

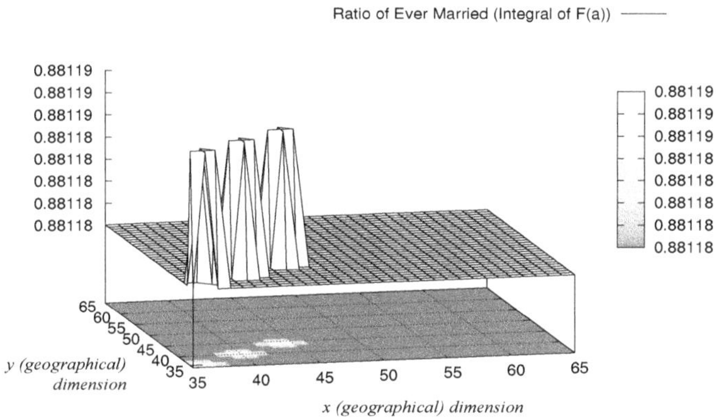

Figure 6.9: Initial bulge of the hypersurface of the marriage function at a theoretical age of 100 (corresponding to 30 years old).

sion to the geographical space conceives the possibility of a loss of control. Reaction-diffusion systems generally have these types of risks (blow-up, explosion) [84].

In the numerical simulation, the values can only be out of one; the likely practical situation is that unmarried (males and) females will be exhausted spatially and between generations.

6.3.2 Quest for a Relatively More Effective Convex Surface

Let us consider the most effective approach for raising the hypersurface when beginning from fixed points. It is already clear that the marriage probabilities of teenagers must be increased. However, the most suitable shape and size of the area are still unknown.

I consider that for a specific area, a distributed bulge is superior to a centralized one. Although the diffusion effect of the marriage function is insufficient, the diffusion interaction is expected to a certain degree. The single initial bulge is numerically compared with the four bulges separated from each other, as shown in Figures 6.10 and 6.11.

Following the passage of 250 theoretical time steps (Figures 6.12 and 6.13), the total marriage probability of all the realms of the space became almost the same. On calculating the probabilities of the median time (100 theoretical times), almost no differences were found.

The four initial bulges had a very slight advantage over the single bulge. Thus, if the intervals of the bulges are adjusted, a greater advantage can be attained, of which few remain.

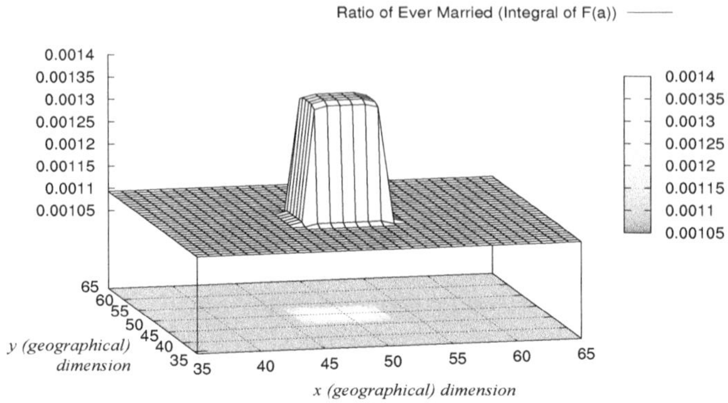

Figure 6.10: Single initial bulge.

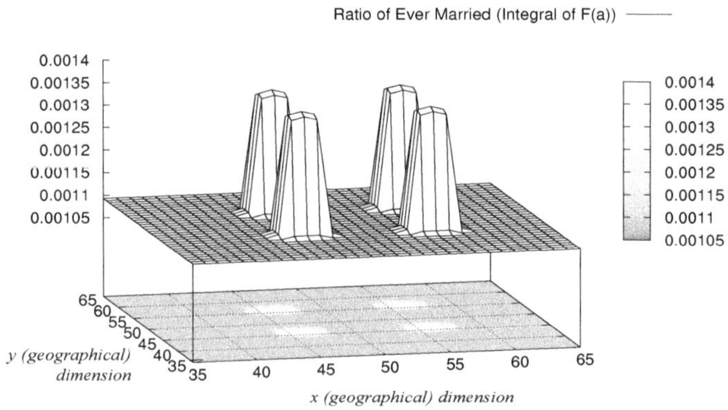

Figure 6.11: Four initial bulges.

6.3.2.1 No Progressive Wave in the Geographical Dimensions

To conclude, my initial expectation was false. There may be no effective approach to adjust the bulge in order to increase marriage probabilities. Regarding the marriage function, unlike that for the number of children, there is no progressive (FKPP) wave in the geographical space. Specifically, the late marriage tendency does not diffuse towards the end of the space. Figures 9.9 to 9.11 (p. 260) show the geographical distribution of the child–woman ratios[3] of middle Japan, whose center is Nagoya. Nagoya is the palest spot on the maps, indicating that it has the lowest fertility. This low fertility is brought about by the late-marriage trend in the center of Nagoya city.

It can be observed that the palest spot has not diffused into its peripherals. Further, the extremely low marriage rate of the center of Tokyo has not diffused into the peripheral prefectures. These observations coincide with

[3]The child–woman ratio is an index of the fertility of a local area.

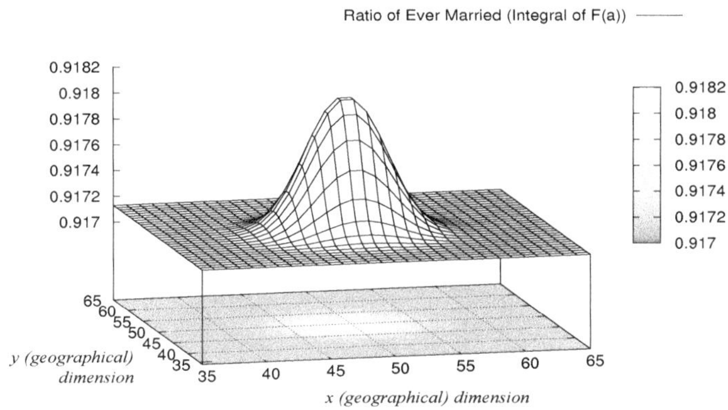

Figure 6.12: Single initial bulge after 200 (theoretical) time steps.

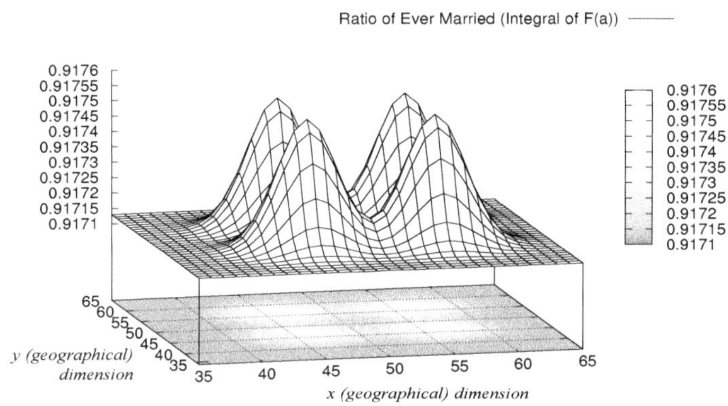

Figure 6.13: Four initial bulges after 200 (theoretical) time steps.

the low diffusivity suggested by the numerical simulations. It is reassuring to know that the low marriage rate has lower diffusivity.

6.3.2.2 Progressive Wave in the Age Dimension

The marriage function does not diffuse well in the x, y dimensions; however, it diffuses swiftly in the *age* dimension. Almost the same patterns are replicated from generation to generation.

Thus, creating a convex hypersurface has an indissoluble effect on the marriage function. As time passes, the progressive wave of marriage probabilities diffuses in the age dimension.

The next problem is the amount of interest yielded by the initial increase.

Evaluating the effectiveness of the marriage function is itself embarrassing, because simple values cannot be used to express the effect on fertility. In this case, the proportion of ever-married individuals at the end of the ovu-

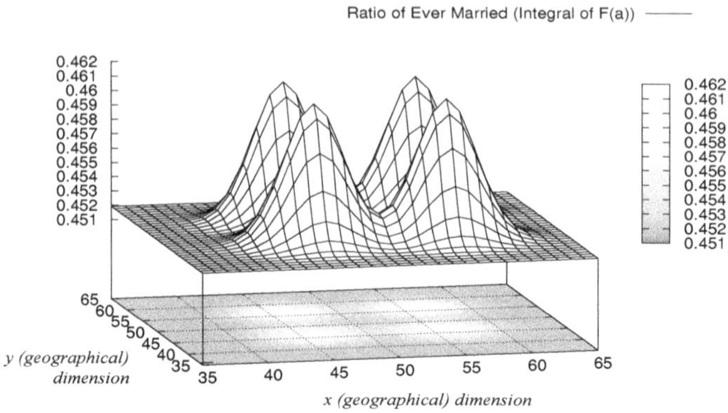

Figure 6.14: Four initial bulges after 150 time steps.

latory phase is not crucial. The early increases in marriage probabilities are significant.

The height of the four bulges was raised (multiplied by 1.2). As a result, the increase in the total marriage probabilities throughout the space was low compared with the original four bulges. Provided that the marriage function has fixed points, their gravity cannot be effectively avoided. The inertia of the total age–space is too high for the artificial convex surface to be effective. Therefore, the timing of the marriage promotion policy must be considered.

6.4 Revisiting the Baby Boom from the Perspective of the Extended Marriage Function

Consider the "baby boom" from the perspective of the extended marriage function. It is understood that there is no simple approach to effectively alter the marriage function. It is empirically evident that the baby boom increased the fertility of several countries. From this experience, we may learn how to effectively increase marriage probabilities.

As mentioned earlier, there is a possibility that the baby boom was a genuine chain reaction driven by the massive influx of soldiers returning from World War II. However, the extended marriage function is required to understand the persistence and massiveness of the American baby boom. In the previous chapter, the temporal exothermic reaction-diffusion characteristic of the (extended) marriage function was recognized. I notice that it is significant to finally instigate a "boom."

6.4.1 American Baby Boom—not the Kuznets Cycle

As an economist, R.A. Easterline [21] insisted that the undulation of American fertility was analogous to the Kuznets cycle. He inferred that the relatively small cohorts born between the 1920s and the 1930s had an advantaged course of life, and therefore, their fertility was higher than before. In contrast, baby boomers were faced with severe competition in entrance examinations and the labor market, and consequently, their fertility was relatively low. Therefore, the fertility and size of cohorts swing.

This famous Easterline hypothesis is no longer considered valid, because in several developed countries, the cohorts' fertility cycles have not occurred since the latter half of the twentieth century. In Japan, the diminishment of the cohort size has persisted for over 40 years.

At this time, it should be understood that Easterline's inference is false; the parallelism he observed between the Kuznets cycle and the fertility undulations was a coincidence. However, he recognized the fact that fertility undulations have a long-term trend, which we should consider. Specifically, the theory regarding the baby boom must be built in accordance with this peculiarity.

From this perspective, the household economics-based explanation of the baby boom is false. For the long-term trend, to ensure that an individual's fertility remains highly active, the socioeconomic conditions, being the background of fertility, must be kept as such. Otherwise, the individual's preferences might change to conform to the alterations in their socioeconomic conditions. In fact, the conditions of the women's labor market changed from 1940 to 1960.

In the mid-1950s, IBM's mainframe (general-purpose) computer was put to use. Females were generally employed for their punching and operating functions. In the United States, female participation in the labor force had already begun by the end of 1950s at the latest.

Generally, it is difficult to keep the background of fertility invariable. Therefore, it is necessary and natural for the theory to possess a long-term component. The marriage function must be built considering this property.

6.4.2 Geographical Extension for Radicalizing the Chain Reaction

In Chapter 5 (p. 112), it was suggested that the marriage function has a chain effect and that it is the cause of the baby boom. The geographical extension can radicalize this chain reaction.

Compare Figure 5.38 with Figure 5.49. The range of probabilities of Figure 5.38 is 0.00009, while the range of probability of Figure 5.49 is 0.0045. Thus, the initial undulations developed by 50 times. This radical development is mainly caused by the exothermic reaction-diffusion characteristic of the extended marriage function.

6.4.2.1 Simulation Settings

The baby boom was simulated in a high-dimensional space to understand how these characteristics alter marriage probabilities. The assumptions of the simulation were as follows:

1. Of the total population of 130.0 million, 10.0 million people were mobilized for war between 1941 to 1945. This was done randomly from all areas of the United States, with a probability of 0.8.

2. Males between the ages of 21 to 30 were mobilized. Therefore, in the United States, 14 cohorts were mobilized during World War II.

3. Each cohort size was about 2.3 million. Therefore, the number of males was 1.18 million. The mobilization rate of each cohort was approximately 0.6 (60 percent). This mobilization was performed randomly from all areas of the United States with a probability of 0.8.

4. All mobilized soldiers returned to their countries.

5. The marriage probabilities of the 1900s cohort obtained from GSS data were assumed initially. The inflection point method was adopted to estimate the marriage probabilities adjusted to the theoretical time interval.

6. 100×100 (geographical) $\times 250$ (age) lattice space.

The marriage probabilities were set to increase over 80 percent of the geographical surface based on the assumption that the mobilized rate of each cohort was approximately 60%. The marriage probabilities were altered from the theoretical age of 2 to 50 over five theoretical time steps (corresponding to two years).

6.4.2.2 How Marriage Probabilities Increase

For the first 20 theoretical time steps, the probabilities of the theoretical ages 20 to 80 were multiplied by 0.8. For the next five theoretical time steps (corresponding to one year), the female marriage probabilities were increased. The method of increasing them is critical. When the increase was assumed using geometric progression, a baby boom could not be achieved. Under the assumption of a geometric progression of marriage probabilities, high fertility remains a relatively lengthy period in the high-dimensional space. Therefore, we assumed a monotonic decrease in the arithmetic progression of female marriage probabilities. At every time interval, a certain increase in probabilities occurred; however, it decreased by a certain amount.

I conjecture that, quantitatively, the mobilized North American soldiers returned at a constant rate at fixed intervals.

6.4.2.3 Two-Sex Problem—Marriage Squeeze Hypothesis

I have exclusively considered the female marriage function; however, the mobilized soldiers were male. Provided that the so-called "two-sex problem" cannot be solved, we must assume an effect on female marriages. See next Figure 6.15. It can be seen that the 1920–1929 American cohort exhibited little decrease in teenage marriage probabilities compared to that of the 1900–1909 cohort about GSS[4] cumulative sample. Therefore, it can be assumed that there was no significant decline in marriage probabilities due to the mobilization of soldiers during World War II.

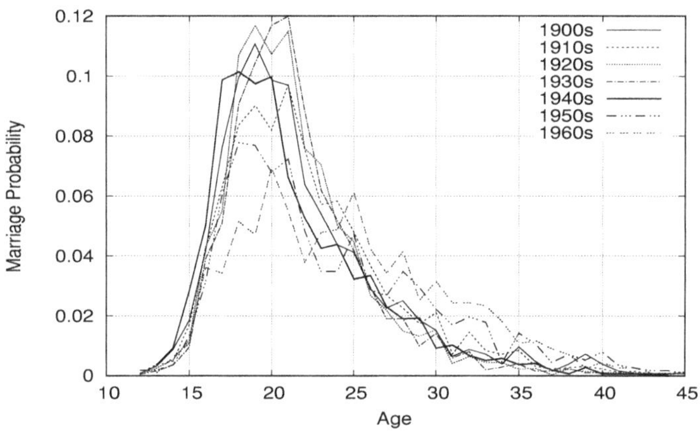

Figure 6.15: Marriage function for each cohorts calculated by GSS data.

The mathematician A.G. Fredrickson presented the model of pair formation in an age-structured population. Following his work, Starverov, Hadeler, and Inaba proposed some mathematical models for pair formation. These models are completely mathematical formalizations that define $\rho(t, a, b)$ as the pair density of female age a and male age b per unit of time and advance further. They easily define these functions as abstract objects. However, the pair density itself is the unidentified object that should be discovered. The possibility of mathematical formalization should be explored considering an arbitrary approach. However, the real law of marriage remains unknown. As a realistic demographer, I attempt to discover how $\rho(t, a)$[5] is determined. Further, I consider that the substantial and concrete two-sex pair density function will not be solved in the near future. This book is intended to explore not the

[4]"The General Social Survey (GSS) is a project by the independent research organization NORC at the University of Chicago, with principal funding from the National Science Foundation."

[5]I positively disregard b. The feasibility of a male's marriage independent of his age negates the two-sex problem.

formalization, but the discovery of the real laws of marriage occurrences, even if they are imperfect or false.

The "two-sex problem" implicitly postulates monogamy, and therefore, it may be a fundamentally false problem formation. When polygamy is considered, the marriage squeeze problem becomes inconsistent. During the period of mobilization, females who desired to marry were not deficient in the resources of males in terms of numbers. Therefore, I presume that there was no decrease in marriage probabilities during World War II; if any, it was negligible.

In the computation, when a large decrease in teenage marriage probabilities is postulated, a baby boom cannot be obtained. To obtain a baby boom, under the valid increase in marriage probabilities following World War II, it is very important to consider the global tendency of the marriage function. If the global decline tendency of the marriage function is potent, the baby boom does not occur.

See Figure 6.15 repeatedly. Between 1900 and 1940, American female cohorts did not display significant late-marriage tendencies in their young ages. It is unnecessary to assume a decline in marriage due to the mobilization during World War II. However, it is necessary to assume a global late-marriage trend nowadays in U.S.

The American baby boom lasted for over 20 years, or 100 theoretical time nowadays in U.S. steps. The temporal increase in the marriage function due to the return of soldiers may have enabled this prolonged period via a chain reaction. I conjecture that these two assumptions of the global decline tendency and the quantity of temporal increase are necessary.

6.4.3 Artificial Baby Boom in High-Dimensional Space

Artificially creating an American baby boom in a high-dimensional space was a challenging task. This is because TFR and other fertility indices provide no information about age-specific marriage probabilities by region. However, to analyze the exothermic property of the marriage function, the geographical dimensions of the marriage function must be provided.

In fact, a baby boom can easily be obtained when extreme values are submitted to the coefficients. Raising the marriage probabilities of 20–30 year old females the year following the war resulted in a baby boom for the next 20 years or longer. It is a far more delicate task to generate a real boom that lasts for about twenty years. In general, it is easier to remain out of control than to regain stability.

On the other hand, this provides an opportunity to estimate the values of the coefficients. The coefficients can be selected to obtain a baby boom that lasts for about twenty years. Of course, the possible combinations are infinite, but we limit their range to valid values. The constraint conditions are as follows:

1. A slight late-marriage trend before the baby boom.
2. The baby boom continues for about 25 years.
3. Following the baby boom, the late-marriage trend continues.

I believe that the slight late-marriage trend is fundamental to the twentieth century United States. The combinations of coefficients that satisfy the second and third conditions are valid coefficients.

6.4.3.1 Simulation Results

At this point, let me consider the features of the North American baby boom afresh. From Figure 5.11 (p. 113), it can be seen that the TFR of the United States began to pick up around 1938. This was not the beginning of the baby boom; it was its rebound following its fall during the Great Depression. Therefore, the outset of the baby boom cannot be clearly distinguished. The decisive peculiarities of the American baby boom are its peak and duration.

It is evident that following the end of demobilization (after 25 theoretical time steps), the "marriage boom" grew even more strongly. For a long time, several demographers and researchers doubted that demobilized soldiers could cause the continuing baby boom. However, the extended marriage function demonstrates this possibility via numerical simulation. The relatively long continuity of the marriage boom and the birth boom linked to it indicate that demobilization caused a continuous baby boom. Without other factors, this may possibly explain the baby boom.

From Figure 5.11 (p. 113), it can be seen that for several countries, the duration of the baby boom was almost the same. Namely, it can be supposed that alterations in the marriage function occurred simultaneously, and the duration was also the same; however, the impact on the function (the hypersurface) differed. A meaningful theory must simultaneously explain these similarities and deviations. I altered the rate of increase in female marriage probabilities, but not the duration of the increases added to the function.

If the strength of the impact on the hypersurface is excluded, the duration of the rise in marriage probabilities is almost the same. The exothermic effect may certainly occur because the increments of the probabilities are significantly larger than the aggregations of additions. I estimate that the additive impact will increase by 2.5 times.

However, the duration of the marriage boom is highly approximate. This is an inevitable consequence of the extension to the geographical (multi-dimensional) space, which is accompanied by a diffusion mechanism. The diffusion occurs based on the flux, and it is proportional to the difference in elevation. Therefore, the higher the convex, the more rapidly it diffuses. Except for the marriage boom cohorts bearing the baby boom, several cohorts in the space remains unchanged, and the marriage boom eventually disappears. An individual's marriage probability is affected by neighboring cohorts and geographical contemporary generation. The true state of the baby boom can

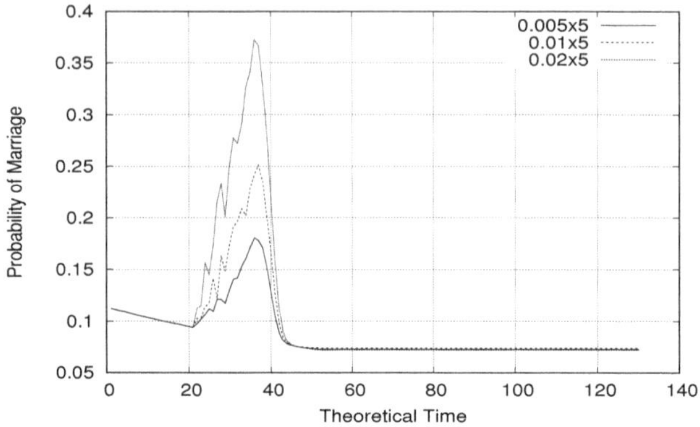

Figure 6.16: Probabilities of marriage at 23 based on the significant increase due to the impact of the returned soldiers.

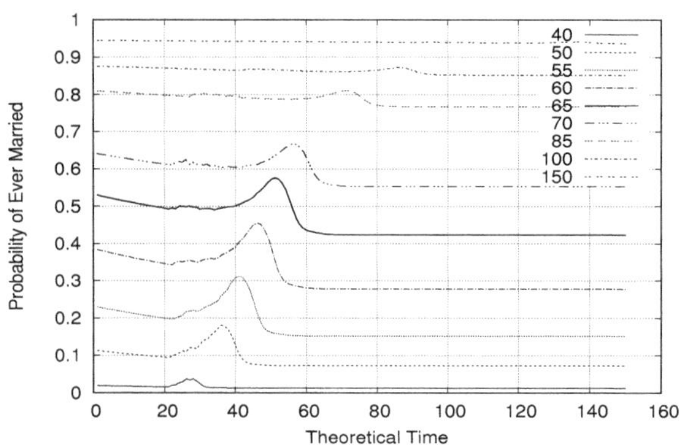

Figure 6.17: Longitudinal probabilities of marriage based on the increase due to the impact of the returned soldiers (the scatter probability is 0.8).

only be understood when we consider marriage in a high-dimensional space and its interactions.

Meanwhile, it should be noted that the duration of the marriage boom is longer than that exhibited in Figure 6.17. This is because the marriage probabilities of the cohorts that experienced the marriage boom developed over time; the probabilities were not as large as that at age 23. This durability of the marriage boom can be understood well from Figure 6.17.

In this simulation, the marriage boom continued longitudinally for almost 15 years. Incremented marriage probabilities proceed on the age dimension because cohorts with incremented probabilities age with the passage of time.

The duration of baby booms of multiple countries made Easterline imagine their synchronicity with the Kuznets cycle. The fact that the prominence of the United States baby boom coincided with the overwhelming economic prosperity of the country made him more confident of this synchronicity. However, in 2020, this synchronicity was lost, and the cyclic undulation of TFR disappeared. The marriage function in a high-dimensional space can yield reasonable explanations for all these phenomena. Specifically, the extended marriage function comprehends the baby boom as a temporal impact on the hypersurface and its exothermic reaction-diffusion. Therefore, my explanation is background independent and offers falsifiability in the predictions.

G.S. Becker insisted that the gender pay gap was the cause of both the beginning and the end of the baby boom. Following World War II, labor compensation for males was relatively higher than that for females, and therefore, females preferred to marry and bear children. On the contrary, labor compensation for females increased relatively compared to that for males at the end of the baby boom; thus, the opportunity costs of marriage and births increased, and fertility declined. Then, was the gender pay gap maximum in 1955? This theory is frequently contradicted by statistical data.

The most vital failure of the household economics theory is that it cannot establish a quantitative relationship between the gender pay gap and the marriage rate increase (or decrease). This shows that it is not a scientific statement in any sense. In Japan, since 1970, citizens have willingly and voluntarily adapted to the reduction of the gender pay gap, as well as the high opportunity cost of marriage and birth. In those days, nobody complained against it.

In 2020, the gender pay gap was held as a hindrance to marriage and births for females in Japan. Public assistance was proposed for female marriage and births to reduce opportunity costs and increase fertility. Compared to the situation ten years ago, hindrances to marriage and birth have reduced, but the TFR has remained stable. Clearly, this intuitive inference of the gender pay gap is not scientific. It lacks a falsifiable prediction. How much assistance to female marriage and births will increase TFR at a quantitative rate?

Social scientists tend to attribute observed phenomena to human intentions ad hoc. Scattered and apparent human motivations are considered. These are not consistent with historical undulations.

A scientific theory of marriage occurrences would consistently and longitudinally explain the undulations of marriage probabilities and offer falsifiable predictions. I believe that there have been no scientific theories of marriage occurrence up to this point. All existing theories of marriage appear to be entirely unsubstantiated.

6.4.4 Effectiveness of the Artificial Baby Boom

The true accuracy of this artificial baby boom should be measured based on the birth function accompanied by the marriage function. This is because the undulations of the marriage function intrinsically affect the birth function in a chain reaction. However, because this is beyond the scope of this chapter, I will confine my analysis to the quantum of marriage probabilities.

Table 6.1: Artificial marriage booms using some additive values.

Ever-married age 23	0.005×5	0.01×5	0.02×5
Initial Prob	0.112463	0.112463	0.112463
Max Increment	0.067748	0.137874	0.259638
Sum of Increment	1.360545	2.101636	3.538595
Effectiveness	1.478853	1.142119	0.961575

I added some values to the marriage probabilities of females between 20 and 29 years over a one-year period (five theoretical time steps); this is the period required for the mobilized soldiers to return. Using this approach, I was able to successfully create artificial marriage booms.

It is surprising that the smallest value is the most effective in increasing the marriage probabilities. The part $(1 - \int F(a)(da))^2$ of the marriage function impedes its development. The effectiveness was measured using the value equal to the sum of the increment divided by the increment $\times 5 \times 46/0.8$ (*ex.* $0.005 \times 5 \times 46/0.8$)[6]. Regardless, an artificial and effective increase can be made to the marriage function. Without the theory of the birth function, we can make an approximate estimation at best. However, at worst, the TFR will increase based on this increment.

It should be mentioned that increments were provided throughout the space $\times$ 0.8. The mobilization could probably be done for a similar area of the United States, and therefore, I adopted this computation. This is an enormous program that cannot be conducted during times of peace. Considering that the marriage function has an insufficient diffusion property on the geographical surface, in order to expect an exothermic effect, the increments must be scattered on the geographical surface, and they should be sufficiently close to each other.

6.4.4.1 What Increases the Sparsity of Increments?

What is the quantitative relationship between the level of proximity and the increase in the hypersurface? The increments were randomly scattered on the hypersurface with a probability of 0.5. The marriage boom became smaller than that with a probability of 0.8. However, the decreasing rate of 0.89 of marriage probability is relatively good compared to the scatter probability ratio 0.5/0.8.

Next, the scatter probability was altered to 0.25. Even with this low scatter probability, a few increases occurred on the hypersurface, as shown in Figure 6.18. However, the increase itself was reduced. This effect should be attributed to the impartiality of the increments in a space. The hypersurface must be raised impartially to achieve a global increase in the marriage probability in the space.

[6] Five is the time for the soldiers to return. The incremented theoretical age is 46. The rate of the scattered area is 0.8.

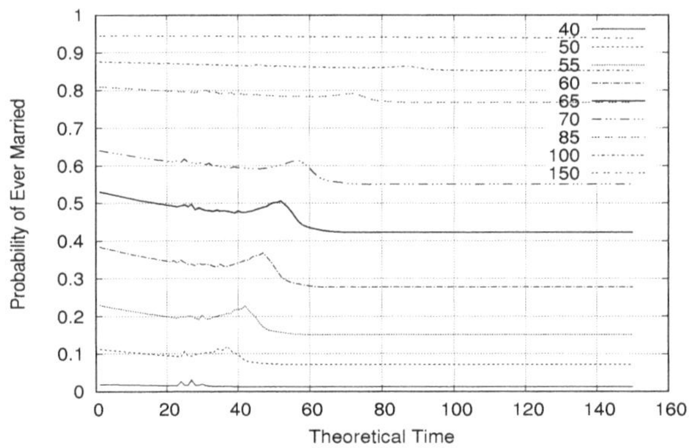

Figure 6.18: Longitudinal probabilities of marriage based on the increase due to the impact of the returned soldiers (the scatter probability is 0.25).

This universal increment requisite suggests that the pronatalist approach of marriage promotion is actually impossible. The increments of marriage probabilities cannot be scattered impartially, as shown in the above example. In particular, it is not effective for promoting marriages in older age groups.

I finally realized that reaction-diffusion systems are not necessarily confined to equations that have a reaction term, such as Equation (2.1) on p. 23. If it has a self-development property with neighboring development, an exothermic reaction effect may be expected.

The effect of returned soldiers causes a continuous increase in marriages for the next five years. This effect is definite. Nevertheless, it is exceptional that marriage probabilities can be increased by 1.2 times through an almost global impact on the hypersurface. The enormous terrorism and destruction of World War II brought about this impact because of the unprecedented mobilization. I do not believe that a similar demographic policy can be artificially created. The effective control of the marriage function continues to be challenging. The occurrence of the real "American baby boom" is linked to the phenomenon of the chain reaction of births. Of course, it is evident that the temporal increase in marriages played a role in the baby boom.

6.5 To End This Chapter

I have not been able to find any effective method to alter the marriage function in a high-dimensional space. I have not yet discussed the development of a convex surface as a pronatalist policy. Of course, this can be performed using background dependent approaches such as economic incentives. Even if it was possible to make someone get married using economic incentives (I do not agree with this idea), effective alteration of the marriage function is nearly impossible.

A genuine countermove against the late-marriage trend would be to enable the freedom of teenagers' sexual behaviors. Modern society has wrongfully suppressed their sexuality for a long time. In particular, it has repressed it under the facade of guardianship and educational development, despite the fact that the imposed education does not actually support their development. Teenagers' natural and erotic functions should be allowed, and society should provide protection for them and their circumstances. In order to overcome the late-marriage trend, support should be provided for teenage pregnancies, and their lives should be assured, even though this will take a considerably long time.

III

BIRTH FUNCTION IN HIGH DIMENSIONAL SPACE

Chapter 7

No Individual Birth Functions Exist

> Prop. XXIX. No individual thing, which is entirely different from our own nature, can help or check our power of activity, and absolutely nothing can do us good or harm, unless it has something in common with our nature.

Spinoza, *Ethics (Ethica Ordine Geometrico Demonstrata)* [16], p. 170.

Are birth behaviors actually products of individualistic decisions? Household economists assume this thesis with simplicity, as do liberals, despite having views on fertility (pronatalism) policies that are contrary to theirs. However, as the marriage function, birth behaviors depend on both time and space neighboring behaviors. This is a scientific fact. Individualism is an illusion that should be recognized as such. "*Free will*" cannot be defined a deterministic manner; it conceals itself in stochastic processes.

It is easily postulated that parents make decisions individually and depending on their socioeconomic conditions. Is this postulation true? Our intuitive understanding of ourselves is often false; self-recognition methods are not reliable at all.

7.1 Quasi Linearity of the Expected Number of Children Based on Marriage Duration

Birth probabilities specific by duration of marriage have quasi-linearity. This serves as indirect but powerful evidence that human births are stochastic and not deterministic. Whether an event is stochastic or deterministic must be judged based on empirical data rather than introspection. The problem that should be solved is the cause of our birth behaviors, not the manner in which we consciously decide to bear children.

7.1.1 Stochastic Variable

The number of children in a family is approximately proportional to the duration of the parents' marriage. This regularity was found by the renowned demographer L. Henry.

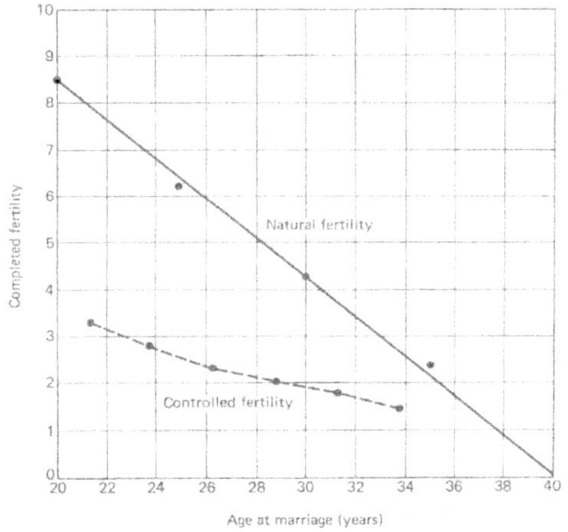

Figure 7.1: Completed fertility and age at marriage: natural and controlled fertility.

Source: L. Henry, *Population Analysis and Models* [32], p. 94.
Originally Figure 7.5 Completed fertility and age at marriage

Henry calculated "natural fertility" from the data of Hindu villages in Bengal, India, and "controlled fertility" from the completed fertility of women married in 1919 in Great Britain (see Figure 7.1). The development of contraception and the difference in socioeconomic settings do not affect this linearity. Henry described this specific peculiarity as follows:

> As in the preceding case, we have a decline in completed fertility with the woman's age at marriage, but it does not follow a straight line; if we fit a line to the straight part of the curve, the intersection with the horizontal axis will be far beyond age 40 years. The decline is often interpreted without taking into account the influence of psychological and social factors on the age at marriage and on the desired number of children. It is then attributed to:

> 1) increasing sterility with age: a woman married at age 25 years has a higher probability of not achieving the number of children she wants to have than a woman married at age 20 years.
>
> 2) contraceptive failures: A woman married at age 25 years who desires, say, 2 children, is exposed for a shorter time to the risk of having one or several more children than she desires than a woman married at age 20 who also wants 2 children.
>
> These two factors do not seem to be sufficient to account for the decline in completed fertility, that is actually observed as age at marriage increases. It is likely that the desired number of children is correlated with age at marriage through the influence of psychological and social factors; women who marry earlier also desire or accept more children. There are other reasons as well to believe that the factors enumerated in (1) and (2) are not a sufficient explanation. An overall change in nuptiality in the direction of earlier marriage, such as that which has been occurring in northern Europe over the last thirty years, does not result in increases in the number of children as it should if factors 1 and 2 were the only ones operating.
>
> Source: L. Henry, *Population Analysis and Models* [32], p. 95.

This is not to support the postulation of "psychological and social factors." Rather, it is to emphasize that the linearity of birth probability cannot be reduced to physiological effects. What Henry called "psychological and social factors" is nothing but neighbor-dependent birth behavior.

Hisashi Inaba also pointed out this linearity in his book [39] in 2003. He proposed a common birth probability p to explain it. If p is adopted as an instantaneous birth probability, the following relation can be obtained:

$$\frac{dC}{dt} = p, \quad f(t) = p \cdot t + C \tag{7.1}$$

Inaba suggested that a common birth probability p is the cause of the linearity. However, birth probabilities may not be a value, but a function or an operator. It is found that assuming birth as a stochastic variable is promising.

It may be noted that in 1937, Teizo Toda, a Japanese family sociologist, noticed the same relation between the duration of marriage and complete fertility [72].

7.1.1.1 Indistinguishability Based on Socioeconomic Conditions

It is generally believed that human birth behaviors are based on socioeconomic conditions. Do more educated or wealthier women tend to execute higher control over their fertility?

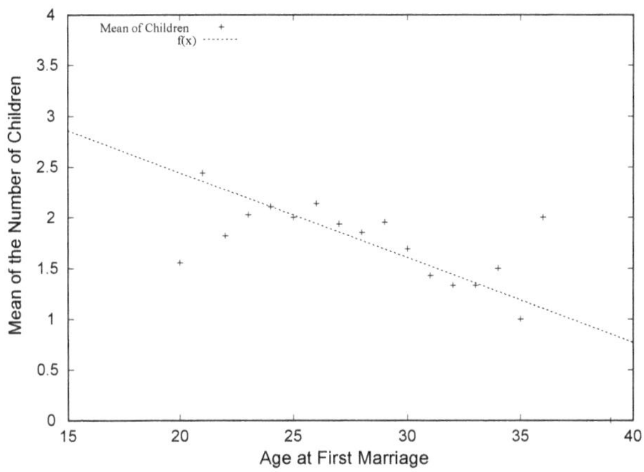

Figure 7.2: Linearity between number of children and age at first marriage (JGSS): educated $R^2 = .6064, n = 543$.

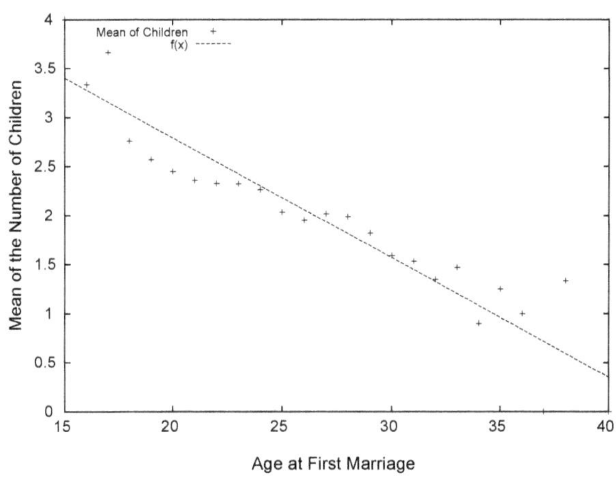

Figure 7.3: Linearity between number of children and age at first marriage (JGSS): less educated $R^2 = .8844, n = 2758$.

Educational Background

The Japanese General Social Survey (JGSS) collected data on the number of children of respondents from 2000 to 2003. In the present work, women are classified into two groups: educated and less educated. Accounting for the fact that the sample size of the educated group is significantly smaller than that of the less educated group, the linearity is the same for both groups. Education level does not seem to alter the dependency of birth behaviors on time.

This indistinguishability of the linearity based on educational difference also appears to be valid for American females. Here, the correlation between their age at marriage and number of children was analyzed using GSS[1] data.

[1] http://gss.norc.org/About-The-GSS.

The linearity between the number of children and age at first marriage is universally applicable to unmarried American females.

It is evident that education does not alter quasi-linearity. Thus, it may be supposed that birth is a stochastic event.

Economic Conditions

Now, the JGSS sample is divided into two groups based on spouse's income to evaluate whether the linearity changes.

For a small sample size of the higher-income group, R^2 decreases compared to the lower-income group. However, it can be visually understood that the linearity is common to both groups. Specifically, it can be concluded that the economic difference does not affect the quasi-linearity of birth probability.

The observed data does not support the general preconception that highly educated and wealthy individuals intellectually control their birth behaviors. Therefore, birth behaviors should be observed as a stochastic phenomenon.

7.1.2 (Class) Differential Fertility

There is a Japanese proverb that says, "Children are poor men's riches." In general, it is assumed that the poorer the parents, the more children they bear. This assumption is not true. The poor tend to marry earlier. This simple fact and the quasi-linearity associate this spurious correlation. Differential fertility is not an exact concept.

I believe that human birth behaviors are stochastically dependent on neighboring behaviors. However, household economists believe that they are dependent on individual conscious choice. They proposed a function of the number of children in a family based on their belief in conscious choice. Their functions have the fatal defect that it may return a not unique number as its formulation.

7.1.2.1 Number of Children—Indefinite Solution

Household economists have not explicitly defined the function of the number of children. The number of children is one of the variables explaining the total household utility U. The parents are assumed to maximize U by optimally adjusting those variables. However, even under certain budget constraints, this problem is indefinite in many cases. Clearly, as the number of children itself cannot be determined as a unique value, their definition is false.

Moreover, their definitions are not concrete. The functions are defined abstractly; they cannot be used to calculate the exact number of children. Thus, the theory of household economics is not falsifiable. The functions are stereotypical expressions that trace human consciousness, and thus, they are obscure. An example of this is the utility function provided by G.S. Becker in his book *A Treatise on the Family* [3]:

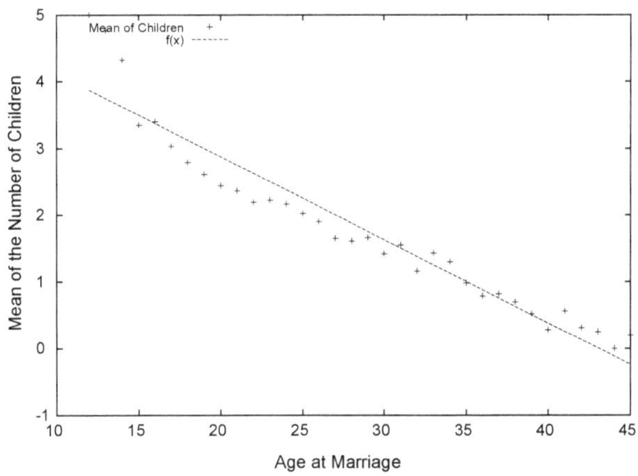

Figure 7.4: Linearity between number of children and age at first marriage: GSS 1972–2008 $R^2 = .9166$.

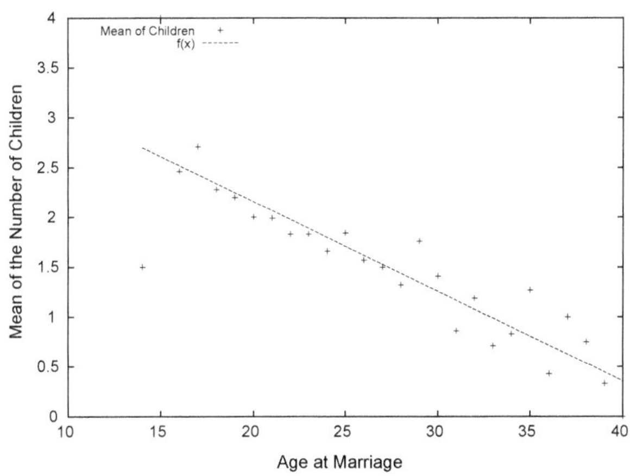

Figure 7.5: Linearity between number of children and age at first marriage: GSS 1972–2008, more than a bachelor, $R^2 = .7442, n = 2002$.

Therefore, these theories can be combined and generalized by assuming that each family maximizes a utility function of the quantity of children, n: the expenditure on each child, called the quality of children, q, and the quantities of the commodities.

$$U = U(n, q, Z_1, \ldots, Z_m) \qquad (5.1)\text{*original}, \qquad (7.2)$$

The Malthusian theory ignores quality and assumes that the demand for births (or number of children) is highly responsive to changes in income (hence, the demand for other commodities may be negatively related to income). ...

Source: G.S. Becker [3], p. 137.

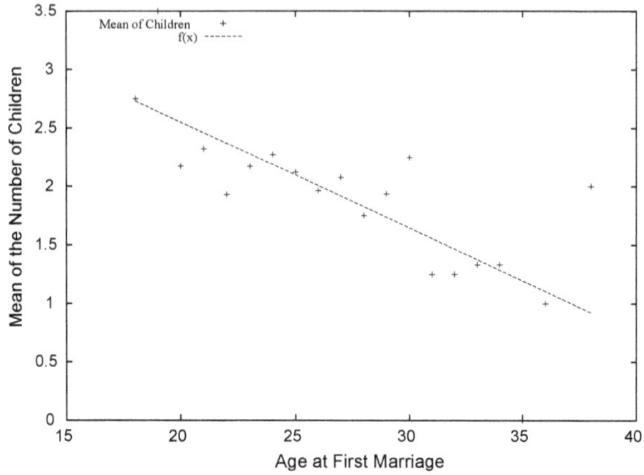

Figure 7.6: Linearity between the number of children and age at first marriage: JGSS, over 6.5 million yen per year, $R^2 = .566, n = 529$.

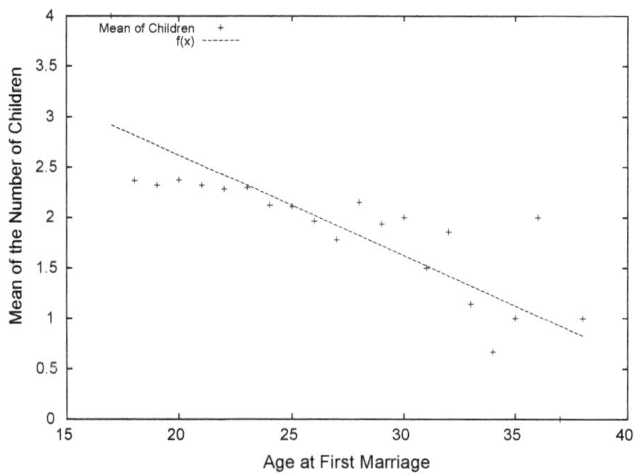

Figure 7.7: Linearity between the number of children and age at first marriage: JGSS, less than 6.5 million yen per year, $R^2 = .668$.

Another example is a modified version of the Becker, Murphy, and Tamura (BMT) model, introduced by Chu:

> Specifically, BMT considered a young/old overlapping-generation model and assumed that at period t, a representative parent has the following utility function:
>
> $$V_t = U(c_t) + \delta(m_t) m_t V_{t+1}, \qquad (11.1)\text{*original}, \qquad (7.3)$$
>
> where V_t and V_{t+1} are the utilities of parents and each child, $U(c_t)$ is the parental utility flow from consumption c_t, and m_t is

> the number of children. $\delta(m_t)$ is the between-generation discount rate determined by parental altruism toward each child. ...
>
> Source: C.Y.C. Chu [9], p. 134.

Both equations are essentially indefinite in the same manner. The maximization of the utility does not change this indefiniteness. Household economists insist that they solve the equation under several constraints. Their insistence is arbitrary and has not been strictly proven.

Coefficients such as δ cannot be determined or estimated. It is exactly questionable that there are common interpersonal coefficients. Therefore, these models are arbitrary and unmathematical.

Here, the nonexistence of such a function is proved.

7.2 Negation of the Existence of the Function of the Number of Children by the Diagonal Method

Over the years, household economists have made several revisions to the void function of the number of children. They readily assume the existence of a function. In other words, functions of the number of children are defined to demonstrate their existence. However, assuming the number of children as a stochastic variable, as Kolmogorov & Chaitin defined, no genuine random variable can be a mapping by the function. Thus, under this assumption, the number of children is not decided by a function.

In addition, by the diagonal method, evidence can be provided showing that there exist no mappings from socioeconomic conditions to the number of children in a family.

Now, the existence of a function to decide the number of children in a family is assumed. A contradiction occurs, and the first assumption is denied. Namely, a proof by contradiction is provided.

7.2.1 Contradiction by the Diagonal Method

The indefinite solution problem and obscurity are sufficient to demonstrate these models to be unscientific objects. However, if their functions are not disproved, household economists may continue to consume and disperse false birth models. For this reason, the function of the number of children is explicitly disproved here.

The definition of the function itself does not demonstrate its existence. For example, Cantor's continuum hypothesis is not proven in Zermelo-Fraenkel set theory. Although the function of the number of children appears to be defined, the definition itself is not necessarily valid.

Here, the function of the number of children is assumed to exist. Nevertheless, the number of children, provided that it is decided, is a single component variable of the household utility. The other variables x_i are adjusted and input such that the number of children is returned. Namely, household economists insist that mappings exist from other variables of socioeconomic and individual conditions to the number of children[2].

The mapping (function) to decide the number of children in each family has infinite independent variables (arguments) as follows:

Table 7.1: Function to decide the number of children in each family.

$$
\begin{array}{ccllll}
c_1 & = & f(x_{11}, & x_{12} & x_{13}, & x_{1j} & \cdots) \\
c_2 & = & f(x_{21}, & x_{22} & x_{23}, & x_{2j} & \cdots) \\
c_3 & = & f(x_{31}, & x_{32} & x_{33} & x_{3j} & \cdots) \\
\vdots & \vdots & \vdots & \vdots & \vdots & \vdots & \cdots \\
c_i & = & f(x_{i1}, & x_{i2} & x_{i3}, & x_{ij} & \cdots) \\
\vdots & \vdots & \vdots & \vdots & \vdots & \vdots & \cdots
\end{array}
$$

First, these mappings are assumed to be to the number of children in each family. Therefore, f can be mapped to all the infinite series of the number of children in a family from prehistoric times to the distant future. When x_j is set, the number of children is determined. This is the insistence of household economists.

A dictionary whose infinite indices are x_j is theoretically defined. **All** the numbers of children in families are completely determined in this dictionary.

Then, the diagonal elements of the left side of Table 7.1 are altered as follows. Now, a new $c_* = f(x_j)$ is obtained for the first time. Thus, the number

Table 7.2: Function to decide the number of children in each family.

$$
\begin{array}{ccllll}
c_1 & = & f(\not{x}_{11}, & x_{12} & x_{13}, & x_{1j} & \cdots) \\
c_2 & = & f(x_{21}, & \not{x}_{22} & x_{23}, & x_{2j} & \cdots) \\
c_3 & = & f(x_{31}, & x_{32} & \not{x}_{33}, & x_{3j} & \cdots) \\
\vdots & \vdots & \vdots & \vdots & \vdots & \vdots & \cdots \\
c_i & = & f(x_{i1}, & x_{i2} & x_{i3}, & \not{x}_{ii} & \cdots) \\
\vdots & \vdots & \vdots & \vdots & \vdots & \vdots & \cdots
\end{array}
$$

of children in a family has been explicitly defined. However, c_* must not represent the number of children. This contradicts the first assumption. Thus, there is no function that calculates the number of children of each parent. Q.E.D.

[2]Naturally, I do not agree with this point of view.

Further, another theorem is supposed here.

> ***Theorem 7.1 Incomputability of the Number of Children***
> *The number of children is an incomputable stochastic number.*

7.2.1.1 Function as an Assumption

As discussed earlier, it is commonly assumed that the number of children is determined by the socioeconomic conditions of parents. Namely, when independent variables are input as arguments of the function, the number of children is returned. The function is assumed to be equal to conscious and unconscious human choices. It is an unconditional assumption that our behaviors are described as a return of an individual's function. This assumption is entirely false.

Social scientists should know that the function cannot be defined in some cases. Intuition is not equal to defining the function. The definition of a utility function tracing conscious human thought patterns and its maximization have long been unclear. The conscious model of human behavior is always false because it is roughly defined *ex post facto* to make ourselves consent.

The reason for this fallacy is that these functions are not defined in geographic or temporal dimensions. In terms of human thought patterns, conscious consideration of our actions is always virtual and is generally disconnected from the time and place of its occurrence. On the other hand, actual behaviors depend on time and space. "Opportunity cost" cannot compensate for this limitation.

In terms of the number of children, the value itself is not returned as a result of calculation defined by a function. It is merely the cumulation of each birth. Therefore, the theory of birth occurrences should be developed in temporal and geographical dimensional space.

In high-dimensional time–space, individuals tend to begin to marry and bear children. This is the process that should be studied. An individual's location in high-dimensional space governs the number of children they bear.

Prior to this, here, the number of children is demonstrated to be a stochastic variable by means of several statistical tests.

7.3 Statistical Test of the Randomness of the Number of Children

The number of children of parents is a stochastic variable because birth is a stochastic trial in an interval of time. It cannot be calculated *a priori*. When millions of values of the number of children in families are known, the subsequent number of children of the million and first of parent cannot be deterministically predicted. Namely, the number of children occur in terms of "Kolmogorov randomness."

The number of children c_i appear as a random sequence of numbers.

$$c_i = 0, 1, 2, 3, \cdots, c_{\max}, \tag{7.4}$$

as in the below sequence:

$$0, 1, 4, 2, 2, 0, 2, 2, 2, 2, 3, 2, 4, 3, 3, 3, 2, 2, 0, 4, 3, \cdots$$

Large-scale social surveys such as GSS inquire about the number of children born to each female respondent. The above sequence is a partial sequence from GSS data. It is regarded as a partial sequence of the infinite sequence c_i. It is checked whether it is statistically a random sequence or not.

7.3.1 Sorting by Independent Variables

If the number of children mapped by an unknown function, c_i varies depending on x_j. This dependence allows c_i to resemble x_j, allowing the sequence to have some idiosyncratic peculiarity. Otherwise, the sequence is random. Thus, there exist no functions that can calculate the number of children.

There is a problem in that the data collected by the social survey is a random sample, and therefore, the partial sequence c_i is already randomized. Therefore, $c_i = f(x_{ij})$ (a record of the dataset) was sorted using several x_j as sorting keys (here, four keys were used: spouse's occupational prestige score (OPS)[3], respondent's education level[4], respondent's OPS[5], and race of respondent[6]). As a result of sorting in ascending order, the sequence c_i queues up as the socioeconomic conditions resembles nearer than before. Therefore, under the function assumption, the sequence c_i may have some idiosyncratic peculiarity (the next c_i may be similar). In other words, adjacent numbers represent the number of children of females from similar socioeconomic backgrounds[7].

7.3.1.1 List of Statistical Tests

[Statistical Tests] Several statistical tests were performed to evaluate the randomness of sequences. These included the following:

1. Test of equal probabilities[8]
    (a) Goodness-of-fit test—χ^2 test
2. Tests of randomness
    (a) Serial correlation test

[3]Variable name of GSS: SPRES80.
[4]EDUC.
[5]PRESTG80.
[6]RACE.
[7]When another key was added for sorting, the sequential order did not change.
[8]The probabilities are not exactly equal. This title was adopted for ease of understanding.

(b) Combination of goodness-of-fit test—χ^2 test

(c) Runs test

i. Length-of-runs test

ii. Runs-up test

iii. Runs-down test

The goodness-of-fit test, serial correlation test, combination of goodness-of-fit test, and length-of-runs test were performed. The tests were designed as follows. As a null hypothesis, the sequence is assumed to be random. This test to adopt the null hypothesis is strict for the random hypothesis. In social science fields, the null hypothesis is generally defined in order to reject it and adopt an alternative. In this fashion, the prior probability of rejecting the null hypothesis is large.

Thus, the probabilities of the number of children are not equal. Let p_i represent the probabilities of the number of children i. Based on the definition of Kolmogorov randomness, if a sequence is truly random, it is irreducible, and nothing else can be calculated in terms of relative frequencies. Let f_i be the frequency of the number of children and n be the total number of children of the sequence. Then, the probabilities p_i must be calculated as $\frac{f_i}{n}$.

Goodness-of-fit test

First, the GSS 1972–2016 cumulative dataset, downloaded from the GSS website http://gss.norc.org/, was used. Here, females who have been married were selected from the 1950–59 cohort. GSS samples were selected from the year 2000 onwards to constrain the number of children of completed fertility. As a result of this procedure, $n = 2349$ was obtained.

The goodness of fit was observed for every 100 numbers of these GSS data. Following this, the number of test intervals was changed, because the results of statistical tests are considerably dependent on the sample size.

The degree of freedom was eight. Thus, $\chi^2_{.05}(8) = 15.50731$ and 19/23 intervals (82.6%) passed (four intervals failed) the χ^2 test. The failure of the first two and the 13th and 14th intervals does not suggest the fallacy of H_0. As Figure 7.4 expresses, the number of children and the marriage duration are strongly correlated. Sorting by the keys (spouse's OPS and respondent's education level, OPS, and race) has the tendency to sort the sequence in ascending order of socioeconomic status. This sorting concentrated females in early marriages in the initial part of the sequence and those in late marriages in the later part.

The first 300 numbers were compared with the last 300 numbers in terms of time of marriage. A significant difference was detected in the age when the first child was born[9].

[9] In terms of age when first married, several of the samples are inapplicable (IAP) to this question. Therefore, it is not possible to calculate the time of marriage.

Table 7.3: Frequency of the Number of Children of the 1950s cohort as per the GSS.

Number of Children	Frequency	Probability
0	259	0.110260
1	464	0.197531
2	809	0.344402
3	498	0.212005
4	182	0.077480
5	69	0.029374
6	28	0.011920
7	22	0.009366
8 or more	18	0.007663
	2349	1

Table 7.4: χ^2 of every twenty-third interval.

interval	1—100	101-200	201-300	301-400	401-500
χ^2	72.7464	42.74243	9.178412	14.86441	5.11596
interval	501-600	601-700	701-800	801-900	901-1000
χ^2	1.990801	3.699001	9.116858	5.671002	6.738515
interval	1001-1100	1101-1200	1201-1300	1301-1400	1401-15000
χ^2	13.95228	15.0869	25.62631	16.89334	3.946327
interval	1501-1600	1601-1700	1701-1800	1801-1900	1901-2000
χ^2	14.45495	2.87552	6.665909	4.231914	8.679055
interval	2001-2100	2101-2200	2201-2349		
χ^2	3.463038	7.313305	13.20605		

A sociologist may insist that the females in the initial portion of the sequence desired several children, or a household economist may say that they have such a preference. These statements are mere explanations that cannot predict the prospective number of children. They are not scientific theories, but an *ex post facto* accounts. Time decisively governs animals' actions and behavior. Humans have no privileges in this regard.

The "differential fertility between social classes" is brought about by "time dependence of birth behavior." If the sequence is randomized in terms of the time of marriage, a perfect random sequence may be obtained. This suggests that "time" is an important factor affecting social behaviors. A legitimate scientific theory of human behavior must regard them in the dimension of time.

Even though the sequence depends on the time of marriage, the results of the tests of randomness demonstrate that it is irrefutably random. The following result of the test of randomness are highly precise.

Serial correlation test

Delayed one to five serial correlation coefficients were calculated for the GSS data used above. z_{rk} was calculated using Equation (7.5) under the assump-

Table 7.5: Delayed k serial correlation coefficients.

	Delayed 1	Delayed 2	Delayed 3	Delayed 4	Delayed 5
r	0.113857	0.082061	0.082559	0.051563	0.0966
z_{rk}	1.530487	1.103079	1.109771	0.693118	1.29851

tion that $\rho = 0$.

$$z_{rk} = \frac{\sqrt{n}\, r_{xy}^{k}}{\sqrt{13}} \tag{7.5}$$

For all $z_{rk} < 1.96$, which is the only conventional standard, we can regard the sequence random.

Compared with the actual quasi-random sequence generated by the Monte Carlo method, the correlation coefficients obtained herein are slightly higher. When the initial portion of the sequence is excluded, the correlation coefficients become lower. Therefore, the time dependence of this sequence causes this result. However, this does not contradict the assessment that there is no regularity in the sequence.

Combination of goodness-of-fit test

From Theorem 7.1, the observed sequence is a genuine sequence that has no regularity and presents itself based on the stochastic principle. Thus, the sequence is a combination of $0,1,2,3,4,5,6,7,8$ by the probabilities of p_i.

The frequencies of f_i are calculated for every nine numbers. To simplify the problem and employ the binomial theorem, let us classify them as the number 2 and "others" (for details, see Table 7.6). Therefore, $p_2 = .0.3444018731$ and $p_{\neq 2} = .0.655598127$ are obtained. Further, for each of the nine numbers, the frequency of k is calculated using the binomial theorem as follows:

$$f_k = {}_9C_k\, p^k (1-p)^{9-k} \qquad (k = 0,1,\cdots,9) \tag{7.6}$$

Table 7.6: Probabilities of f_k of 2 among ten numbers: combination test data.

2	Others	Probability	Expected	Observed
0	9	0.0223738	6	8
1	8	0.1057815	28	26
2	7	0.2222785	58	59
3	6	0.2724595	71	70
4	5	0.2146946	56	52
5	4	0.1127844	29	31
6	3	0.0394989	10	13
7	2	0.0088928	2	2
8	1	0.0011679	0	0
9	0	6.816941E-05	0	0

The result of the chi square test is $\chi^2 = 2.164495, \chi^2_{.05}(8) = 15.50731$. The null hypothesis that the sequence is random cannot be rejected[10]. For the cases $p_0, p_{\neq 0}$, $p_1, p_{\neq 1}$, $p_3, p_{\neq 3}$, H_0, and $p_5, p_{\neq 5}$, H_0 is not rejected. However, for $p_4, p_{\neq 4}$, H_0 is rejected.

Length-of-runs test

The runs test evaluates the randomness of the arrangement of the sequence. If the number of children emerges solely dependent on probabilities, it is easy to compute the length of runs of the number. For example, the probabilities of the length of runs of 2 can be computed as shown in Table 7.7.

Table 7.7: Length of runs of 2 (where the number of children is 2).

$$
\begin{array}{rcl}
2 & \rightarrow & (1-p_2)p_2\,(1-p_2) \doteq .148027 \\
2,2 & \rightarrow & {p_2}^2(1-p_2)^2 \doteq .050981 \\
2,2,2 & \rightarrow & {p_2}^3(1-p_2)^2 \doteq .017558 \\
2,2,2,2 & \rightarrow & {p_2}^4(1-p_2)^2 \doteq .006047 \\
2,2,2,2,2 & \rightarrow & {p_2}^5(1-p_2)^2 \doteq .002083 \\
2,2,2,2,2,2 & \rightarrow & {p_2}^6(1-p_2)^2 \doteq .000717 \\
2,2,2,2,2,2,2 & \rightarrow & {p_2}^7(1-p_2)^2 \doteq .000247 \\
2,2,2,2,2,2,2,\cdots & \rightarrow & \vdots\ (1-p_2)^2 \doteq \ \vdots
\end{array}
\tag{7.7}
$$

First, the expected frequencies were calculated using the probabilities, as shown in Table 7.7. The value of each run length was 2. The expected and observed frequency were compared. The goodness-of-fit test is $\chi^2 = 7.390338$, $\chi^2_{.05}(7) = 14.06714$. H_0 could not be rejected.

The same test was performed for the length of runs of $0,1,3,4,5$. For the runs of $1,3,4,5$, H_0 could not be rejected, but for the runs of 0, H_0 was rejected. I counted triples 0 five times for two expected. This deviance is the cause of the large χ^2 value.

7.3.1.2 Result of the Statistical Test of GSS 1950s Data

Here, the results of the statistical tests were rearranged. Surprisingly, it was found that the observed frequencies precisely matched the expected ones. It can be anticipated that human reproductive behaviors are simply stochastic phenomena.

1. Test of equal probabilities
   (a) —Goodness-of-fit test—χ^2 test: 19 intervals passed, while four failed.

[10]Some may say that expected frequencies such as 1 are too small to test. However, only the null hypothesis is being considered. H_0 is proposed, where the sequence is random; for this, small expected frequencies provide more rigidity. The a priori probability that H_0 is rejected is larger than that in the case where the expected frequencies are much larger.

2. Tests of randomness
   (a) Serial correlation test
   Delayed 1, 2, 3, 4, 5 $r < .12$. The delayed 1 is somewhat near the critical region; however, it is not significant in the normal standard.
   (b) Combination of goodness-of-fit test
      i. Number 0 $\chi^2 = 10.72987$, $\chi^2_{.05}(5) = 11.07050$. Passed.
      ii. Number 1 $\chi^2 = 4.767846$, $\chi^2_{.05}(7) = 14.06714$. Passed.
      iii. Number 2 $\chi^2 = 2.164495$, $\chi^2_{.05}(8) = 15.50731$. Passed.
      iv. Number 3 $\chi^2 = 5.413546$, $\chi^2_{.05}(7) = 14.06714$. Passed.
      v. Number 4 $\chi^2 = 14.34069$, $\chi^2_{.05}(5) = 11.07050$. Failed.
      vi. Number 5 $\chi^2 = 0.329524$, $\chi^2_{.05}(3) = 7.814728$. Passed.
   (c) Runs test
      i. Length-of-runs test
         A. runs of 0 $\chi^2 = 14.08367$, $\chi^2_{.05}(3) = 7.814728$. Failed.
         B. runs of 1 $\chi^2 = 3.210707$, $\chi^2_{.05}(4) = 9.487729$. Passed.
         C. runs of 2 $\chi^2 = 7.390338$, $\chi^2_{.05}(7) = 14.06714$. Passed.
         D. runs of 3 $\chi^2 = 5.33264$, $\chi^2_{.05}(5) = 11.0705$. Passed.
         E. runs of 4 $\chi^2 = 2.840323$, $\chi^2_{.05}(3) = 7.814728$. Passed.
         F. runs of 5 $\chi^2 = 0.561538$, $\chi^2_{.05}(2) = 5.991465$. Passed.

The above result suggests that the null hypothesis that the sequence is random cannot be rejected by means of GSS data in the case of the 1950s American female cohort.

7.3.1.3 Sequence by Another Sorting Method

A sequence sorted based on other keys is also considered. The keys are as follows: family income, spouse's prestige score, respondent income, spouse's education level, and race. The result of the statistical test for this sequence are following.

1. Test of equal probabilities
   (a) —Goodness-of-fit test—χ^2 test: 18 intervals passed, while five failed.
2. Tests of randomness
   (a) serial correlation test
   Delayed 1, 2, 3, 4, 5 $r < .126$. Delayed 2 is near the critical region; however, it is not significant to the normal standard of the two sides.
   (b) Combination of goodness-of-fit test
      i. Number 0 $\chi^2 = 4.706533$, $\chi^2_{.05}(5) = 11.07050$. Passed.
      ii. Number 1 $\chi^2 = 11.70658$, $\chi^2_{.05}(7) = 14.06714$. Passed.
      iii. Number 2 $\chi^2 = 6.02594$, $\chi^2_{.05}(8) = 15.50731$. Passed.
      iv. Number 3 $\chi^2 = 4.607304$, $\chi^2_{.05}(7) = 14.06714$. Passed.

v. Number 4 $\chi^2 = 6.06714$, $\chi^2_{.05}(5) = 11.07050$. Passed.
vi. Number 5 $\chi^2 = 0.743095$, $\chi^2_{.05}(3) = 7.814728$. Passed.

(c) Runs test

i. Length-of-runs test

A. runs of 0 $\chi^2 = 2.020042$, $\chi^2_{.05}(3) = 7.814728$. Failed.
B. runs of 1 $\chi^2 = 2.625423$, $\chi^2_{.05}(4) = 9.487729$. Passed.
C. runs of 2 $\chi^2 = 4.421956$, $\chi^2_{.05}(7) = 14.06714$. Passed.
D. runs of 3 $\chi^2 = 2.398226$, $\chi^2_{.05}(5) = 11.0705$. Passed.
E. runs of 4 $\chi^2 = 7.423656$, $\chi^2_{.05}(3) = 7.814728$. Passed.
F. runs of 5 $\chi^2 = 0.561538$, $\chi^2_{.05}(2) = 5.991465$. Passed.

Family income is not the key to distinguishing the number of children; the resulting sequence appears to have more randomness than that obtained by sorting by the key of OPS. Therefore, the effectiveness of the pronatalism income support system for childbearing and education is questionable.

As an empirical law, it appears that for any chosen set of keys, the order of the queue does not change significantly. The fundamental randomness may also remain unchanged.

7.3.2 Statistical Tests for Number of Japanese Children

The randomness of the number of children is a universal principle not limited to the United States. Here, this principle is verified for Japan using Social Stratification and Social Mobility (SSM) and JGSS data. In addition, the time dependence of the sequence of the number of children is also recognized. This is also a universal principle.

7.3.2.1 SSM Survey Data

> The SSM project has surveyed Japanese people's social status, social mobility, and consciousness of class and inequality, with national representative random samples.
>
> Source: Sigeto Tanaka, http://tsigeto.info/ssm/e.html.

This survey collected the number of children of respondents. Here, a sequence of the number of children was extracted from the 1995 SSM survey and statistically tested. The procedure to develop this sequence is presented below.

1. Concatenate the data of Sheet A and Sheet B for the number of children.
2. Exclude respondents who have never been married.
3. Extract respondents between the ages of 40–59.
4. Exclude IAP, NA, and DK.

5. Recode IAP, NA, and DK of the respondents' education level, their spouse's education level, and their income, which are the sorting keys used, to the code of the missing value.
6. Invert the order of the variable for the question "Is good education necessary to get ahead?" such that its value is directly related to the number of positive responses.

Thus, the higher educated, higher income, and higher education–oriented parents' number of children form the later portion of the sequence. As a result of the above procedure, the total number of children was found to be 2434. This sequence was statistically tested.

Results of SSM sequence

The sequence of the number of children is definitively random. Theorem 7.1 and the proof by the diagonal method merely negate the deterministic function of the number of children. There may be a stronger theorem that certifies its stochastic feature.

1. Test of equal probabilities
   (a) Goodness-of-fit test—χ^2 test. The data were divided into 300 numbers. Seven intervals passed, while one (601–900) failed.
2. Tests of randomness
   (a) Serial correlation test — Delay 1 to 5 $r_k < .07$. All passed.
   (b) Combination of goodness-of-fit test
   — For number 0, $\chi^2 = 10.08263$, $\chi^2_{.05}(4) = 9.48773$ Failed.
   — For number 2, $\chi^2 = 4.08266$, $\chi^2_{.05}(8) = 15.50731$ Passed. From numbers 1 to 5, this test was passed.
   (c) Length-of-runs test
      i. Length of runs of 0 $\chi^2 = 2.995575$, $\chi^2_{.05}(2) = 5.991465$. Passed.
      ii. Length of runs of 1 $\chi^2 = 0.58584$, $\chi^2_{.05}(3) = 7.814728$. Passed.
      iii. Length of runs of 2 $\chi^2 = 12.72127$, $\chi^2_{.05}(11) = 19.67514$. Passed.
      iv. Length of runs of 3 $\chi^2 = 7.437607$, $\chi^2_{.05}(5) = 11.0705$. Passed.
      v. Length of runs of 4 $\chi^2 = 0.016667$, $\chi^2_{.05}(2) = 5.991465$. Passed.
      vi. Length of runs of 5 $\chi^2 = 0.0$, $\chi^2_{.05}(1) = 3.841459$. Passed.

In Japan, low fertility has considerably eliminated the time dependence of the sequence of the number of children. Therefore, Japanese data can more easily be regarded in terms of a random sequence compared to American data.

7.3.2.2 JGSS Data

A procedure similar to the SSM data test was adopted to create a partial sequence of the number of children from JGSS data.

Procedure to develop a sequence from JGSS data

1. Exclude respondents who have never been married.
2. Extract respondents between the ages of 40–59.
3. Exclude IAP, NA, and DK.
4. Recode IAP, NA, and DK of the respondent's education level, their spouse's education level, and their income, which are the sorting keys used, to the code of the missing value.
5. Invert the order of the variable for the question "Is good education necessary to get ahead?" such that its value is directly related to the number of positive responses.

Thus, the higher educated, higher income, and higher education–oriented parents' number of children form the later portion of the sequence. As a result of the above procedure, the total number of children was found to be 4325. This sequence was statistically tested.

Results of JGSS sequence

1. Test of equal probabilities
   (a) Goodness-of-fit test—χ^2 test. The data were divided into 500 numbers. Eight intervals passed, while one (1–500) failed.
2. Tests of randomness
   (a) Serial correlation test — Delay 1 to 5 $r_k < .03$. All passed.
   (b) Combination of goodness-of-fit test
   — Each number from 0 to 4 passed this test.
   (c) Length-of-runs test
      i. Length of runs of 0 $\chi^2 = 0.104641$, $\chi^2_{.05}(3) = 7.814728$. Passed.
      ii. Length of runs of 1 $\chi^2 = 2.973626$, $\chi^2_{.05}(4) = 9.487729$. Passed.
      iii. Length of runs of 2 $\chi^2 = 3.384109$, $\chi^2_{.05}(11) = 19.67514$. Passed.
      iv. Length of runs of 3 $\chi^2 = 0.063736$, $\chi^2_{.05}(5) = 11.0705$. Passed.
      v. Length of runs of 4 $\chi^2 = 0.343042$, $\chi^2_{.05}(2) = 5.991465$. Passed.

For the various lengths of intervals, the goodness-of-fit test similarly suggests that the sequence of the number of children is random.

7.3.2.3 Comparison with the Hierarchical Differential Sequence

The proof by the diagonal method only indicates that there is no deterministic function of the number of children by individual parents. The proof and statistical tests suggest that the number of children is not consciously and individually determined. However, there is a level of randomness.

In terms of the level of randomness, the number of children is a more random sequence than the number of books read in a month. To understand the level of randomness of the number of children, a statistical test of the sequence of the number of books read is presented using the same sorted records.

Sequence of the number of books read in a month

JGSS included a question on the number of books the respondent reads in a month. The sequence is as in the below sequence.

$$4,1,0,2,0,0,1,2,2,4,9,1,8,0,0,8,0,\cdots$$

The numbers 1, 2, and 3 indicate the respective number of books, 4 indicates 4 or more books, 8 indicates IAP, and 9 indicates "no answer."

1. Test of equal probabilities
   (a) Goodness-of-fit test—χ^2 test. I divided the data into 500 numbers. Three intervals passed, while six (1–1000, 1501–2000, 3001–4235) failed. Zero readers were concentrated in the initial portion.
2. Tests of randomness
   (a) Serial correlation test — Delay 1 to 5 $r_k < .05$. All passed.
   (b) Combination of goodness-of-fit test
   For the number 0, $\chi^2 = 161.7183, \chi^2_{.05}(8) = 15.50731$. Failed.
   For the number 1, $\chi^2 = 0.17576$, $\chi^2_{.05}(7) = 14.06714$. Passed.
   For the number 2, $\chi^2 = 10.57678$, $\chi^2_{.05}(5) = 11.0705$. Passed.
   For the number 3, $\chi^2 = 6.006616$, $\chi^2_{.05}(4) = 9.487729$. Passed.
   For the number 4, $\chi^2 = 12.14069$, $\chi^2_{.05}(4) = 9.487729$. Failed.
   For the number 8, $\chi^2 = 15.90927$, $\chi^2_{.05}(5) = 11.0705$. Failed.
   (c) Length-of-runs test
      i. Length of runs of 0 $\chi^2 = 18.81354$, $\chi^2_{.05}(9) = 16.91898$. Failed.
      ii. Length of runs of 1 $\chi^2 = 4.47564$, $\chi^2_{.05}(5) = 11.0705$. Passed.
      iii. Length of runs of 2 $\chi^2 = 4.68534$, $\chi^2_{.05}(3) = 7.814728$. Passed.
      iv. Length of runs of 3 $\chi^2 = 9.889474$, $\chi^2_{.05}(2) = 5.991465$. Failed.
      v. Length of runs of 4 $\chi^2 = 2.25350$, $\chi^2_{.05}(3) = 7.814728$. Passed.
      vi. Length of runs of 8 $\chi^2 = 75.40905$, $\chi^2_{.05}(4) = 9.487729$. Failed.

The expected hierarchical difference is readily observable. Clearly, not all variables construct completely random sequences.

7.3.3 Older (Partial) Sequences

The randomness of the sequence of the number of children of parents is a ubiquitous and general law that is independent of time and space. Namely, the manner in which the stochastic variables occur is the problem to be solved.

The birth function will be argued in the following chapter. However, prior to this, the ubiquitous nature of the randomness of the sequence of the number of children is verified.

7.3.3.1 U.S. 1930–1939 Cohort

The 1930s cohort is explored because it experienced a "baby boom," the massive reproductive period in post-war fertility history. This large-scale event might affect the randomness of the sequence of the number of children.

1. Test of equal probabilities
   (a) Goodness-of-fit test—χ^2 test. The data were divided into 200 numbers. 11 intervals passed, while seven (1–600, 1001–1600, 3400–3761) failed.
   The numbers 6, 7, 8 were concentrated in the initial portion.
2. Tests of randomness
   (a) Serial correlation test — Delay 1 to 5 $r_k < .071$. All passed.
   (b) Combination of goodness-of-fit test
   For the number 0, $\chi^2 = 14.13177, \chi^2_{.05}(5) = 11.0705$. Failed. Each number from 1 to 4 passed this test.
   (c) Length-of-runs test
      i. Length of runs of 0 $\chi^2 = 8.53026$, $\chi^2_{.05}(3) = 7.814728$. Failed.
      ii. Length of runs of 1 $\chi^2 = 2.746829$, $\chi^2_{.05}(3) = 7.814728$. Passed.
      iii. Length of runs of 2 $\chi^2 = 1.416971$, $\chi^2_{.05}(5) = 11.0705$. Passed.
      iv. Length of runs of 3 $\chi^2 = 13.15891$, $\chi^2_{.05}(5) = 11.0705$. Failed.
      v. Length of runs of 4 $\chi^2 = 3.166014$, $\chi^2_{.05}(4) = 9.487729$. Passed.
      vi. Length of runs of 5 $\chi^2 = 0.672576$, $\chi^2_{.05}(3) = 7.814728$. Passed.

The null hypothesis is still not rejected, although the probabilities of the number of children are altered.

7.3.3.2 U.S. 1910–1919 Cohort

Next, the 1910s cohort is explored because it experienced the "Great Depression" and World War II, two significant disastrous events in history.

1. Test of equal probabilities
   (a) Goodness-of-fit test—χ^2 test. The data were divided into 200 numbers. Seven intervals passed, while three (1–200, 1201–1400, 2001–2340) failed.
   The numbers 6, 7, 8 concentrate in the initial portion. This feature is the same as in the 1930s.
2. Tests of randomness
   (a) Serial correlation test — Delay 1 to 5 $r_k < .082$. All passed.
   (b) Combination of goodness-of-fit test
   — From 0 to 4, each number passed this test.

(c) Length-of-runs test

i. Length of runs of 0 $\chi^2 = 1.870744$, $\chi^2_{.05}(4) = 9.487729$. Passed.
ii. Length of runs of 1 $\chi^2 = 5.020445$, $\chi^2_{.05}(4) = 9.487729$. Passed.
iii. Length of runs of 2 $\chi^2 = 2.880237$, $\chi^2_{.05}(5) = 11.0705$. Passed.
iv. Length of runs of 3 $\chi^2 = 3.430477$, $\chi^2_{.05}(4) = 9.487729$. Passed.
v. Length of runs of 4 $\chi^2 = 1.792839$, $\chi^2_{.05}(3) = 7.814728$. Passed.
vi. Length of runs of 5 $\chi^2 = 1.116208$, $\chi^2_{.05}(3) = 5.991465$. Passed.

The significant historical disasters did not alter the randomness of the sequence of the number of children. From the observer's perspective, the numbers of children are random; however, our intentions are considered to be conscious. From the past to the present, the sequence of the number of children is random. However, as shown, the probabilities change. The problem that should be solved is the manner in which this alteration occurs. Conscious free will does not alter probability.

7.4 Irreducibility of the Sequence of the Number of Children

Do functions of the number of children, such as Becker's (Equation (7.2)) or BMT's (Equation (7.3) on p. 172) have informational value? The sequence of the number of children that can be regarded as random has already been examined. "Kolmogorov complexity" tells us that the sequence itself may be the shortest length.

Individual birth functions such as Becker's or BMT's imply that future utility is expected, and the number of children is consciously calculated through rational reasoning. They require enormous amounts of information to determine the numbers of children. Contrastingly, the birth function proposed in the following chapter calculates the probability of birth based entirely on the previous neighboring state (sequentially and iteratively). It has been insisted that a more reducible approach to decide the number of children is possible.

When considering the problems of randomness of sequence and geographical dimension, I noticed that the necessity of two-dimensional space is similar to a two-dimensional cellular automata. The birth function had already been enhanced to multi-dimensional space in order to acquire the dynamics and superior features of the birth function independently from the cellular automata. However, while G. Chaitin [8] indicated algorithmic irreducibility in a cellular automata universe, the randomness of the sequence of the number of children coincides with the irreducibility he pointed out.

7.4.1 Evaluation of the Function Using Information Theory

It is important to evaluate the value of the birth function as an information quantity. The sequence of the number of children is composed of 2^4 bits at most. As the sequence is infinite, using Cantor's ω, the information quantity of the sequence is

$$2^4\omega \tag{7.8}$$

Then, assuming that the equations that compute the number of children have concrete programs, what is the length of their programs? They must have a minimum length of 2^4, because they must return the number of children for given inputs. Let C be the minimum amount of the program. If the input-independent variables are infinite integers (such as 8-bit integers)

$$(2^8\omega + 2^4 + C)\omega \tag{7.9}$$

If independent variables are finite, say $2^8 = 1\text{byte}$,

$$(2^8 + 2^4 + C)\omega \tag{7.10}$$

When the independent variables contain real numbers, the information quantity is larger than that in Equation (7.10). Assume that an individual birth function is an extravagant waste of information. The function's significance as information is completely lost. Despite this waste of information, the individual birth functions cannot concretely predict the number of children.

In contrast to this wastage, the birth function proposed herein is parsimonious.

7.5 Conscious Choice of Birth—Distribution of the Number of Children

Almost all disciplines that consider birth behaviors assume that females or parents consciously choose the number of their children overtly or implicitly. However, they cannot propose a theory that can concretely compute the distribution of the number of children. From this perspective, this assumption is entirely unscientific. Those who believe it should answer relevant following questions; nevertheless, I will consider them here.

The sequence of the number of children of parents (females) has no regularities, despite being sorted based on socioeconomic keys. Therefore, it is likely that females under similar conditions have no tendency to choose the same number of children. Namely, it is a logical consequence that conscious choices are not relevant to the coordination of the number of children.

7.5.0.1 Free Choices are Random

Utilitarians insist that human choices are made freely and consciously. This assertion is a preconception, not a logical argument. According to them, the number of children depends upon external and internal conditions. However, it has been shown that similar external conditions bring about a random number of children. Thus, internal free and conscious choice is arbitrary. Namely, "free and conscious choice" is equivalent to a random choice.

Human consciousness resists this conclusion. However, sociologist T. Parsons analyzed the ends of utilitarians as follows.

> Hobbes' system of social theory is almost a pure cause of utilitarianism, according to the definition of the preceding chapter. The basis of human action lies in the "passions." These are discrete, randomly variant ends of action: "There is no common rule of good and evil to be taken from the nature of the objects themselves." In the pursuit of these ends, men act rationally, choosing, within the limitations of the situation, the most efficient means. But this rationality is strictly limited, reason is "servant of the passions" it is concerned only with questions of ways and means.
>
> T. Parson [61], p. 90.

If a utilitarian postulates that humans make a free conscious choice when bearing children, this choice may be random. Otherwise, human "passion" is not synchronous or sympathetic. Namely, utilitarianism and free conscious choice must accept random birth as an ends of their behaviors. It is impossible to compute the future number of children of parents with random passion.

It is somewhat strange that free choices, based on conscious will, result in random choices. However, this is an inevitable logical consequence. Random choices are relevant to the postulation of the single birth probability p and the binomial distribution of the number of children. Expectedly, this contradicts the empirical distribution of the number of children. Another empirical contradiction will also be proposed.

7.5.0.2 Ideal Number of Children and the Number of Children in America

The most common and probable statement is that humans will consciously attempt to achieve their ideal number of children. This hypothesis is no more than a primitive preconception; the data collected by social surveys do not support it.

To begin with, using GSS data, the relation between the number of children and the ideal number of children was analyzed. The sample of the 1910s American cohort produced the following results. See Figure 7.8. If every female attempts to achieve her ideal number of children, these different distributions of the number of children might not have occurred. There is no indication that the ideal number of children of respondents was the cause of their real number of children. The correlation coefficient is .152; though it is significant statistically, there appears to be no correlation. The causality between the two variables is not discernible from the scatter diagram (Figure 7.9).

For the younger generation (1970s cohort), the effect of the ideal number of children on reality is lost in a maze. It is surprising that the distribution of the ideal number of children is the largely similar. It is difficult to imagine

Table 7.8: Number of children and ideal number of children of the 1910s American cohort.

	Number of Children	Number of Ideal Number of Children
0	13.14	1.20
1	17.87	1.06
2	24.94	41.04
3	18.62	26.16
4	11.28	18.33
5	5.26	2.59
6	3.14	1.13
7	1.81	0.73
EIGHT OR MORE / As many as what	3.94	7.77
Frequency (Missing)	2264 (3)	2264 (758)

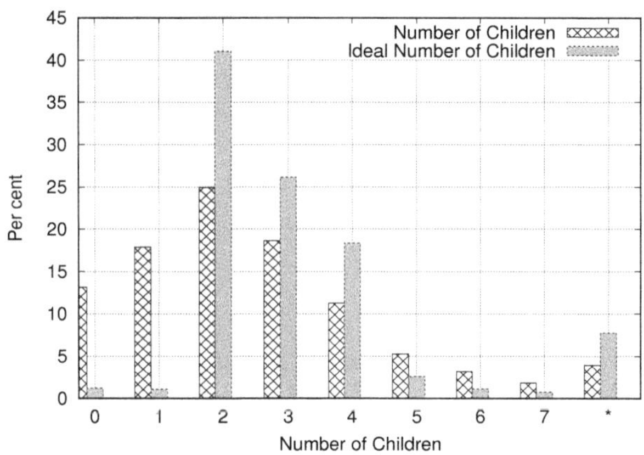

*Eight or more/as many as given.

Figure 7.8: Number of children and ideal number of children of 1910s American cohort.

that every female would achieve their ideal number of children. The correlation coefficient was higher than that in the 1910s, but it remained at .199.

7.5.0.3 Ideal Versus Actual Number of Children in Japan

On analyzing the Japanese distribution of the number of children, it appears that the ideal number has no causal effect on the distribution of the actual number of children. The correlation coefficient is also low (.168). Like Americans, Japanese people did not generally bear their ideal number of children. The scatter diagram (Figure 7.13) shows no relation between the two variables.

Observing the younger generation in Japan, it can be said that no relationship is universal. It can be seen that the correlation coefficient increases

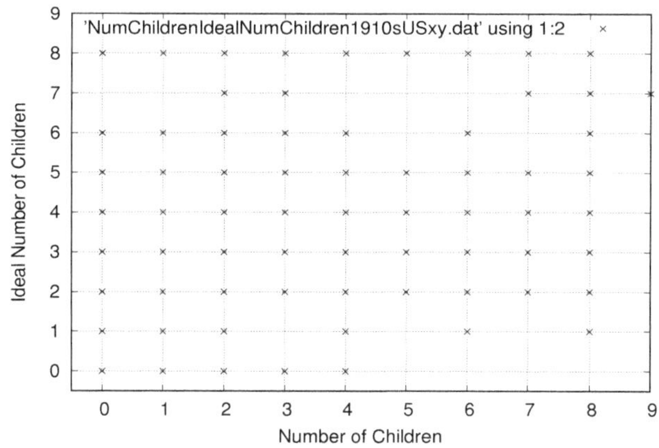

Figure 7.9: Scatter diagram of the number of children and the ideal number of children of the 1910s American cohort.

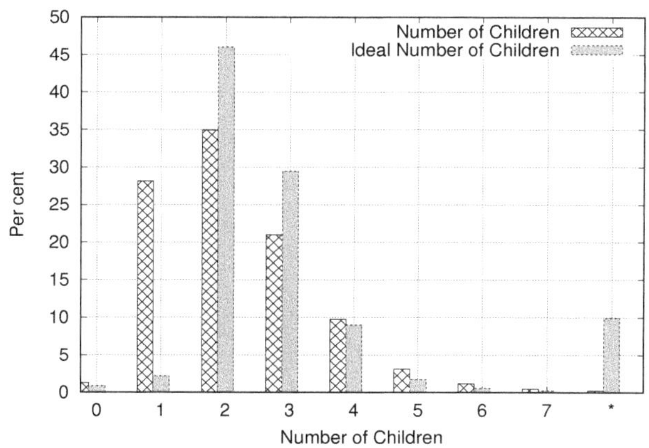

*Eight or more/as many as given.

Figure 7.10: Number of children and ideal number of children of the 1970s American cohort.

to .277. However, this is not because the younger Japanese generation began to achieve their ideal number of children. The tendency to converge to the number 2 brought about these computational effects. It is difficult to find a relationship between every phenomenon. Although humans are programmed to believe that our behaviors are consciously decided, theories based on empirical data must be considered. The ideal number of children is not considered in birth function theory; the distribution of the number of children is calculated without it.

On rephrasing the ideal number of children to "the desired number of children" or "the planned number of children" no definite relationship is observed.

Table 7.9: Number of children and ideal number of children of the 1970s American cohort.

	Number of Children	Number of Ideal Number of Children
0	1.27	0.86
1	28.12	2.19
2	34.91	46.01
3	20.98	29.50
4	9.77	9.00
5	3.11	1.72
6	1.14	0.55
7	0.48	0.23
EIGHT OR MORE	0.22	
/ As many as given		9.94
Frequency(Missing)	2283(5)	2283(1005)

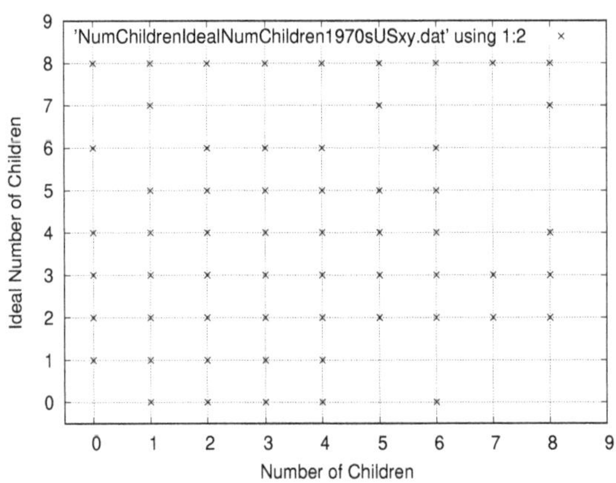

Figure 7.11: Scatter diagram of the number of children and the ideal number of children of the 1970s American cohort.

There is no valid model of the number of children based on the ideal number of children. This relationship is an illusion based merely on intuition.

The distribution of the number of children is not a result of free and conscious choice. If one was to say that other animals intend to bear a certain number of children, the idea would be ridiculed. Then, why is mankind considered to have such privileges?

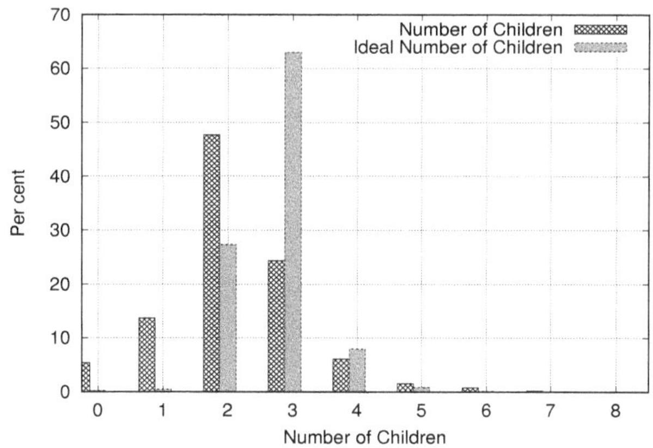

Figure 7.12: Number of children and ideal number of children before the 1950s Japanese cohort.

7.5.0.4 Ideal Number of Children is a Response to the Number of Children in High Dimensional Space

Utilitarians and adherents of free conscious choice take it for granted that humans freely and arbitrarily hold in mind their ideal or desired number of children. This is a superficial preconception of observed data.

See Figure 7.16. The GSS data present the averages of the ideal number of children, which change by cohort. The baby boom in the U.S. was a massive demographic event, but it was not sufficiently large to change the structure of the global high-dimensional space. Except during the baby boom, the mean ideal number of children undulates in sync with the mean number of children. The ideal number of children is essentially a shadow of reality.

Our conscious ideal number of children varies based on reality; however, the manner in which this ideal is envisioned does not vary.

This relationship between reality and consciousness will be similar in Japan. From the 1970s to the present, the number of children per family has very gradually and slightly declined. The ideal and planned number of children have followed this decline. Why do we always assume that conscious will is the cause of human actions? It is a result in every way.

7.6 Necessity of Multi-Dimensional Space

Although the sequence of the number of children is random, the probability does change and undulate. Considering only one-dimensional space—a simple sequence of the number of children—these various undulations cannot be described. The degree of dimensions of space must be extended in order to satisfactorily cxplain the dynamics of the random sequence.

Table 7.10: Number of children and ideal number of children before the 1950s Japanese cohort.

	Number of Children	Number of Ideal Number of Children
0	5.42	0.21
1	13.75	0.46
2	47.67	27.32
3	24.33	62.97
4	6.13	7.95
5	1.58	0.88
6	0.79	0.17
7	0.25	0.04
8	0.08	
	2405(5)	2390(15)

Table 7.11: Number of children and ideal number of children before the 1950s Japanese cohort.

	Number of Children	Number of Ideal Number of Children
0	10.30	
1	19.29	1.20
2	46.90	36.18
3	20.83	57.78
4	2.26	4.25
5	0.36	0.54
6		
7		0.06
	1680(7)	1672(15)

Sorting by the key of any socioeconomic variable cannot break the random property of the sequence. The only way to arrange them in clusters based on specific properties is to arrange them in the order of time and space. Moreover, it is not possible to continue to arrange them in simpler sequences.

7.6.1 Necessity of Geographical Dimensions

Fertility ($\doteqdot$number of children per family) has geographical differences. Namely, the number of children can be uniquely arranged on the geographical surface. Regardless of the number of independent socioeconomic variables that are added, the sequence of the number of children cannot be arranged in a regular manner. This suggests that socioeconomic hypereuclidean space is not relevant to the theory of the number of children. When geographical

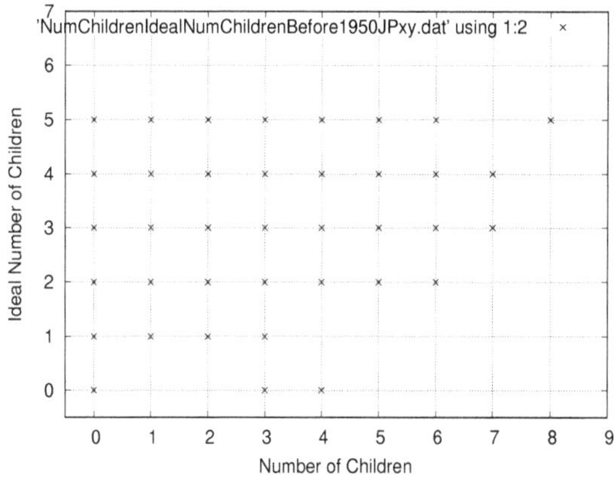

Figure 7.13: Scatter diagram of the number of children and the ideal number of children before the 1950s Japanese cohort.

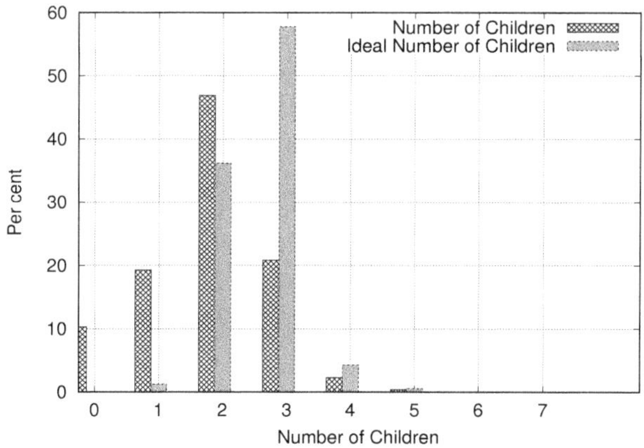

Figure 7.14: Number of children and ideal number of children following the 1950s Japanese cohort.

x, y dimensions are adopted, the number of children can be arranged using a stochastic and theoretical formulation.

7.6.1.1 Consideration of Geographical Dimensions

Now, the geographical difference in fertility (the number of children of parents) can easily be found, as shown in Figures 2.1—2.4 (pp. 21–22). Namely, in the two-dimensional space, the numbers of children in families can be arranged in a unique order of significance for the first time. I have realized that two-dimensional space is necessary to meaningfully arrange them.

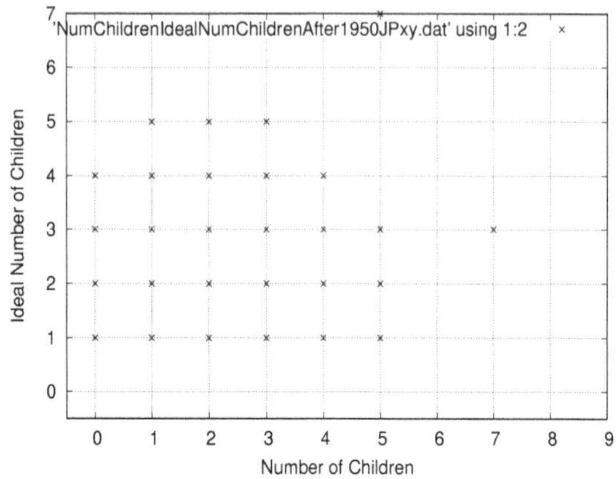

Figure 7.15: Scatter diagram of the number of children and the ideal number of children following the 1950s Japanese cohort.

The number of children of parents is countable, but no meaningful mapping can be defined to them in the hypereuclidean space of socioeconomic variables. The mapping of birth can only be defined in geographical space.

For a long time, the social sciences have wrongly neglected the importance of geographical dimensions. It is time to recognize their fundamental necessity in order to describe human actions. Humans cannot transcend geographical distance even when utilizing networks with perfect freedom. We live in local spaces with neighboring people. This chapter has been titled "No Individual Birth Functions Exist" with these connotations. Humans, like all animals, are limited by geographical constraints.

7.6.2 Necessity of Time Dimension for Observation

The number of children of parents changes over time. It begins from zero and consistently increases by one[11]. Objectively observing this phenomenon, the number of children as a quantum alters over time, not based on human conscious will. The development of this quantum and the realization of this regularity must be investigated. Therefore, a time dimension is necessary to effectively observe the birth function.

Sociologists such as M. Weber foolishly assumed human conscious will as a cause of behaviors. However, the causal relationship has significance solely in the time dimension. Time is necessary to describe the alteration of human birth behaviors. When the relationship between preceding and present actions is accurately formulated, the causality is naturally constituted.

[11] There are exceptions,however.

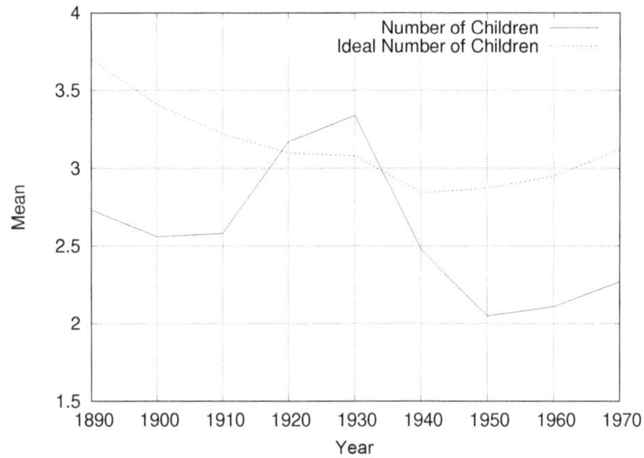

Figure 7.16: Undulation of the mean number of children and mean ideal number of children (GSS).

Assuming only the two-dimensional space, the numbers of children of parents cannot be meaningfully arranged. Additionally, to trace dynamic alterations, the third dimension is necessary–the time dimension.

Human behaviors are entirely dependent on time and space; we are naturally constrained by them.

7.7 Conjecture

Principle 7.7.1 (Causes of behaviors)
The causes of current behaviors are limited to previous behaviors. Human conscious will does not cause our behavior. The causal relation should be based simply on the relation of behavior.

Chapter 8

Birth Function for a Cohort

> PROP. XXVIII. Every individual thing, or everything which is finite and has a conditioned existence, cannot exist or be conditioned to act, unless it be conditioned for existence and action by a cause other than itself, which also is finite, and has a conditioned existence; and likewise this cause cannot in its turn exist, or be conditioned to act, unless it be conditioned for existence and action by another cause, which also is finite, and has a conditioned existence, and so on to infinity.

Spinoza, *Ethics (Ethica Ordine Geometrico Demonstrata)* [16], p. 27.

8.1 Only One Birth Function Exists

There is no function that can determine the number of children. However, there certainly exists one that can determine the probability of birth. Using this approach, all the behaviors starting from the era before fertility decline started to the present day as well as those in the future should be explained.

The *demographic transition theory* and *second demographic transition theory* set simple premises to explain the alteration of human birth behaviors. These development stage theories provide mere explanations and terminologies at most and cannot produce any theoretical predictions. The *second demographic transition theory* by van de Kaa [75] or Lesthaeghe [49] claims that the very low fertility of Mediterranean and far-east countries is partly caused by the alteration of the value of marriage. However, when fertility increases in these countries again, they should build the *third demographic transition theory* based on the fact that the value of marriage will change once more. Clearly, the development stage theory is an *ad hoc* illustration that cannot produce any deductive reasoning. What should be pursued is not a likely explanation, but a predictable and falsifiable theory.

In this chapter, a single birth function that can universally describe human birth behaviors longitudinally and cross-culturally is investigated. Such a birth function is possible because through the reaction-diffusion theory, it was confirmed that in both Western countries and Eastern Japan, the wave of the decline in the number of children travelled with similar velocities. Therefore, I assume that birth function is identical for all of humanity.

8.1.1 Birth Process

Consider the simplest case of a single cohort.

A female in a cohort acts out her reproductive behaviors as follows:

1. Physiologically effective fecundate: reaching sexual maturity to bear children.
2. Marriage—include common law marriage and exclude married individuals who have no possibility of intercourse.
3. Birth.

The birth pattern of a cohort can be observed based on the age groups of married individuals, as shown in Table 8.3. This is arranged normally, with the marriage time of zero taken as the starting point. This listing method is standard but flawed because it ignores the chronological order.

Time is the main factor affecting every event in social phenomena. It is the central aspect of causality. The event of birth should be considered in terms of time. According to this law, Table 8.3 must be rearranged chronologically, as shown in Table 8.4. Figure 8.1 shows this chronological rearrangement graphically. **When they are chronologically ordered, the causal relation between events can be effectively understood.**

8.2 Calculation of Birth Probabilities by Age and Specific Age at Marriage

Here, birth probabilities are calculated by age and specific age at marriage, by availing the vital statistical data based on the work by Ito & Bando [40]. It must be noted that their calculations are for the period birth probabilities which is measured in a calendar year. Here, these data are interpreted as virtual cohort data[1].

According to the vital statistics of Japan [55], in 1985, 12262 babies were borne by women who were twenty years old and within one year of marriage in

[1] This is a standard virtual method for theoretical considerations.

Japan. Expectedly, these were women were aged between 19–20 and married[2] in 1984 or 1985. However, their birth probabilities are not equal.

Table 8.1: Population likely to bear children at age 20 within a year of marriage in 1985.

	1984	1985
Age 19	13654	11984
Age 20	25277	25191

For these calculations, I assume that births and marriages occur equally in a period.

Females married in December 1985 have nearly equal zero probabilities. Females married under the age of 20 have high probabilities of bearing children. Those who were married in 1984 at age 20 had high probabilities of bearing children. Namely, the population who married in 1984 and 1985 must be carefully considered based on the age at marriage.

Let me consider the birth probability of a female who married in 1984 under the age of 20. A woman who married on January 1st, having recently turned 19 years old, had nearly the entire year to bear a child, but had a small probability to bear at 20 years old. On the other hand, a 19 years and 11 months old female married on December 31 had nearly zero probability of bearing a child at age 19.

Contrarily, on December 31, 1984, a female aged 19 years, 11 months, and 30 days had the entire year to bear a child at age 20. A marriage with a 19-year-old female in June could not produce a child until the following June, when she turned 20 years old. Therefore, the probability of bearing a child at the age of 20 is proportional to the month in which the marriage occurs.

Let x be the age (as a continuous time variable) at which the marriage took place in a year. Thus, the domain of x is $0 \leq x \leq 1$.

In 1985, females who married in 1984 at age 19 make up the population that had not borne a child in 1984 and bore one in 1985 within a year of marriage. Thus, their birth probability can be expressed as

$$(1-x)^2 = x^2 - 2x + 1 \tag{8.1}$$

This is integrated from 0 to 1 (for one year).

$$\left[\frac{1}{3}x^3 - x^2 + x\right]_0^1 = \frac{1}{3} \tag{8.2}$$

The weight of birth probability for the combination of 19-year-old females married in 1984 was $\frac{1}{3}$.

[2]Common law marriage is included in this definition of marriage.

In 1985, females who married in 1984 at age 20 make up the population who did not give birth to a child in 1984 but bore one in 1985 at the age of 20. Thus, their birth probability can be expressed as

$$(1-x)x = x - x^2 \tag{8.3}$$

This is integrated from 0 to 1 (for a year).

$$\left[\frac{1}{2}x^2 - \frac{1}{3}x^3\right]_0^1 = \frac{1}{6} \tag{8.4}$$

The weight of birth probability for the combination of 20-year-old females married in 1984 was $\frac{1}{6}$.

Females who married in 1985 at age 19 make up the population that did not bore a child at age 19 and bore one in 1985 at age 20. Thus, their birth probability can be expressed as

$$x(1-x) = x - x^2 \tag{8.5}$$

This is integrated from 0 to 1 (for a year).

$$\left[\frac{1}{2}x^2 - \frac{1}{3}x^3\right]_0^1 = \frac{1}{6} \tag{8.6}$$

The weight of birth probability for the combination of 19-year-old females married in 1985 was $\frac{1}{6}$.

In 1985, females who married at age 20 make up the population who had borne a child at age 20 and did not bear one at age 21. Thus, its probability can be expressed as

$$(1-x)^2 = x^2 - 2x + 1 \tag{8.7}$$

This is integrated from 0 to 1 (for a year).

$$\left[\frac{1}{3}x^3 - x^2 + x\right]_0^1 = \frac{1}{3} \tag{8.8}$$

The weight of birth probability for the combination of 20-year-old females married in 1985 was $\frac{1}{3}$.

Thus, the following weighting matrix can be obtained for the populations of Table 8.1. The latent population that can bear a child in 1985 within a year of marriage is computed by the following equation:

$$\begin{aligned} & M* \\ = \ & \frac{1}{6}\left(2M_{1984}^{19} + M_{1985}^{20} + M_{1984}^{20} + 2M_{1985}^{20}\right) \end{aligned} \tag{8.9}$$

Thus, the latent population who can bear a child in 1985 is computed as follows:

Table 8.2: Weights of birth probabilities.

	1984	1985
Age 19	1/3	1/6
Age 20	1/6	1/3

$$\frac{1}{6}\left(2\times 13654+11984+25277+2\times 25191\right)=19158.5$$

Further, 12262, which is the number of children borne by 20-year-old females within a year of marriage, is divided by the value obtained the latent population above. Thus birth probability of females who married at age 20 in 1985 in Japan is

$$\frac{12262}{19158.5}=0.64002923.$$

The result[3] of computation is the probability of birth based on age and specific age at marriage. By repeatedly performing the same calculation for a retrospective marriage year, a set of birth probabilities is obtained (Table 8.3).

Compared Figure 8.1 with Figure 8.2 which presents next year birth probability of marriage age specific, we can realize the similar patterns are replicated.

8.2.1 Birth Probabilities in 1939 in Japan

In 1941, Ayanori Okazaki presented the 1939 birth probabilities in Japan [59]. Here, they have been rearranged in chronological order, as shown in Table 8.5. This probability is not measured in the same manner as Table 8.4. Okazaki's data counted the occurrences of births in a calendar year. Therefore, I recomputed them in accordance with actual time as accurately as possible. The results of the re-computation are shown in Table 8.6.

Figure 8.3 and Figure 8.4 illustrate these data. Comparing Figure 8.4 with Figure 8.1, birth probabilities in Japan have changed significantly with the passage of time. Moreover, in 2010, birth probabilities have transformed dramatically to reflect very low fertility, as shown in Figure 8.5.

The birth function must naturally and independently explain these alterations. The hypotheses of household economics do not explain this dynamic variation in birth probabilities. Existing explanations always depend on subjective and conscious alterations. The theory depending on human consciousness cannot yield the morphological dynamics of birth and marriage functions. That is, human will cannot explain the curved lines constructed by birth probabilities. As Spinoza said, human conscious will is not the true cause at all. A reasonable and natural theory must transcend human conscious will. They are just explanations which every one can understand. Intelligibleness is not enough to be scientific.

[3]The values of Table 8.3 are multiplied by 1000.

Table 8.3: Birth probabilities of Japanese females in 1985.

Age at Marriage	Birth Rate specific Duration of Marriage (in 1000 pair)															
	0	1	2	3	4	5	6	7	8	9	10	11	12	13	14	15-19
15	34.9	1808.2	767.8	1141.1	974.0	876.9	913.0	731.7	468.8	865.4	878.0	487.5	417.7	850.4	345.3	252.2
16	772.8	788.2	517.1	412.3	342.4	284.5	211.2	149.6	173.5	154.4	188.1	96.8	83.3	119.1	63.4	119.0
17	830.2	583.5	388.5	346.3	287.7	212.7	179.0	161.5	118.6	114.5	113.7	53.8	48.9	39.9	39.5	65.1
18	744.9	397.4	352.6	294.6	221.0	192.0	142.7	115.4	100.3	82.7	71.2	39.6	34.8	25.9	18.7	47.2
19	738.4	365.3	339.8	291.7	218.3	183.7	141.7	118.6	90.4	73.5	55.3	38.8	29.0	20.2	16.8	32.3
20	640.0	347.0	316.7	286.3	225.2	181.5	139.0	105.7	78.4	61.0	47.5	30.1	23.6	16.4	11.8	21.8
21	509.9	355.7	310.8	293.1	232.1	185.4	134.4	97.3	77.6	55.6	40.5	25.4	17.6	11.7	7.5	14.8
22	435.5	380.2	308.5	305.8	250.1	187.0	137.2	102.5	72.9	50.4	35.4	23.4	15.0	9.4	6.2	10.6
23	384.3	394.9	305.9	317.2	250.7	186.6	134.3	98.4	67.2	47.1	32.0	20.2	12.2	8.2	4.9	7.4
24	357.2	395.9	299.3	313.7	244.0	186.1	132.3	92.1	64.8	41.6	28.3	17.1	11.0	6.4	3.7	4.9
25	357.1	393.2	294.6	305.1	242.6	179.1	124.2	84.8	57.9	37.5	23.9	14.4	8.3	5.9	2.9	3.3
26	366.0	366.4	283.3	297.9	224.1	162.0	114.2	77.8	50.0	31.0	20.5	12.0	7.3	3.7	2.1	2.2
27	375.3	350.0	280.9	289.3	203.1	151.1	103.6	67.9	42.3	27.2	15.7	9.7	4.6	2.9	1.9	1.4
28	397.0	350.7	283.6	264.0	195.2	142.1	92.6	60.8	36.2	23.5	14.1	6.8	4.1	2.0	1.2	0.9
29	416.0	333.8	266.8	251.4	185.9	128.0	80.3	49.6	30.5	18.8	9.6	6.2	3.0	1.5	1.3	0.5
30	420.4	316.3	271.4	231.6	166.1	111.7	69.5	42.1	24.1	13.2	8.3	3.9	1.7	0.7	0.6	0.3
31	395.2	314.5	259.9	215.9	149.9	98.8	56.0	32.9	17.1	10.0	5.0	4.5	0.9	0.6	0.7	0.2
32	383.7	308.6	239.5	196.8	125.6	82.7	44.0	23.4	14.6	9.7	5.1	3.1	1.0	0.3	0.0	0.1
33	361.1	296.4	217.7	166.9	112.4	65.4	38.6	20.8	11.1	6.5	2.5	1.1	0.7	0.0	0.0	0.0
34	362.8	256.4	198.2	144.2	88.1	46.7	28.0	15.9	8.0	2.6	2.2	1.0	0.0	0.0	0.0	0.0
35	320.6	239.3	177.0	118.2	65.1	36.5	20.7	8.3	3.4	2.2	0.7	0.0	0.2	0.0	0.0	
36	295.2	219.4	136.9	89.0	47.3	23.8	13.1	5.5	3.2	0.3	0.3	0.3	0.0	0.0		
37	253.3	183.7	95.9	71.0	35.3	18.6	5.8	1.2	2.1	0.7	0.3	0.0	0.3			
38	213.4	135.4	80.1	45.1	27.1	10.1	5.4	1.2	0.4	0.0	0.0	0.0				
39	157.9	115.5	63.9	35.8	15.3	7.0	1.4	1.0	0.0	0.5	0.0					
40	118.1	88.4	37.5	16.2	9.3	2.1	0.6	0.6	0.5	0.5						
41	80.3	65.4	19.7	8.4	3.0	1.3	0.6	0.0								
42	55.3	32.5	10.0	5.8	1.4	0.7	0.0									
43	23.2	14.6	4.8	1.5	0.8	0.0										
44	18.8	4.4	2.5	0.8	0.0											
45	8.5	2.6	0.8	0.0												
46	3.6	0.0	0.0													
47	1.0	0.0														
48	0.0															
49	0.0															

Source: ITO&BANDO [40], p. 61. Re-tabulated and translated by author. *Original last four columns about the expected ever born children are omitted.

Table 8.4: Rearranged birth probabilities in Chronological Order.

Marriage Duration	1	2	3	4	5	6	7	8	9	10	11	12	13	14	15
Age 16	.77279	.78824	.51712	.41234	.34243	.28451	.21121	.14963	.17346	.15438	.1881	.09677	.08326	.11908	.06344
17		.83017	.58353	.38855	.34630	.28768	.21268	.17902	.16146	.11860	.11446	.11369	.05384	.048932	.039905
18			.74495	.39739	.35263	.29463	.22095	.19196	.14271	.11541	.10026	.08266	.07125	.03959	.03480
19				.73840	.365254	.33979	.29172	.21834	.18366	.14168	.11858	.090435	.07355	.05527	.03883
20					**.64005**	.34702	.31674	.28629	.22517	.18154	.13898	.10572	.07844	.06104	.047509
21						.50986	.35565	.31080	.29312	.23211	.18543	.13443	.09728	.07757	.05557
22							.43549	.38021	.30849	.30576	.25012	.18700	.13717	.10249	.07291
23								.38430	.39487	.30590	.31718	.25069	.186630	.13432	.09880
24									.35719	.39591	.29929	.31370	.24401	.18615	.132260
25										.35712	.39324	.29464	.30509	.24263	.17908
26											.36602	.366449	.28329	.29791	.22408
27												.37528	.349951	.28092	.28928
28													.39703	.35070	.28361
29														.41596	.33384
30															.42039
⋮															

* I recomputed each values using data from census and vital statistics. And I rearranged chronologically.

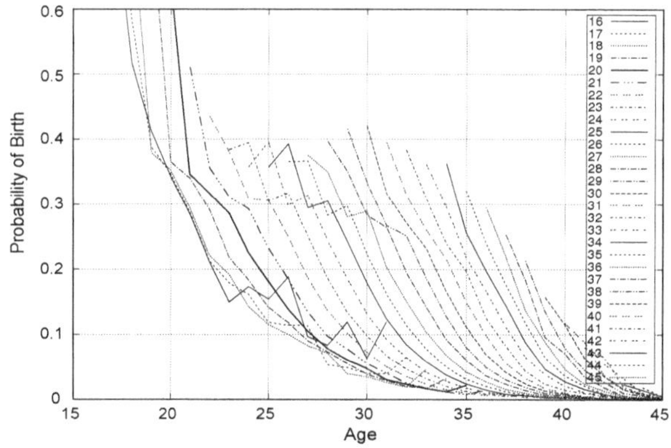

Figure 8.1: Rearranged birth probabilities of Japanese females in 1985.

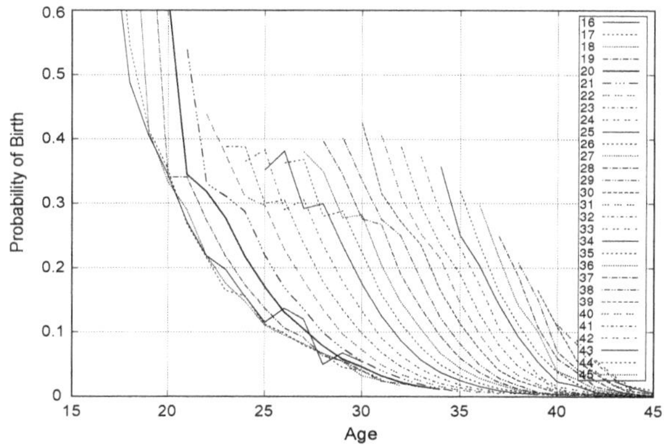

Figure 8.2: Birth probabilities of Japanese females in 1986.

8.2.1.1 England–Wales in 1939

In his article [10], Colin Clark provided a record of marriage age and duration–specific birth probabilities of England–Wales in 1939. If the place differs, the birth probabilities also differ. Therefore, it can be assumed that the law is not the birth probability itself, but possibly the law of variation. Clark wrote:

> ... The probability of a birth to any particular marriage this year is very much a function of the number of births which have already taken place in this marriage during recent years.
>
> Colin Clark, Age at Marriage and Marital Fertility [10], p. 414.

Namely, not sufficiently, he recognized the probable causal chains of birth probabilities. In Japan, in 1939, 1985, and 2010, the birth probabilities of a particular marriage age were replications of other marriage ages. Like the marriage function, births are neighboring (time) dependent. The dynamism of the birth function resides in this neighboring dependence.

Table 8.5: Birth probability by age and specific age at marriage in 1939 (calendar year) in Japan.

Marriage Age	15	16	17	18	19	20	21	22	23	24	25	26	27	28	29	30	...
15	0.06	0.31	0.24	0.22	0.28	0.24	0.23	0.29	0.25	0.26	0.22	0.27	0.24	0.23	0.23	0.23	...
16		0.11	0.36	0.32	0.32	0.29	0.28	0.29	0.28	0.30	0.27	0.27	0.25	0.26	0.26	0.25	...
17			0.13	0.45	0.28	0.30	0.29	0.31	0.31	0.29	0.27	0.29	0.29	0.26	0.27	0.26	...
18				0.14	0.44	0.26	0.29	0.31	0.29	0.27	0.27	0.28	0.27	0.27	0.26	0.24	...
19					0.15	0.49	0.27	0.31	0.32	0.29	0.29	0.30	0.26	0.28	0.29	0.27	...
20						0.16	0.47	0.28	0.32	0.29	0.30	0.26	0.29	0.27	0.28	0.27	...
21							0.17	0.46	0.29	0.33	0.27	0.29	0.28	0.27	0.27	0.27	...
22								0.18	0.43	0.27	0.31	0.17	0.40	0.27	0.26	0.27	...
23									0.19	0.44	0.25	0.31	0.27	0.26	0.25	0.26	...
24										0.17	0.43	0.26	0.30	0.25	0.26	0.27	...
25											0.20	0.39	0.21	0.29	0.25	0.23	...
26												0.16	0.37	0.28	0.25	0.26	...
27													0.20	0.35	0.26	0.24	...
28														0.16	0.37	0.26	...
29															0.12	0.37	...
30																0.14	...

The *Marriage and Population* table was rearranged here [59], pp. 232–233. *Chronologically rearranged by author.

Table 8.6: Birth probability by age and specific age at marriage in 1939 in Japan.

Marriage Duration	1	2	3	4	5	6	7	8	9	10	11	12	13	14	15	···	
15	0.22	0.28	0.23	0.25	0.26	0.24	0.26	0.27	0.26	0.24	0.25	0.26	0.24	0.23	0.23	0.24	···
16		0.29	0.34	0.32	0.29	0.29	0.29	0.29	0.29	0.29	0.27	0.26	0.26	0.26	0.26	0.24	···
17			0.30	0.36	0.29	0.30	0.30	0.31	0.30	0.28	0.28	0.29	0.28	0.27	0.27	0.26	···
18				0.36	0.35	0.28	0.30	0.30	0.28	0.27	0.28	0.28	0.27	0.27	0.25	0.24	···
19					0.40	0.38	0.29	0.32	0.31	0.29	0.30	0.28	0.27	0.29	0.28	0.27	···
20						0.36	0.38	0.30	0.31	0.30	0.28	0.28	0.28	0.28	0.28	0.27	···
21							0.33	0.38	0.31	0.30	0.28	0.29	0.28	0.27	0.27	0.26	···
22								0.34	0.35	0.29	0.24	0.29	0.34	0.27	0.27	0.26	···
23									0.35	0.35	0.28	0.29	0.27	0.26	0.26	0.27	···
24										0.39	0.35	0.28	0.28	0.26	0.27	0.26	···
25											0.40	0.30	0.25	0.27	0.24	0.23	···
26												0.35	0.33	0.27	0.26	0.25	···
27													0.38	0.31	0.25	0.27	···
28														0.35	0.32	0.25	···
29															0.31	0.28	···
30																0.32	···

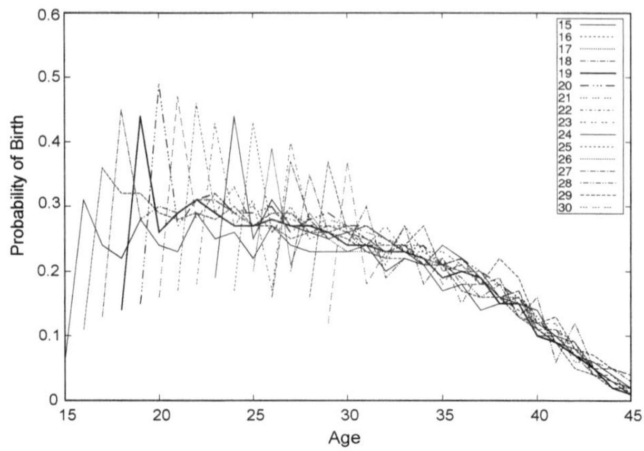

Figure 8.3: Birth probabilities in 1939 (calendar year) in Japan.

Drawn using the data of Okazaki, *Marriage and Population* [59], pp. 232–233.

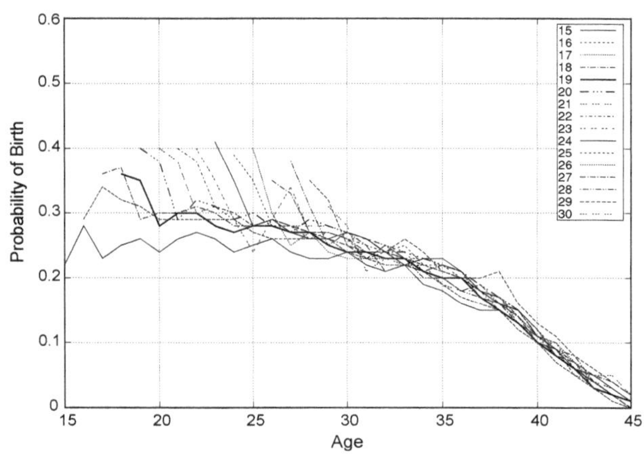

Figure 8.4: Adjusted birth probabilities at 1939 in Japan.

8.3 Regularity of Birth Probabilities

Therefore, comparing Figure 8.1 with Figure 8.2, for a very small time period (such as a year), the birth probability between cohorts was highly similar. On comparing the ages at marriage in the cohort, it can be seen that neighboring ages at marriage have similar birth probabilities.

A law of birth behaviors that can explain both these historical alterations and the pseudo-stationarity must be found. This clearly inconsistent question cannot be answered immediately in this chapter. Here, the objectives that will be accomplished are discussed.

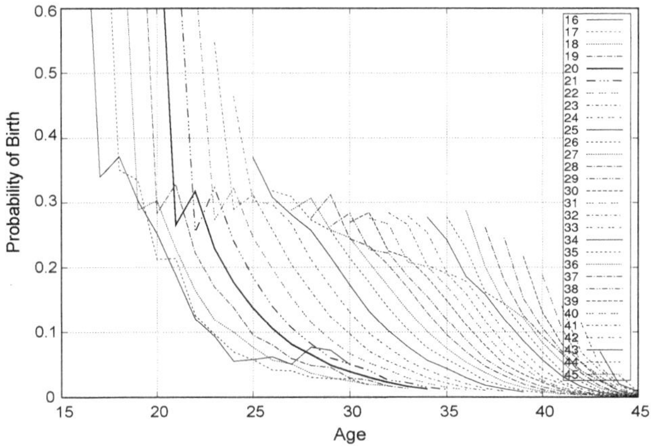

Figure 8.5: Birth probabilities of Japanese females in 2010.

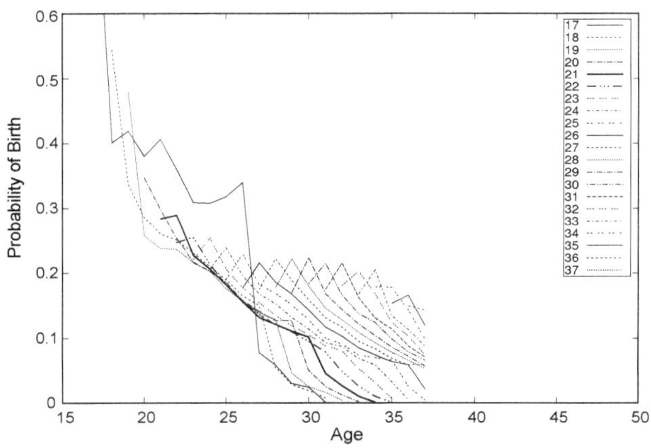

Figure 8.6: Birth probabilities based on age at marriage in England-Wales in 1939.

8.3.1 Precise Computation of Birth Probability by the Birth Function

Let us review the explanation that human conscious selection has changed, and these changes have brought about these historical alterations. Consider whether this effectively serves as a theoretical explanation for historical modification.

The explanation cannot produce any predictions. Household economists and sociologists do not compute birth probabilities. They merely adjust the explanation to fit the situation. As Spinoza said, human conscious will cannot cause these alterations; rather, the cause of the alteration of human conscious selections should be investigated. To do this, the birth function must be formulated.

It is important to observe the regularity of the birth function with respect to time order—that is, its causality. In Table 8.4 on page 204, the horizon of time = line of the present moves horizontally from left to right. Causality accrues on this horizon, not in the human mind.

8.3.1.1 Quasi-Replication

One remarkable feature is that the birth probabilities of females married in a given year approximate those of females married in the previous year. The previous dependence of birth behaviors is anticipated; therefore, the probability of birth for each marriage age may be expressed according to the equation. Let $b(a_i, m_j)$ be the probability of a birth for a female of age a_i who married at age m_j. Let ψ be the coefficient of the effect of previously married females.

$$b(a_{i+1}, m_{j+1}) = \psi\, b(a_i, m_j), \tag{8.10}$$

Note that Equation (8.10) cannot completely describe the birth probabilities of the cohort. Regardless of the chosen value of ψ, birth probabilities are not traced by Equation (8.10). Namely, the probability of "fecundability" is needed.

8.4 Fecundability

Describing the probabilities of birth in terms of the physiological concept of fecundability is essential. Here, the primary source of information related to fecundability is the study by J. Wood and M. Weinstein [81].

They explain fecundability as follows:

> One of the most important parameters in the study of human reproduction is *fecundability*, customarily defined (following Gini)[1] as the probability that a married woman in a susceptible state[2] will conceive during a month of exposure to unprotected intercourse. Fecundability affects achieved fertility through its relation to the average time required to conceive: in a homogeneous population, the mean waiting time to conception is equal to the reciprocal of the mean fecundability.[3] Thus, fecundability determines a major segment of the interval between births as well as contributing to the interval between marriage and first birth. Fecundability can also be thought of as the transition probability for passage from the susceptible state to pregnancy.[4] As such, it plays much the same fundamental role in fertility studies as Mendelian transmission ratios play in population genetics.

A Model of Age-specific Fecundability,
James W. Wood and Maxine Weinstein [81], p. 85.

Nonetheless, as per their definition, the concept of fecundability has some ambiguity. According to Wood & Weinstein, fecundability is classified into two definitions ([81], p. 87) *effective fecundability* is constructed by three stochastic variables. The concepts of fecundabilities described in their article ([81], p. 113) have been summarized below.

1. *total fecundability*
2. *effective fecundability*
   (a) ζ_j: probability that a woman aged j is post-menarcheal and pre-menopausal.
   (b) $\Pr_j(o)$: probability that a woman aged j is undergoing an ovulatory cycle.
   (c) γ_j: probability of embryonic or fetal loss at age j.

Effective fecundability is equal to $\zeta_j \cdot \Pr_j(o) \cdot (1-\gamma_j)$. Here, this *effective fecundability* is adopted in place of physiological fecundability. The concrete values estimated by Wood & Weinstein are presented in Table 8.7. They have been visualized in Figure 8.8 and Figure 8.9. Birth probabilities in 1939 in Japan appear to exhibit effective fecundability.

Although Wood and Weinstein defined *total fecundability*, their definition has not been adopted here. They estimated it by multiplying the daily probability of intercourse at age j with the *effective fecundability*. However, the daily probability of intercourse and the usage of various contraceptive methods are social phenomena that are not individually decided. Because these behavioral factors related to fecundability have undergone historical alterations, I do not consider them as Wood & Weinstein did. The theory proposed herein contributes a novel idea to this point.

8.5 Regularity of Birth

Let the age specific *effective fecundability* at age a_i be $Fe(a_i)$. Effective fecundability is assumed to be common between Western and Eastern countries because it is relevant to all *Homo sapiens*[4]. Therefore, Equation (8.10) evolves.

$$b(a_{i+1}, m_{j+1}) = Fe(a_{i+1})\, \psi\, b(a_i, m_j). \tag{8.11}$$

When ψ is appropriately selected, it may be $1.4 < \psi < 1.5$, and the values computed by Equation (8.11) are precisely approximated to the observed values between the ages of 25–45. As previously mentioned, ψ is a sociological constant, referred to herein as the *reproductive coefficient*. Here, several cohorts in Japan were selected. A detailed explanation of each cohort's regularity is presented.

[4]This assumption has not caused any issues.

Table 8.7: Reference values of the components of fecundability for contemporary US women, listed by single year of age.

Age	ζ	$Pr_j(o)$	γ	$Fe(a) = \zeta \cdot \Pr_j(o) \cdot (1-\gamma)$	Probability of Daily intercourse
10	0.0062	0.4647	0.3675	0.0018	0.3803
11	0.0586	0.4691	0.3530	0.0178	0.3755
12	0.2269	0.4697	0.3397	0.0704	0.3709
13	0.5093	0.4873	0.3276	0.1669	0.3652
14	0.7623	0.5294	0.3167	0.2758	0.3585
15	0.9043	0.5900	0.3070	0.3697	0.3511
16	0.9707	0.6596	0.2984	0.4492	0.3434
17	0.9954	0.7221	0.2911	0.5095	0.3356
18	0.9985	0.7681	0.2850	0.5484	0.3277
19	1.0000	0.8060	0.2801	0.5802	0.3201
20	1.0000	0.8426	0.2763	0.6098	0.3128
21	1.0000	0.8784	0.2738	0.6379	0.3058
22	1.0000	0.9068	0.2725	0.6597	0.2992
23	1.0000	0.9283	0.2723	0.6755	0.2930
24	1.0000	0.9427	0.2734	0.6850	0.2870
25	1.0000	0.9505	0.2756	0.6885	0.2815
26	1.0000	0.9514	0.2791	0.6859	0.2763
27	1.0000	0.9519	0.2837	0.6818	0.2715
28	1.0000	0.9548	0.2896	0.6783	0.2670
29	1.0000	0.9589	0.2967	0.6744	0.2627
30	1.0000	0.9617	0.3049	0.6685	0.2588
31	1.0000	0.9622	0.3143	0.6598	0.2550
32	1.0000	0.9622	0.3250	0.6495	0.2516
33	1.0000	0.9635	0.3368	0.6390	0.2483
34	1.0000	0.9670	0.3499	0.6286	0.2452
35	1.0000	0.9710	0.3641	0.6175	0.2424
36	1.0000	0.9740	0.3796	0.6043	0.2396
37	0.9969	0.9752	0.3962	0.5870	0.2371
38	0.9938	0.9762	0.4140	0.5685	0.2347
39	0.9938	0.9769	0.4331	0.5504	0.2324
40	0.9923	0.9778	0.4533	0.5304	0.2303
41	0.9892	0.9786	0.4747	0.5085	0.2282
42	0.9846	0.9784	0.4973	0.4843	0.2263
43	0.9722	0.9753	0.5212	0.4540	0.2245
44	0.9460	0.9697	0.5462	0.4163	0.2227
45	0.9043	0.9625	0.5724	0.3722	0.2211
46	0.8503	0.9568	0.5998	0.3256	0.2195
47	0.7685	0.9414	0.6284	0.2688	0.2180
48	0.6512	0.9199	0.6582	0.2048	0.2166
49	0.5139	0.8896	0.6893	0.1420	0.2152
50	0.3812	0.8558	0.7215	0.0909	0.2139
51	0.2546	0.8248	0.7549	0.0515	0.2126
52	0.1497	0.8013	0.7895	0.0253	0.2114
53	0.0818	0.7805	0.8253	0.0112	0.2103
54	0.0386	0.7537	0.8623	0.0040	0.2092
55	0.0201	0.7132	0.9005	0.0014	0.2081
56	0.0077	0.6808	0.9399	0.0003	0.2071

Source: Wood & Weinstein [81], p. 113,
* The expected total cyclic length at age j has been omitted in this table.
* And the column of $Fe(a)$ (effective fecundability) is added by author.

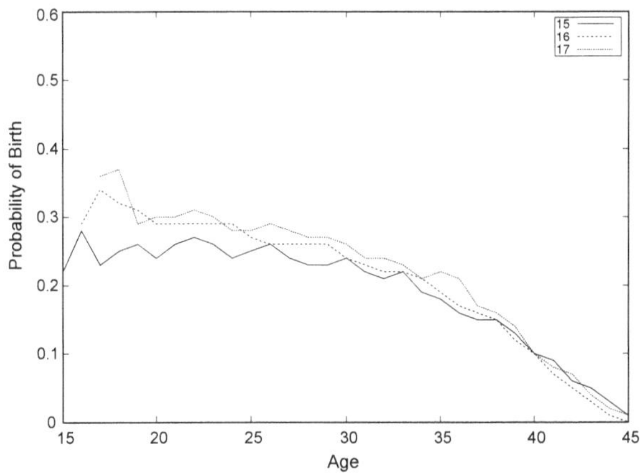

Figure 8.7: Birth probabilities for specific ages at marriage in 1939: 15, 16, and 17 years old.

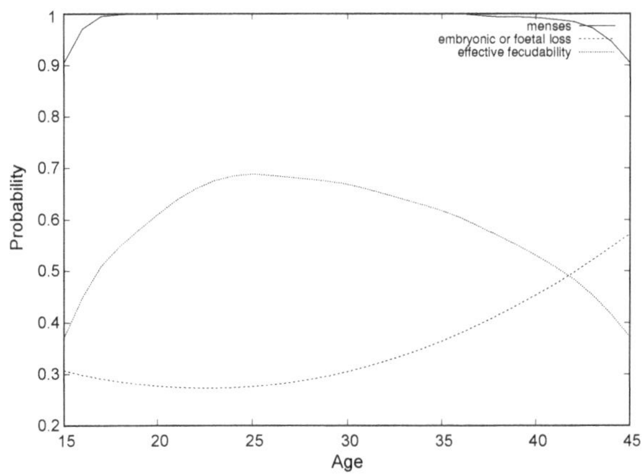

Figure 8.8: Fecundability estimations by Wood & Weinstein: ages 15–45 for comparison.

8.5.1 1965 Cohort in Japan

By 1979, the vital statistics of the population of Japan had documented birth by calendar year. Therefore, the marriage age–specific birth probabilities for the 1965 cohort can be calculated. Equation (8.11) presents a fine fit, as shown in Figure 8.10.

For the first year, the calculated values are slightly high. In other expressions, the observed values are slightly low. The birth probability of the first year of marriage cannot be completely described by Equation (8.11); this decline in birth probabilities for females married at the age of 25 may be brought about by the "burst of Japan's economic bubble." It is one of the few exceptions to the background independence of the birth function.

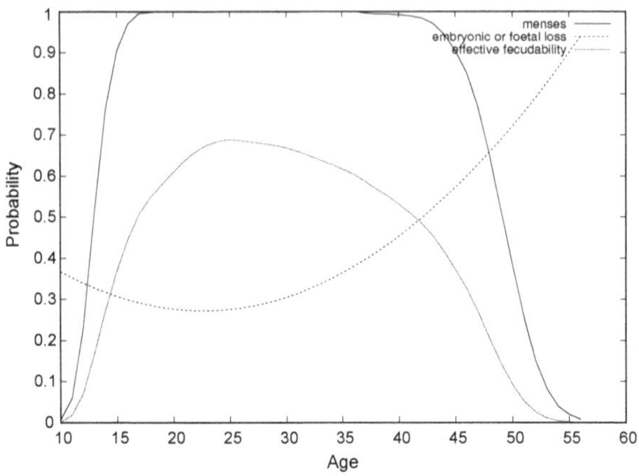

Figure 8.9: Fecundability estimations by Wood & Weinstein: ages 10–60.

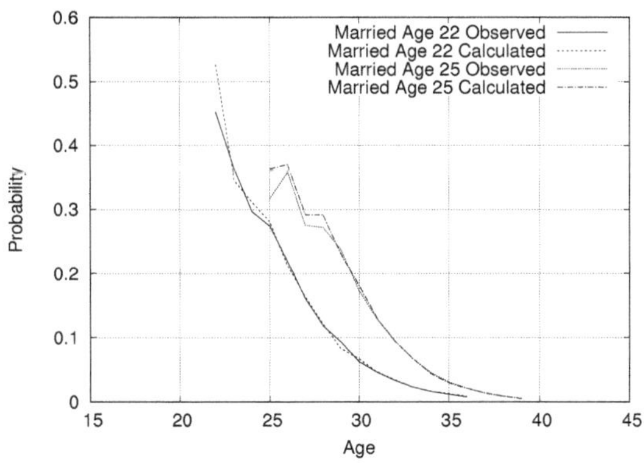

Figure 8.10: Birth probability for females aged between 22–25 in the 1965 cohort, = 1.48.

8.5.1.1 *Idiosyncrasy*

For young married females, such as those married between the ages of 15–21, another fitness coefficient is required. For very young marriages, registered marriages are considered to be fewer than real marriages. Within the scope of the cohort theory, this idiosyncrasy must be dealt with by introducing another coefficient $\rho(a_i)$. I will explain about the idiosyncratic coefficient $\rho(a_i)$ in the following section.

Equation (8.11) calculates the approximate values of birth probabilities of those who married after age 22. In reality, the number of young marriages in Japan have reduced significantly, and Equation (8.11) is valid for a considerable number of births.

8.5.2 1970 Cohort in Japan

The fine fitness of Equation (8.11) is firm. On careful examination, the probability of birth in the 1970 cohort is slightly lower than that in the 1965 cohort. This slight and regulated decline leads to the idea that the decline of birth probability is a stochastic process which may not be brought about by conscious human behavioral changes.

8.5.3 1975 and 1980 Cohorts in Japan

Equation (8.11) remains valid for the 1975 and 1980 cohorts. The probability of birth had been also declining at this time. Here, the total fertility for each marriage age was calculated using the following equation, let m be the marriage age:

$$\int_{m}^{a} b(a)(da) \doteq \sum_{a_i=m}^{a_i=a_n} b(a_i) \tag{8.12}$$

The computed total fertility rates for females married at the ages of 22 and 25 are presented in Table 8.8. Note that these populations are not yet sterilized.

Table 8.8: Computed total fertility rates of married age 22 and 25 of each cohort.

cohort	22	25
1965	2.176	2.047
1970	2.065	1.832
1975	1.985	1.930
1980	2.108	

It may be speculated that Japanese marital fertility reached its lowest point after the 1975 cohort.

8.5.4 1939 and 1985 Cohorts in Japan

Okazaki's data are probably calculated based on period data. Despite the fact that Japan had a high fertility rate at the time, the decline in fertility had begun. The momentum of the decline was concealed by the active birth behaviors (shown in Figure 8.13).

The 1985 birth probabilities for specific ages at marriage computed herein are identical to those calculated by Ito and Bando. Expectedly, the equation is falsifiable.

Moreover, the theoretical significance is that Equation (8.11) is longitudinally valid.

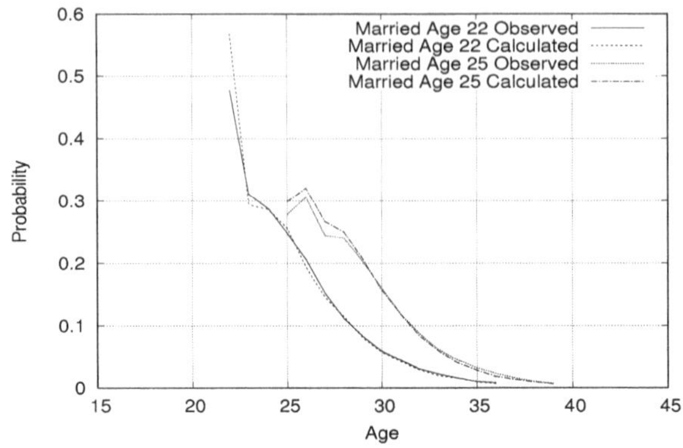

Figure 8.11: 1970 Cohort Marriage Age 22–25 BirthProb, $\psi = 1.48$.

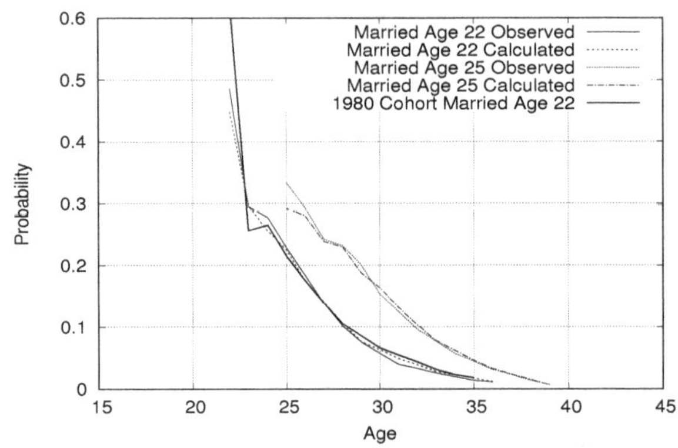

Figure 8.12: Birth probability for females aged between 22–25 in the 1975 cohort, $\psi = 1.48$.

8.5.5 England–Wales—C. Clark's Data

C. Clark's data is not strictly based on cohorts, but for the England–Wales data, the validity of Equation (8.11) is repeatedly recognized. Thus, the birth function may be universal for all *Homo sapiens*.

8.5.6 General Law

The marriage age–specific birth probability for age j is completely dependent on that for age $j-1$. Children are borne to replicate previous births. This is not a conscious imitation. Like many other animals, humans are programmed to bear children in this manner. With the decline in effective fecundability, female fertility is reducing. This is the fate of all animals.

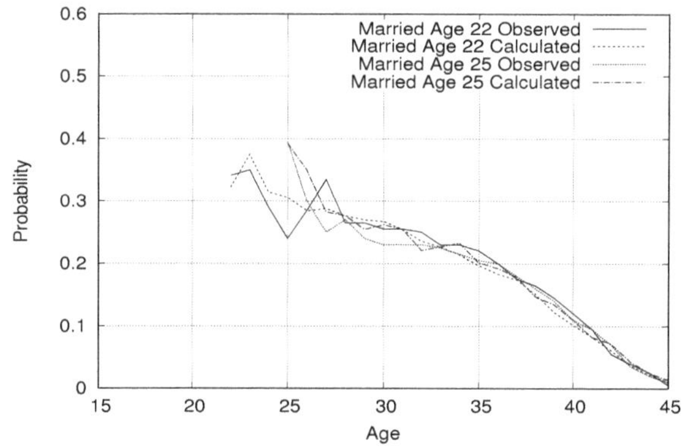

Figure 8.13: Birth probability for females between the ages of 22–25 in the 1939 cohort in Japan, $\psi = 1.48$.

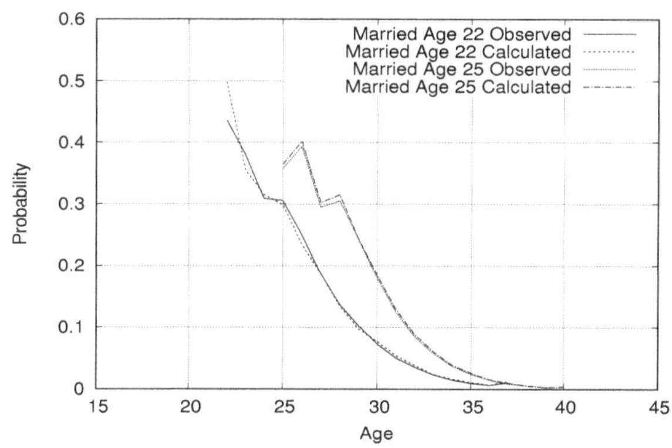

Figure 8.14: Birth probability for females between the ages of 22–25 in the 1985 cohort in Japan, $\psi = 1.48$.

8.6 Birth Function for a Cohort

The birth function essentially represents the replication of previous births. This relation itself is precisely the causality. On accepting this principle, the birth function of a cohort can be formulated as a recurrence relation from initial values. However, this recurrence relation has some unique difficulties. Thus far, the observed values have been used sequentially for each $b(a, m)$. Beginning with an initial $b(a, m)$ and computing successive values from it presents problems.

In terms of the final results of this work, the birth function for cohorts is a superfluous divergence. However, to find the dimensions that are needed, it was necessary to consider this stepwise and construct a birth function in high-dimensional space. This process is described in detail below.

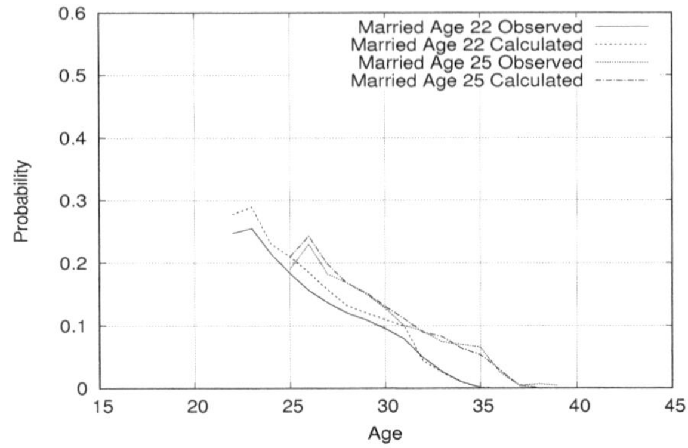

Figure 8.15: Birth probability for females between the ages of 22–25 in 1939 in England–Wales, $\psi = 1.48$.

8.6.1 Virtual 1985 Cohort

8.6.1.1 Model Zero

The birth probabilities must also be considered for the population who married young (such as between the ages of 15–18). Their birth rates are very high because of their low registration rate. Here, these idiosyncrasies are operated by introducing the idiosyncratic coefficient $\rho(a)$. Therefore, Equation (8.11) is altered as follows.

$$b(a_{i+1}, m_{j+1}) = \rho(a_{i+1})\, Fe(a+1)\, \psi\, b(a_i, m_j). \tag{8.13}$$

Compared with the observed model (Figure 8.17), the atmosphere of the marriage age–specific birth function is reproduced fairly well in Model Zero (Figure 8.16). However, some vital defects cannot be avoided. "Model Zero" provides a lower probability of elderly birth (bearing children at around 32 years old) compared to the observed model. Modifying these differences by increasing the reproductive coefficient ψ has an adverse effect: the probabilities of younger birth become much higher. Generally, a large ψ cannot be taken because the birth function grows exponentially, meaning that the population catastrophe heightens. Namely, it is necessary that ψ is neither excessively large nor excessively small. I reason that ψ is a highly specific value, and human reproduction may be elaborately adjusted.

The recurrence equation has a systematic, unmanageable problem: the few initial errors grow exponentially. The marriage age–specific birth function also faces this problem, which may not be applicable to the real world. "Model Zero" does not sufficiently represent reality. Some type of increasing element must be added to Equation (8.13).

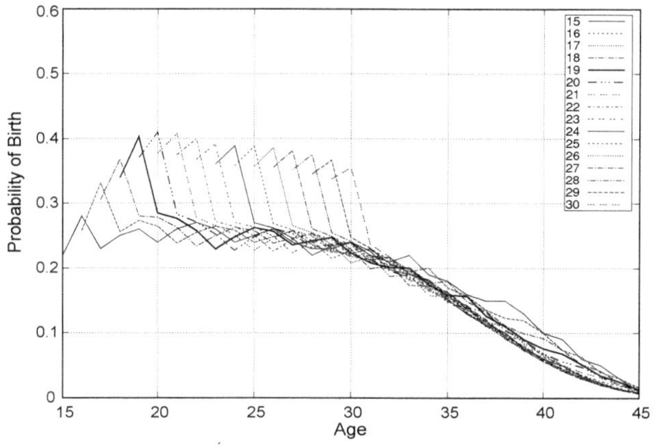

Figure 8.16: Model Zero ($\psi = 1.45$).

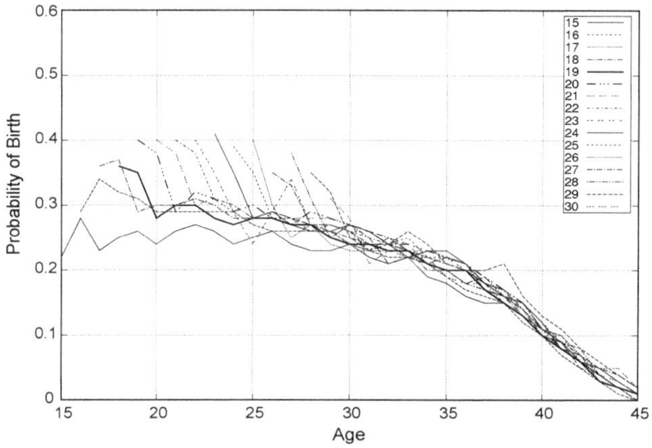

Figure 8.17: Observed in 1939 (for comparision).

8.6.1.2 Model One

Some prior studies have used the proportional hazard model or multinomial logit analysis to demonstrate that females who marry young tend to restrain conception, and those who marry later achieve the mean number of children. Here, the hypothesis that parents bear the mean number of children introduced by comparing the mean number of children in their cohort. Thus I constituted Equation (8.14) as "Model One".

$$\begin{aligned} &b(a_{i+1}, m_{j+1}) \\ &= \rho(a_i)\, \psi\, Fe(a_{i+1}) \left\{ b(a_i, m_j) + \kappa \left(\sum_{j=1}^{m_j} \sum_{i=m_j}^{a} b(a_i, m_j) - \sum_{i=m_1}^{a} b(a_i) \right) \right\}, \end{aligned} \tag{8.14}$$

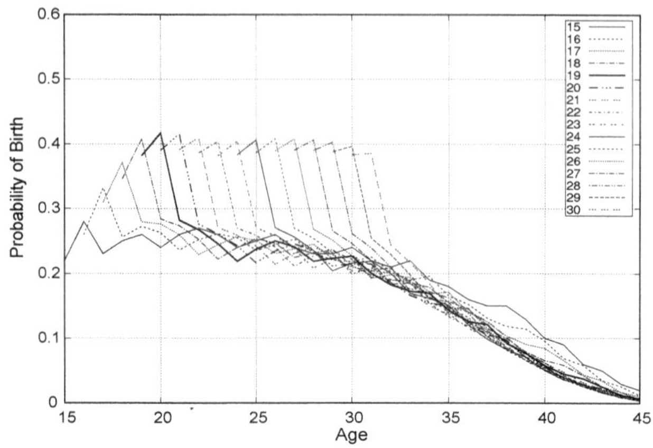

Figure 8.18: Model One ($\psi = 1.46$, $\kappa = 0.003$).

Model One was a greater failure than expected. Although ψ was taken to have a small value, the pattern of the birth function differs significantly from the observed model. As shown in Figure 8.18, the initial birth probabilities for individuals between the ages of 25–29 remain higher; notwithstanding these penalties, the lower probabilitics in oldcr agcs did not improve.

Parents do not compare the number of children they bear with their cohort's mean. If this were to happen, they would immediately bear children such that the mean is attained. It is likely that parents do not perform such comparisons because they cannot calculate it.

8.6.1.3 Model Two

Following the failure of Model One, I was not immediately able to establish a suitable model. In order to negate the decline in the momentum of the function, a term that can compete with the decline of the effective fecundability must be added. For this reason, I added the square of age to the equation that is generally used in statistical analysis.

$$b(a_{i+1}, m_{j+1}) = \rho(a+1)\,\psi\, Fe(a_{i+1})\left\{b(a_i, m_j) + \kappa\, {a_i}^2\right\} \tag{8.15}$$

This "Model Two" can be regarded as a good estimate of reality. However, the square of age lacks tangibility in human social interactions. Sociological theory must be built on tangible social objects rather than statistical techniques. Almost all people cannot calculate the square of their age.

Moreover, strictly speaking, some details of Model Two (Figure 8.19) do not match the observed model. For the most part, the probabilities for the age group 20–29 are slightly lower, and with penalties, the last part of birth behaviors(age 30–35) observed are relatively higher than the model. These minute differences were also not eliminated by changing the square of age to the cube of age.

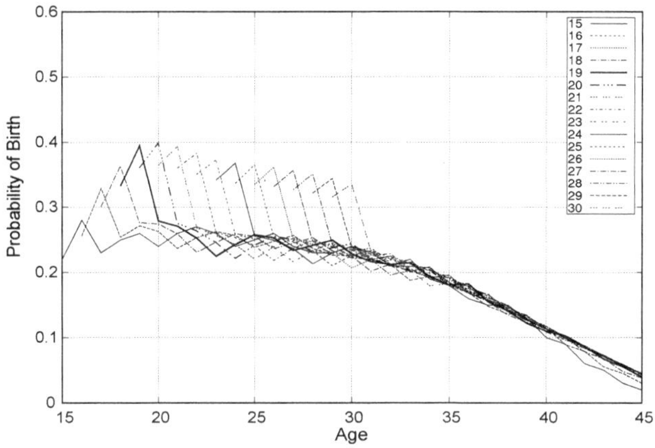

Figure 8.19: Model Two ($\psi_1 = 1.444$, $\kappa_2 = 0.005$).

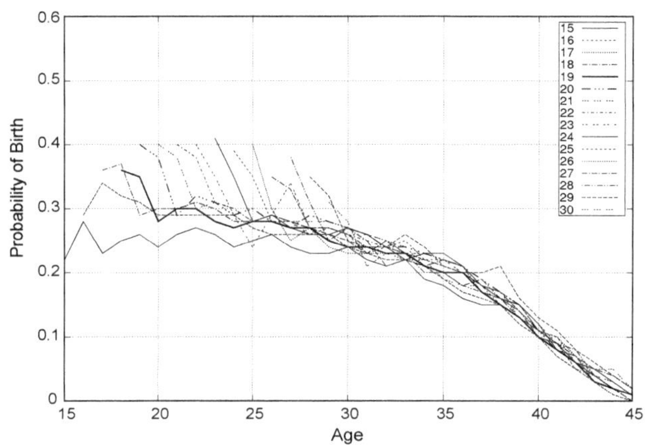

Figure 8.20: Observed in 1939 (for comparison).

8.6.1.4 Model-Three

I attempted to combine Model One and Model Two to create a new model based on the principle that more terms generally achieve higher fitness. However, this attempt was unsuccessful.

8.6.1.5 Model-Four and Model-Five

Some increase in the probabilities of birth is observed from for females between the ages of 25–35 (See Figure 8.2 for an example). This increase is observed for the birth probabilities in 1939 and those of the England–Wales data presented in the articles by C. Clark.

I assumed that for the purpose of describing these increases, the principle of neighboring dependence needed to be strongly incorporated. At this time, I had not abandoned the hypothesis of comparison to the neighboring number of children. Therefore, Model Four and Model Five assumed the term that caused this effect, and both models were unsuccessful.

Following several failures, I finally found that parents do not compare their number of children with the global value of the number of children. Humans are far more illogical than expected. In these terms, it is erroneous to postulate supreme intelligence through "game theory." The inquiry into birth functions is essentially to find the most effective mapping from the present set of births to the successive set of births. Human cognitive capacity needs to be assumed for this mapping, and this would be our true capacity. I explored several possibilities.

8.6.1.6 Model Six

Consider the problem of Model Zero once again. The main divergences are that in the late marriage, the expected probabilities $= \rho(a)\ Fe(a+1)\ \psi\ b(a,m)$ are lower than those observed. Namely, additive catch-up effects to Model Zero are necessary. Birth probabilities are merely replications of previous birth probabilities of previously married females. However, the decline of *Fe* with the increase in age is higher than in previous birth probabilities. The appropriate catch-up mechanism that is not approaching the mean had to be found, because if every birth occurs in order to approach the mean, the first birth probability should be higher than has been observed, and the birth probability would expand exponentially.

Humans do not actually approach the mean number of children. This suggests that no actual norms exist for the number of children. In reality, human birth behaviors have no relation to the so-called "norm."

What type of catch-up is happening? Humans approach the number of children of their predecessors ($\subset$ the neighbors in time-space) as far as possible. Thus, the following term was added to Model Zero.

$$\left(\sum_{i=m_j}^{a_i} b(a_i, m_{j-1}) - \sum_{i=m_j}^{a_i} b(a_i, m_j)\right) \tag{8.16}$$

Thus, the following birth function was obtained.

$$\begin{aligned} & b(a_{i+1}, m_{j+1}) \\ & \quad = \rho(a+1)\ Fe(a_{i+1}) \\ & \qquad \left(\psi\, b(a_i, m_j) + \kappa \left(\sum_{i=m_j}^{a_i} b(a_i, m_{j-1}) - \sum_{i=m_j}^{a_i} b(a_i, m_j)\right)\right) \end{aligned} \tag{8.17}$$

By setting the appropriate coefficient κ in Equation (8.17), a comparatively superior function is obtained (Figure 8.21). However, the divergences are large.

This appeared to have no explanation. I attempted to obtain a solution by examining the function using data from 1985.

8.7 Integration of Birth Function with 1985 (Virtual) Cohort

Let us compare Figure 8.24 with Figure 8.22. It is doubtful that the same function can describe the probabilities of both cohorts. However, it is meaningless to postulate alterations of conscious motivations, because the theory cannot be tested for the predictions that are deducted from itself. It is already clear that in the range of a single cohort, this is a difficult task. To demonstrate the limit of the cohort concept, my work in this area is presented.

8.7.1 Idiosyncrasy Rates of 1985

Primarily owing to the lack of registration, for very young married females, $\rho(a)$ must be multiplied. This is because the married population is higher than that which has been registered. For the birth probabilities in 1939, the following values have been set; other values have been set for 1985.

See Table 8.9. It is clear that $\rho(a_i)$ remained similar from 1939 to 1985 for individuals between the ages of 16–18. I cannot help but admire those very young married parents, who rarely adopted modern methods of contraception. Their birth behaviors are largely longitudinally constant.

8.7.2 Model-Six for 1985

Figure 8.23 presents the results of the calculation of Model Six, where $= 1.44$, $\kappa = 0.02$ was set. Unfortunately, the catch-up effect remained insufficient. The cause of this shortage of catch-ups is that in the process of

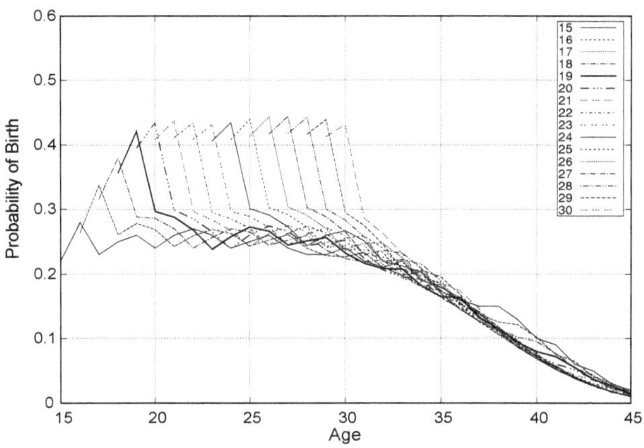

Figure 8.21: Model Six ($\psi = 1.46, \psi_1 = 0.01$).

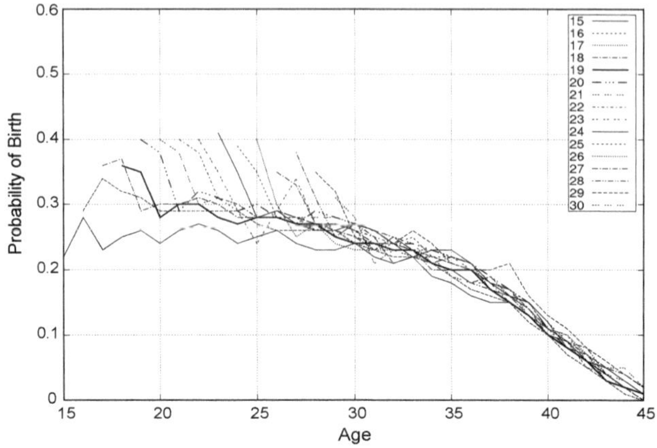

Figure 8.22: Observed data from 1939 (for comparison).

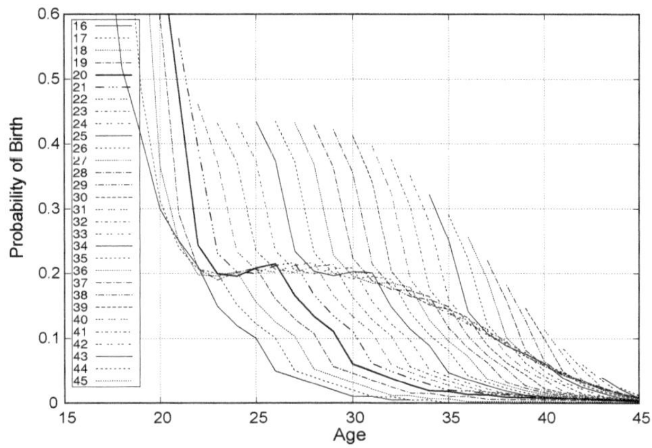

Figure 8.23: Age-specific birth probabilities in 1985 by Model Six: $\psi = 1.44$, ${}_1 = 0.02$.

fertility decline in time–space, neighboring birth probabilities simultaneously decline, and the catch-up also declines with age. Essentially, the factors that increase as age increases must be identified.

Additionally, the catch-up effect must be identified within the system of fertility decline. The conscious agent appears to be apt to explain this. However, Spinoza states (Prop. XXXII), "Will cannot be called a free cause, but only a necessary cause."[5] Conscious catchups conceal the true neighbor-dependent nature of human behavior.

Despite it being a divergence, Spinoza's proof for Prop. XXXII is introduced herein. In addition, the analysis conducted in this work is deepened.

[5] *Ethics*, Spinoza, p. 30.

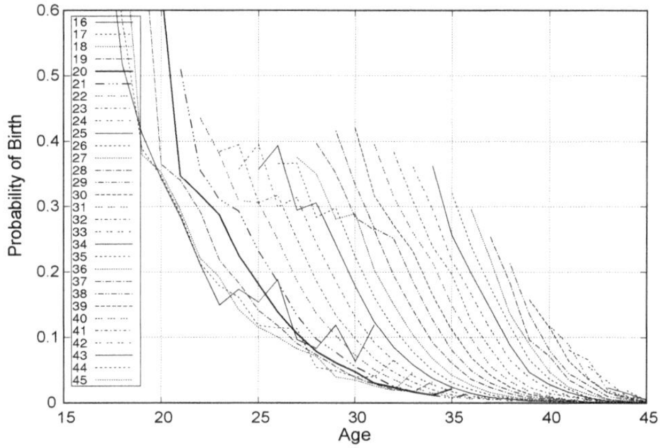

Figure 8.24: Age-specific birth probabilities observed in 1985 (redisplaying).

Table 8.9: Idiosyncratic rates ($\rho(i)$) of birth function.

	1939	1985
15	4.3	1.68
16	1.8	1.8
17	1.6	1.5
18	1.4	1.2
19	1.3	1.15
20	1.15	0.95
21	1.075	0.80
22	1.025	0.85
23	1.0	0.88
24	1.0	0.9
25	1.0	1.0

> Proof. Will is only a particular mode of thinking, like intellect; therefore (by Prop. XXVIII) no volition can exist, nor be conditioned to act, unless it be conditioned by some cause other than itself, which cause is conditioned by a third cause, and so on to infinity. But if will be supposed infinite, it must also be conditioned to exist and act by God, not by virtue of his being substance absolutely infinite, but by virtue of his possessing an attribute which expresses the infinite and eternal essence of thought (by Prop. xxiii.). Thus, however it be conceived, whether as finite or infinite, it requires a cause by which it should be conditioned to exist and act. Thus, (Def. VII) it cannot be called a free cause, but only a necessary or constrained cause. Q.E.D.

Spinoza, *Ethics (Ethica Ordine Geometrico Demonstrata)* [16], p. 30.

Human behaviors occur in an infinite (or perhaps finite) chain. In this chain, the law that governs one sequence to the next is vital and inclusive.

Our conscious wills are inconsequential. The prime objective is to define this law, whose mapping is from one sequence to the next.

Background independence is another crucial and instructive principle in which demographic phenomena are almost entirely decided independent of socioeconomic background. Human behavior is simply a reaction to neighboring situations, not rational action for a future socioeconomic state. This principle of background independence is another factor in the elimination of conscious will as a cause.

8.7.3 Model Eight

[6]What, then, is the cause of human will to level with the average case? The underlying cause was sought herein. Recognizing that the catch-up in 1939 was greater than that in 1985, it can be supposed that the variance of the number of children in the system is a probable cause. The variance in the number of children was investigated here, as shown in Figure 8.25.

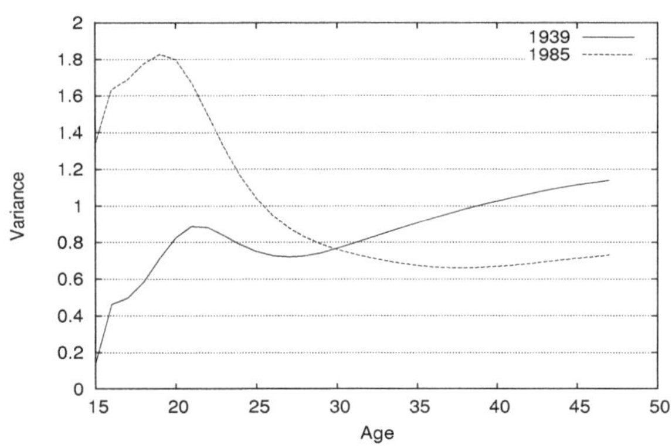

Figure 8.25: Growing of variances in the system in 1939 and 1985, Japan.

However, the variances in the system are also large for the youngest parents, as shown in Figure 8.25. Therefore, the concept of catch-up being the objective of older individuals is not sufficient. Reflectively, statistical values such as variance are not imagined by us. It is not the cause of the catch-up because almost all human cannot calculate it as a statistical value.

The mechanism of catch-up should be simpler than any other. The truth is sometimes unremarkable. From Figure 8.26, it can be seen that both means monotonically grow and moderate in the first half of a, and rapidly grow in the middle. The mean in 1939 was larger than that in 1985. These figures of the mean in the system are identical to those of the catch-up. It is estimated that the mean of the number of children is an index of the catch-up in the system.

[6]"Model Seven" was also examined. This model assumed another type of catch-up to neighboring birth probabilities in vain.

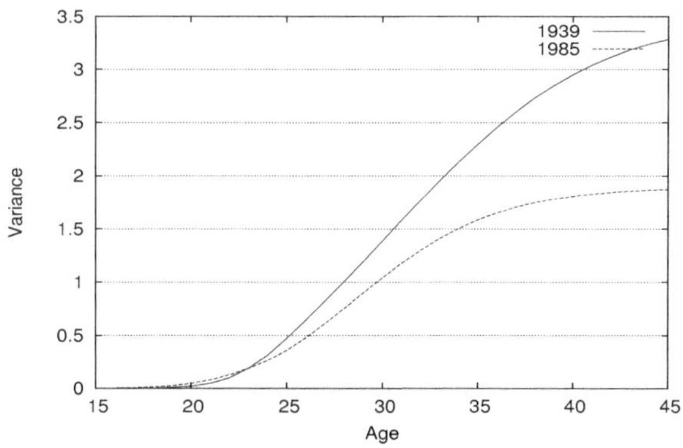

Figure 8.26: Growing of the averages in the system in 1939 and 1985, Japan.

The coefficient of catch-up behavior may be proportional to the mean number of children in the system. The actual cause of human catch-up behaviors is the number of neighboring children. The higher this number, the greater are our catch-up efforts. This situation is compared to that of a marathon race. When running in a marathon race, we attempt to catch up to the runner just ahead of us. However, runners pace themselves based on pacemakers or other factors resulting from the restraint brought about by one another. We run in accordance with the general atmosphere, whether that be lethargic or proactive.

Human birth behavior is similar to this example. Our catch-up depends on the atmosphere created by contemporary birth behaviors. This is the exact cause of the catch-up. I propose that the mean number of children is an accurate index of this atmosphere.[7] However, it should be noted that this does not suggest that we actually calculate the mean number of children in families. We are simple animals.

8.7.3.1 Coefficient of Catch-up

Because this catch-up effect is relatively smaller than the neighboring replication principle, the coefficient is small. The ψ (reproduction coefficient) value for each cohort was estimated using the least square iteration method. It is clear that the birth function has a catch-up effect, which has a maximum value of 0.2. The birth probabilities of very young married females can be ignored because their registrations are fewer than the number of actual common-law marriages. However, because the manner in which ψ is combined with the catch-up effect is unknown, they are assumed to have a linear combination. The coefficient of the catch-up effect is denoted as κ. $\mu(a_*)$ means the mean number of children ever born in the space. Thus, "Model-Eight" is defined.

$$b(a_{i+1}, m_{j+1}) = \rho(a_{i+1})\, Fe(a_{i+1})\, (\psi + \kappa\, \mu(a_*))\, b(a_i, m_j), \tag{8.18}$$

[7] The strange conception, atomosphere is solved by the geographical extention.

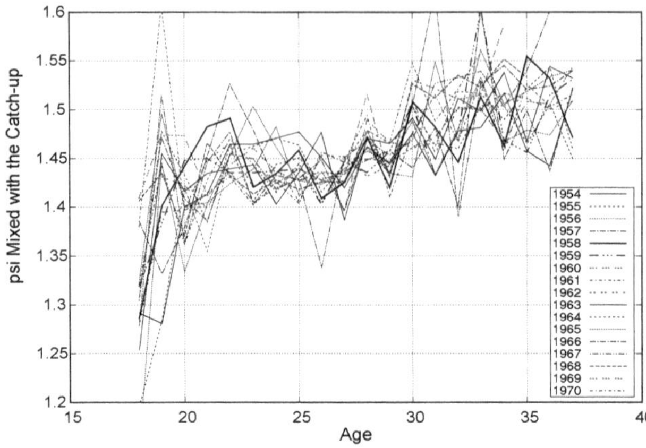

Figure 8.27: ψ combined with the catch-up effect.

Equation (8.18) naturally describes that the catch-up effect is minimal during its initial stages, and increases as age at marriage increases. The cause of catch-up is not human conscious will but neighboring behaviors after all. Spinoza is timelessly right. The conclusive reasoning for this fact is now presented.

8.7.3.2 Analysis of the Derivatives of the Birth Function

When the catch-up effect is assumed to be proportional to the number of children in the system, it means that $\psi + \kappa\, \mu(a_*)$. From Figure 8.26, it can easily be deduced that the catch-up effect in 1939 was larger than that in 1985 as the mathematical model. This appears to contradict my preconception. Living in a time of very low fertility in Japan (2019), I believe that couples in late marriages today ought to be more eager to catch-up than those in 1939. At that time, there was a fear of overpopulation in Japan, and demographers suspect that normal parents did not consciously want to bear more children.

However, contrary to our intuition, the catch-up effect in 1939 was greater than in 1985. This fact can be verified via statistical analysis of the observed data independent from $\mu(a_*)$. First, the polynomial approximation regression is adapted to each marriage-specific probability[8]. Namely, they are regarded as high-dimensional polynomials. The Weierstrass approximation theorem certifies their existences. Second, their second-order derivatives are computed to compare the catch-up effect.

Married at Age 30

The following table and figure represent the second-order derivatives. It is quite clear that $b''(a)$ values in 1939 are greater than those in 1985. The

[8]Every function is nearly perfectly approximated by the fourth-dimension polynomial—$R^2 > .978$.

Table 8.10: Birth probabilities of females married at age 30 in 1939 and 1985.

	30	31	32	33	34	35
30(1939)	0.32	0.29	0.24	0.22	0.24	0.20
30(1985)	0.4204	0.3163	0.2714	0.2316	0.1661	0.1117
	36	37	38	39	40	41
30(1939)	0.16	0.17	0.16	0.15	0.16	0.15
30(1985)	0.1117	0.0695	0.0421	0.0241	0.0132	0.0083
	42	43	44	45		
30(1939)	0.09	0.07	0.04	0.02		
30(1985)	0.0039	0.0017	0.0007	0.0006		

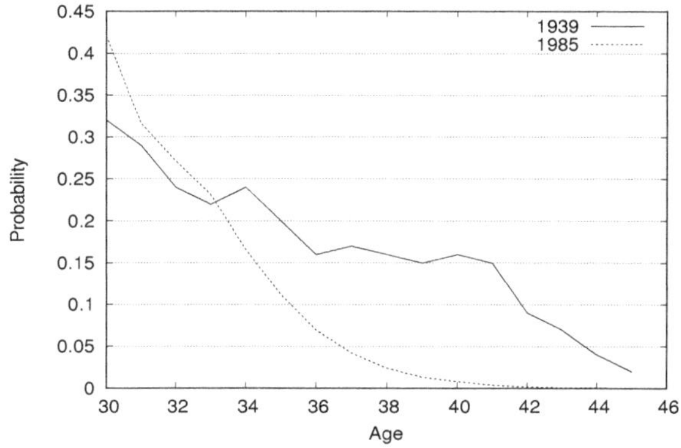

Figure 8.28: Birth probabilities of females married at age 30 in 1939 and 1985.

Table 8.11: Second-order derivatives of the probabilities of females married at age 30 in 1939 and 1985.

	30	31	32	33	34	35
30(1939)	0.04657	0.01817	0.00434	-0.00011	0.00057	0.00306
30(1985)	-0.00001	-0.00049	-0.00145	-0.00289	-0.00481	-0.00721
	36	37	38	39	40	41
30(1939)	0.00497	0.00487	0.00223	-0.00250	-0.00797	-0.01187
30(1985)	-0.01009	-0.01345	-0.01729	-0.02161	-0.02641	-0.03169
	42	43	44	45		
30(1939)	-0.01096	-0.00108	0.02289	0.06700		
30(1985)	-0.03745	-0.04369	-0.05041	-0.05761		

catch-up effect was unarguably greater in 1939. The fact that the theoretical predictions contradict our intuition is an evidence for the birth function theory.

In 1939, the b'' values are almost consistently positive throughout the observed intervals, while in 1985, they became negative.

Some believe that the pronatalist policies of the fascist and militarist Japanese government in those days encouraged parents to bear more children. This idea is unreasonable because the birth rates and their second-order derivatives were decreasing at the time.

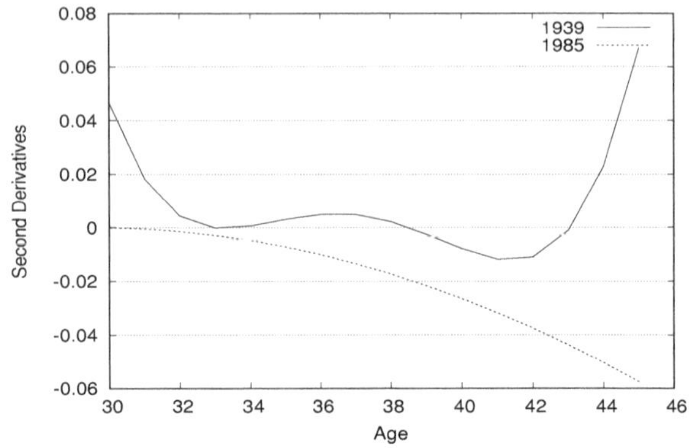

Figure 8.29: Second-order derivatives of the probabilities of females married at age 30 in 1939 and 1985.

The neighboring effect was and continues to be significant. Human conscious thoughts and actions are the results of neighboring behaviors. Our subjective decisions are a stochastic necessity. Contrary to the beliefs of many sociologists, causality is not the relationship between meanings, but between neighboring behaviors.

Married at age 35

To marry at age 30 is no longer considered a late marriage for females in Japan nowadays. Here, the birth probabilities of females married at age 35 are analyzed for catch-up behaviors.

According to expectations, in 1985, the catch-up behavior of females married at age 35 was more prominent than that of females married at age 30. Compare the second-order derivative of Figure 8.31 on p. 231 with that of Figure 8.29 on p. 230.

In 1939, the birth probabilities were computed by the birth function. Therefore, the 1939 values are only for reference; they are insufficient to describe the actual catch-up effect. The birth function for a single cohort is not realistically viable.

8.7.3.3 Effect of Infertility Treatment

Human consciousness mirrors neighboring behaviors. However, I found an odd catch-up that cannot be explained by Equation (8.18). The birth function is largely downward convex. In 1985, strange fluctuations were seen for females who married late. Consider the second-order derivative of females married at the age of 30 in 2000 (Table 8.14 and Figure 8.32).

Derivatives change as $b'' < 0 \rightarrow b'' > 0 \rightarrow b'' < 0$. The birth function unnaturally changes between approximately 35 to 40 years of age. Equation (8.18) does not include this unnatural reactivation. Expectedly, once the protuber-

Table 8.12: Birth probabilities of females married at age 35 in 1939 and 1985.

	35	36	37	38	39	40
35(1939)	0.2899	0.2571	0.1662	0.1468	0.1275	0.0982
35(1985)	0.3206	0.2393	0.1770	0.1182	0.0651	0.0365
	41	42	43	44	45	
35(1939)	-0.0892	0.0751	0.0535	0.0330	0.0203	
35(1985)	0.0207	0.0083	0.0034	0.0022	0.0007	

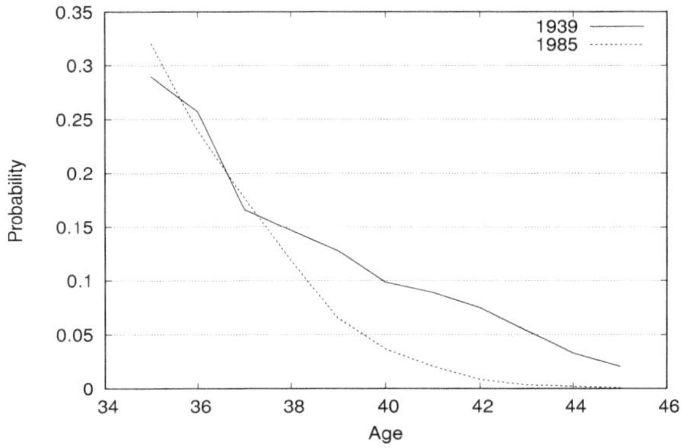

Figure 8.30: Birth probabilities of females married at age 35 in 1939 and 1985.

Table 8.13: Second-order derivatives of the probabilities of females married at age 35 in 1939 and 1985.

	35	36	37	38	39	40
35(1939)	-0.12821	-0.01498	0.02459	0.02417	0.00894	-0.00437
35(1985)	0.00488	0.00932	0.01232	0.01388	0.01400	0.01268
	41	42	43	44	45	
35(1939)	-0.00753	-0.00075	0.00724	-0.00072	-0.05028	
35(1985)	0.00992	0.00572	0.00008	-0.00700	-0.01552	

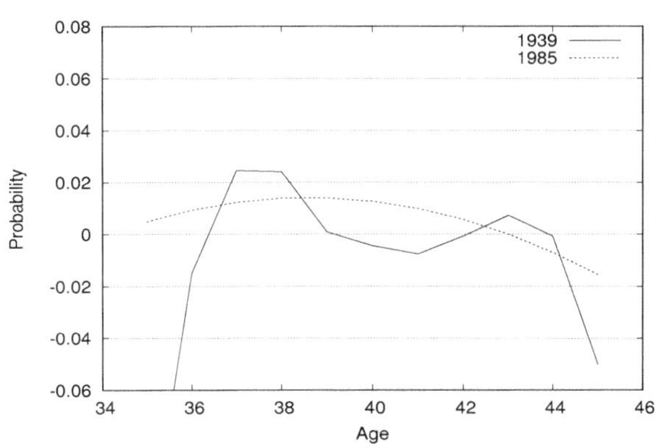

Figure 8.31: Second-order derivatives of the probabilities of females married at age 35 in 1939 and 1985.

Table 8.14: Second-order derivatives of the probabilities of females married at age 30 in 2000 and 2010.

	30	31	32	33	34	35
30(2000)	-0.00802	-0.00484	-0.00207	0.00029	0.00225	0.00380
30(2010)	-0.01761	-0.01231	-0.00748	-0.00312	0.00776	0.00420
	36	37	38	39	40	
30(2000)	0.00567	0.00599	0.00591	0.00542	0.00452	
35(1985)	0.00716	0.00965	0.01167	0.01323	0.01431	

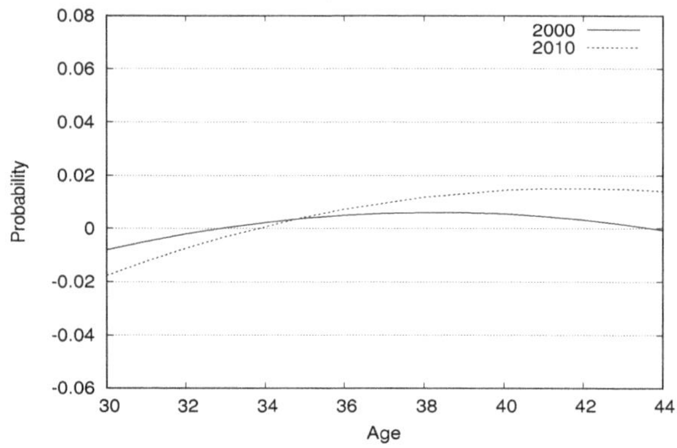

Figure 8.32: Second-order derivatives of the probabilities of females married at age 30 in 2000 and 2010.

ance had been made, it was replicated. I believe that this reactivation was brought about by the popularization of infertility treatment.

In Japan, the birth probabilities of women who married late declined in the latter half of the 20th century and improved slightly in the 21st century. Humans adapted to the very low fertility by means of the revival of full utilization of pregnancy.

8.7.4 Randomness of the Sequence of the Number of Children

Equation (8.18) is a simple mapping as below.

$$b(a_{i+1}, m_{j+1}) = \rho(a_{j+1})\, Fe(a_{i+1})\, ((\psi + \kappa\, \mu(a_*))\, b(a_i, m_j)),$$

Why does this simple mapping generate a random sequence of the number of children?[9] One reason is that as organisms, humans are stochastic converters. An organism cannot behave accurately.

Another more significant reason is that $b(a_{i+1}, m_{j+1})$ and $b(a_i, m_j)$ are arranged in order of the time of marriage in a cohort. In the case where

[9]Exactly, this mapping generates somewhat idiothincratic random sequence. When it is arranged in the socioeconomic order, it truens nearly perfect random sequence.

they are arranged in a causal order, $b(a_{i+1}, m_{j+1})$ can easily be calculated. Without arranging it, the relation between the probabilities of $b(a_{i+1}, m_{j+1})$ and $b(a_i, m_j)$ cannot be recognized. Without a time–space causal universe, the regularity of human birth behaviors cannot be found for the sequence on a single dimension using the data of social surveys. Having already recognized the necessity of enhancing the birth function to high-dimensional space, I understand this reason well.

A third reason is that the number of children is the summation of each result of the stochastic trials. When infinite socioeconomic variables are recognized, their interrelations cannot be deduced without a time–space arrangement.

8.8 Statistical Test of Birth Function for a Cohort

The optimal birth function for a cohort was obtained herein. Assuming that the birth function is constant (which, in reality, it is not), future birth probabilities and expected number of children can naturally be predicted. I intend to verify the birth function.

However, on calculating birth probabilities of a cohort using the vital statistics, I already know the number of births for every age at marriage in a cohort. Generally, these predictions are tautological. Nevertheless, I may test the birth function for 1939 data because Okazaki [59] did not present the birth probabilities for individuals married at over 31 years of age. The birth function can inductively compute the birth probabilities as far as possible. Therefore, here, the expected number of children of females married at age 35 are calculated and tested using data from a large-scale social survey.

The expected number of children for females married at age 35 was found to be 1.3718792 by the birth function. I construct the following hypotheses.

$$\begin{cases} H_0: & \mu = 1.3718792 \\ H_1: & \mu \neq 1.3718792 \end{cases} \tag{8.19}$$

When the sample mean of the number of children is not significant, the birth function is not disproved. From the National Family Research of Japan (NFRJ)[10] and JGSS, which were mentioned above, seven female samples born in 1939 and married at age 35 were obtained. The mean number of children is 1.2857. The calculation of the t-test does not reject H_0.

$$t = \frac{1.2857 - 1.3718792}{\frac{0.4880}{\sqrt{7}}} \doteq -0.467231 \tag{8.20}$$

However, in the actual population, the number of children of women who married at age 35 is not normally distributed. Therefore, the t-test is not valid.

Table 8.15: Distribution of the number of children of females married at age 35 in 1939 in Japan.

Number of Children	Probability
0	0.2197000
1	0.3705012
2	0.2673672
3	0.1090762
4	0.0280025
5	0.0047597
6	0.0005479
7	0.0000429
8	0.0000023
9	0.0000001

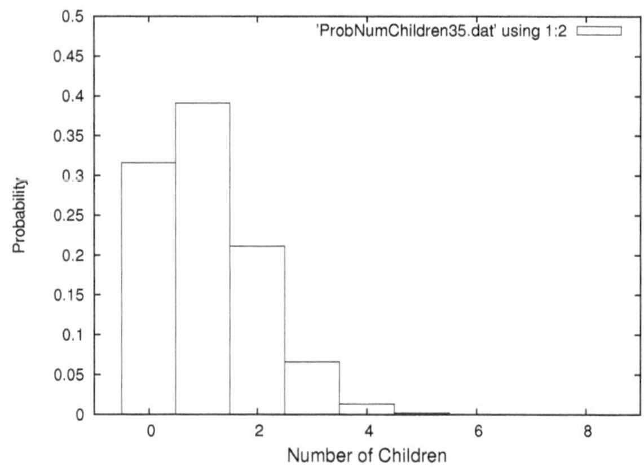

Figure 8.33: Distribution of the number of children of females married at age 35 in 1939 in Japan.

Using the birth function, the distribution of the number of children can be theoretically computed assuming there is sufficient computer memory. This task will be discussed in detail in a later chapter. Regardless, the number of births that follow the birth function for n arbitrarily prescribed females married at age 35 can be calculated.

Let n be seven because the sample we obtain is seven females. A total of 10000 probabilistic trials were conducted. The distribution of the number of children for females married at age 35 in the 1939 cohort in Japan was obtained as shown in Table 8.15. This distribution is then represented as a graph, as shown in Figure 8.33. The point is that $\overline{x} = 1.28571$ is probable as the mean of seven females. As a result of this numerical experiment, the probability of the

[10] http://nfrj.org/nfrj-profile.htm.

mean of n, that is, $\overline{x} < 1.28571$, is $2358/10000 = .2385$. Namely, $\overline{x} = 1.28571$ is sufficient.

Under the commonly considered level of significance, $\overline{x} = 1.28571$, $t = -0.467231$ has no significance for H_0. Although the small sample size makes these tests insufficient to confirm the birth function, the sample characteristic does not contradict the birth function.

8.9 Fixed Point of Birth Function for a Cohort

In Japan, where the fertility rate has dropped below the population replacement level, the birth probabilities appear to have stopped decreasing. See Figure from 8.34 to 8.36. At a glance, they display the same pattern seen in the 1970s birth cohort.

It is self-evident that the birth function for a cohort has fixed points. However, the condition for this is confirmed under the Banach fixed-point theorem.

$$b(a_{i+1}, m_{j+1}) = \rho(a_{i+1})\, Fe(a_{i+1})\, (\psi + \kappa\, \mu(a_*))\, b(a_i, m_j),$$

I use "Browder fixed point theorem" for this case.

Boundedness the product of Fe and b ensure boundedness because they are probability densities defined between the interval $[0, 1]$.

Closedness the product of Fe and b ensure closedness.

Convexity The previous analysis of the second-order derivatives of birth probabilities presented the tendency that they are approaching $0 \geq$. Therefore, it is easy to conjecture for all a, m, $b''(a, m) \geq 0$. However, this is difficult to prove. As the equation of the birth function is a recursion formula, it cannot be concretely analyzed.

However, the equation of the birth function suggests that the convexity of it depends on its element—*effective fecundability* is whether convex or not, because other variables are just unvariable numbers.

I analized it numerically, *effective fecundability* is almost convex but not convex for some intervals. In low dimensional degree, birth function is still unstable. The proof of the convexity might only be possible when the birth function is extended in a high-dimensional space by attaching the diffusion property. If it lacks diffusion property, it expresses unstable trajectories.

However, we can expect convexity for rather large intervals of *effective fecundability*, once the birth function begin to decline for its shape in Figures 8.34, 8.35, and 8.36.

The birth function appears to satisfy the conditions of "**Boundedness, Closedness, Convexity**", and thus, it has fixed points.

Human fertility is controlled by the mixture of physiology and social neighboring effect in order to adapt flexible to changing environment.

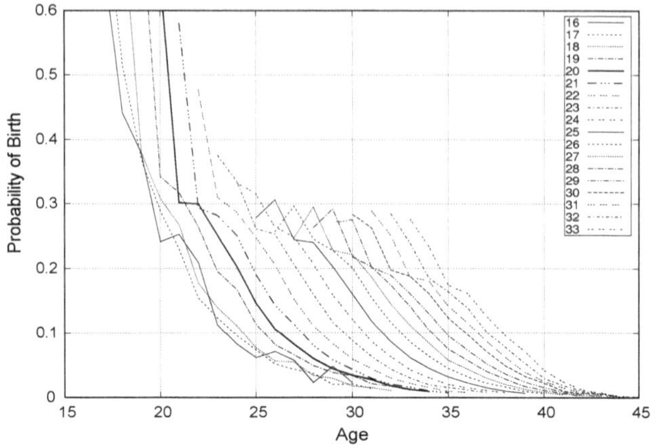

Figure 8.34: Probabilities of birth in the 1970 cohort in Japan.

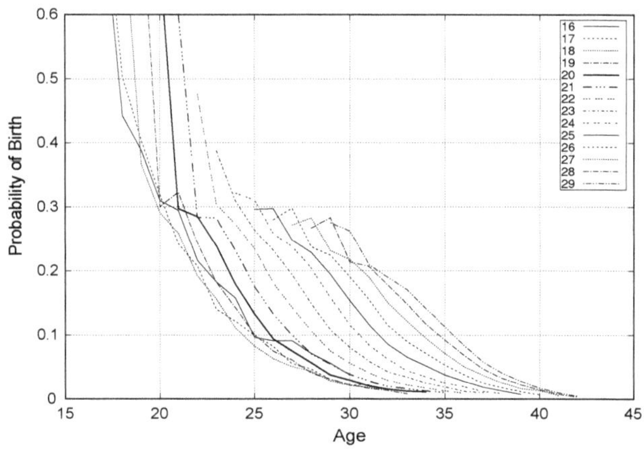

Figure 8.35: Probabilities of birth in the 1973 cohort in Japan.

8.10 Strong Relation between Late Marriage Tendency and Birth Function

On examining the catch-up effect, I perceived that the late marriage tendency inevitably caused the decline of birth probabilities. The delay of marriage slightly decreases $\mu(a_*)$; in the initial stage, this decrease went unnoticed. The slight decrease in $\mu(a_*)$ barely reduced the birth probabilities. These chain effects gradually but steadily developed without being consciously recognized during its initial stages.

In Japan, the late marriage tendency began around 1920. The present-day low fertility in the country was inevitable. In 1920, nobody was conscious of the decline in fertility. It is evident that human consciousness is not the cause of fertility decline.

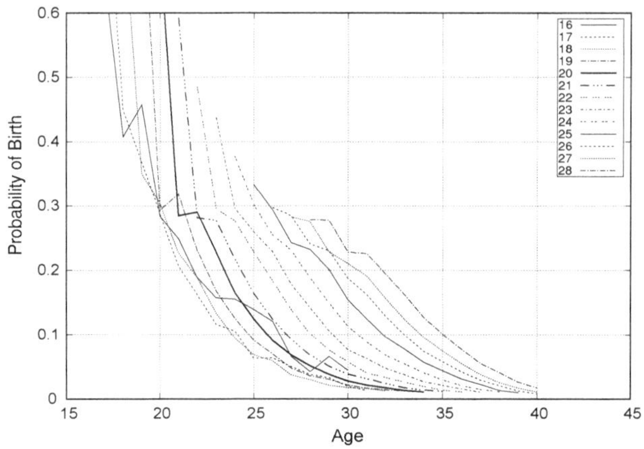

Figure 8.36: Probabilities of birth in the 1975 cohort in Japan.

In 1941, Japanese demographer Ayanori Okazaki made the following (fascist) statement:

> It should be noted that for France in the latter half of the nineteenth century, for other European countries after the beginning of the twentieth century, fertility gradually declined, as Table 1[11] presents. Every country, as a national population policy, desired high fertility, and they adopted some programs to promote births and founded childcare facilities to accomplish the goal of the pronatalist policy. Nevertheless, except for the effect of the Nazi pronatalist policy[12], almost no effect can be observed. We must realize an arduous mission: to prevent fertility decline and increase fertility.
>
> The fertility rate of my country (Japan) is still much higher than that of European countries. However, since 1920, it has shown signs of a slight tendency to decline. "The Guideline to Build the Population Policy" (originally Jinkou Seisaku Kakuritu Yokou) publicized four major objectives: to ensure the eternal growth of the population (of Japan), to be superior to other countries in terms of fertility and capacity, to retain military and work power for the high military state, and to maintain the leadership of Asian people. ...

*Source: *Kazoku to Jinkou* (Family and Population), Ayanori Okazaki [59], pp. 226–227.
Translated from Japanese to English by author.

[11]Table 8.16 at the end of this chapter.

[12]The fertility increase during this period was a temporal reaction to the previous Great Depression and the subsequent confusion and panic. It is quite clear that the Nazi population policy had a negative effect.

Okazaki realized the difficulty in effectively preventing fertility decline, and he obtained the knowledge to observe the pronatalist population policy of European countries empirically. However, he regarded Japanese fertility decline in those days as temporal and a fugitive whim that would be easily remedied. He made a foolish but common mistake: to perceive his own country as an exception and to underestimate the initial signs. Regardless, he did not realize that, without serious intention of Japanese parents as the cause of the diminishing number of children, they decline their number of children as a stochastic process.

In fact, there were no causes of fertility decline instigated by human will in Japan.

8.10.1 The Reverse is not Always True

The delay of marriage inevitably results in a decline in fertility. However, the reverse is not always true; early marriage tendency does not necessarily bring about an increase in fertility. The reason is that $\kappa\mu(a_*)$ is significantly smaller than ψ, so that the neighboring effect is far greater than the temporal increase of $\mu(a_*)$. In the nineteenth century, fertility decline occurred simultaneously with early marriage tendencies in Europe. The marriage function does not have the same diffusibility as the birth function because it is an integral function.

I am inclined to believe that the human population is designed to decline easily. A female can reproduce in surplus, but as a group, this surplus is controlled. This image coincides with evolutionary theory. Moreover, the birth function appears closely and subtly associated with marriage function. The human reproductive mechanism is more delicately designed than we once thought.

8.10.1.1 Effectiveness of Pronatalist Policy Measures

The birth function and its fixed points clearly suggest that fertility decline in Japan is a certainty. This is caused by the inevitable human behavioral dependence on neighboring behaviors. Japan has long been on the journey of population decline, although the end is not very far.

The fear of population decline is not new. It is a recurrent phenomenon in the world. In the section "The Recurrent Fear of Population Decline" of their book *The Fear of Population Decline*, Teitelbaum & Winter wrote the following:

> That history has never been fully explored. To do so is to see how demographic issues have touched the most important political and social questions of the past century, both in Europe and in North America. This is not to say that anxiety about population decline has been constant over the past 100 years. It ebbs and flows, depending both upon demographic realities and perceptions

> of the links between population change and economic, social, and political power.
>
> In the West, the fear of population decline last flourished during the Great Depression of the interwar years. It went into remission during the two postwar decades from 1945 to the mid-1960s. Since that time, anxious voices have been raised again, first in the socialist states of Eastern Europe, then in Western Europe, and most recently in North America. These cries of alarm are less shrill than those of the 1930s, but there are unmistakable signs that concern about population decline is on the rise. Indeed, in some countries, pronatalist policy measures verging on the draconian have been proposed or enacted, and the whole subject is further complicated by the fact that demographic trends in the developed and developing world have diverged to an unprecedented degree.
>
> Source: *The Fear of Population Decline*, M.S. Teitelbaum and J.M. Winter [71], p. 1.

Although they lacked the view of fertility decline in Asia, the decline in fertility can be considered as a universal phenomenon in human history. Its recurrence is brought about by the regularity which involves Equation (8.18) wherein the decline in marriage probability tends to trigger the decline of birth probabilities. This law is upheld by the assumption that humans behave according to neighboring behaviors, which is undeniable.

As Ayanori Okazaki said, the pronatalist policy measures of European countries in the 1930s were not effective. During the next wave of fear of population decline, the almost draconian pronatalist policy measures were unsuccessful. Similar pronatalist policy measures in the 1990s in Japan were also ineffective.

Assuming the birth function theory is accurate, if the process of birth probabilities is decreasing, the pronatalist policy measures are essentially ineffective. A large number of articles insisting that pronatalist policy measures are effective has been published. Recently, several researchers have insisted that maintaining "work-life balance" is assumed as an effective pronatalist policy measure. However, from their inception, these arguments have one common, fatal flaw: an inability to predict the value of fertility without the policy measures. In other words, they have not been able to predict the (numerical) value of the effect of pronatalist policy measures because their statements depend on our conscious states.

From this perspective, every statistical analysis of the pronatalist policy measures ever published is completely unscientific because it lacks the falsifiability. When the TFR coincidentally increases while some pronatalist policy measures are being adopted, they may be estimated to be effective.

8.11 Limitations of the Birth Function for a Cohort

The study of the birth function for a cohort brought about several findings. It demonstrates that human birth behaviors fundamentally replicate previous birth behaviors. Human conscious intuition, including my own, deny this thesis because we rationalize based on our situation and decide whether or not to bear children. However, our consciousness is unreliable. We are incapable of conceiving ourselves adequately; a scientific theory cannot be constructed based on our conscious will.

> PROP. XXVII. The idea of each modification of the human body does not involve adequate knowledge of the human body itself.

Spinoza, *Ethics (Ethica Ordine Geometrico Demonstrata)* [16], p. 67.

A scientific theory must be constructed based on observed data and extrapolation from them. The birth function for a cohort shows that this is the only viable approach.

The birth function for a cohort clearly demonstrates that the catch-up birth behaviors exists. Without a catch-up term, the birth function cannot compute adequate birth probabilities. We simply attempt to replicate previous birth behaviors, but the magnitude is slightly affected by extra-local numbers of children.

The birth function for a cohort precisely estimates the effect of infertility treatment for the first time.

Despite this incredible success, the birth function for a cohort is insufficient and has manifest limitations. The alteration of the birth function depends entirely on its initial values. The birth function for a cohort necessarily lacks actual dynamics. A cohort is only quasi-causal within itself. For actual dynamics, the birth function must be extended to high-dimensional space.

From another point of view, geographical factors must be added to the historical dynamics of fertility undulation based on the results of historical demography. This problem will be explained in the following chapter.

Table 8.16: Fertility rates of Western principal Nations.

Year	France	England -Wales	Germany	Italy
1801	32.9			
1806	31.4			
1811	31.6			
1816	32.9			
1821	31.7			
1826	31.4			
1831	30.3			
1836	29.2			
1841	28.5	32.2	36.4	
1846	27.3	33.8	36.0	
1851	27.1	34.2	36.7	
1856	26.3	34.4	33.4	
1861	26.9	34.6	35.7	
1866	26.4	25.2	37.8	38.7
1871	22.9	35.0	34.5	37.0
1876	26.2	36.3	40.9	39.3
1881	24.9	33.9	37.0	38.1
1886	23.9	32.8	37.1	37.0
1891	22.6	31.4	37.0	37.2
1896	22.5	29.6	36.3	34.8
1901	22.0	28.5	35.7	32.5
1906	20.5	27.2	33.1	32.1
1911	18.7	24.4	28.6	31.5
1916	9.5	21.0	15.2	24.0
1921	20.7	22.4	25.3	30.3
1926	18.8	17.8	19.5	27.7
1931	17.5	15.8	16.0	24.9
1936	14.7	14.9	18.8	22.9

*Source: *Kazoku to Jinkou (Family and Population)*, Ayanori Okazaki [59], pp. 222–224. Re-tabulated and translated from Japanese to English by author.

Chapter 9

Necessity of Extending the Birth Function

> Since it is evident that no simple general model of the relationship between demographic changes and the industrial revolution is likely to "save phenomena,.....

E.A. Wrigley, *Population and History* [83], p. 164.

Even in the 1960s, for those with good judgment and independence from academic trends, it was fairly evident that the demographic transition theory was failing. Why is it not completely denied today? The reason is simple: there has been no alternative theory. The theory of marriage and birth functions is not only an alternative but a more general and sophisticated sociological and demographic theory. Comparatively, "demographic transition" is merely a patchwork of unordered intuitions.

By the previous chapter, it has been established that mathematical and theoretical extensions are inevitable. However, empirical data will also be demonstrated here.

9.1 Need for Geographical Factors

If the reaction-diffusion equation describes the dynamics of the total number of children of parents in 19th century Europe, the birth function must describe diffusion dynamics.

The fact that fertility is geographically distributed confirms the geographical extension of the birth function. At first glance, this contradicts the reaction-diffusion system, because it requires the smoothing of fertility. However, as reaction-diffusion is a large-scale model of overall fertility, undulations of the TFR (approximately 0.1–0.2) are probable in geographic dimensions at a micro scale of time.

The fact that fertility undulates is significant. A meaningful theory of the birth function must systematically explain these undulations. Moreover, the longitudinal and geographical dynamic undulation of the TFR is observed. An effective birth function must explain these dynamics in its own system.

9.1.1 Religious Differential Fertility Explained by Spatial Differentials

It is generally taken for granted that religion has a certain effect on fertility. It is believed that humans marry and bear children based on religious factors. It is an explicit and conventional assumption that we act rationally as utilitarian persons, and our deviant actions are due to religious values.

However, these are entirely outdated preconceptions, because empirical studies do not support these assumptions. Schellekens and Poppel researched the historical registration data of The Hague and came to the following conclusions:

> There are no significant differences between Catholics and Protestants for the period before the transition. Marital fertility for these two groups only started to diverge in the 1880s[1]. Catholics lagging behind other religious groups. After controlling for occupational categories, the coefficient of being Catholic remains almost unaffected. This lends support to the argument that the characteristics hypothesis does not explain the relatively high fertility of Catholics.
>
> J. Schellekens and F. van Poppel, Religious differentials in marital fertility in The Hague (Netherlands) 1860–1909, p. 79, in *Religion and the Decline of Fertility in the Western World* [68].

If religious values had some definite effect, also before the transition, there should have been some significant differences between the two groups. The doctrine of Catholicism did not affect birth behaviors in The Hague. Catholics are a minority in the Netherlands and are distributed uniformly in The Hague. In this situation, as Catholics receive neighboring effects equivalent to those of Protestants, no difference was seen. Namely, the manner of their distribution in space is crucial. Without considering the distribution of the group in the (geographical) dimensional space, every inference is superficial.

> Before and during the transition, Jewish infant and child mortality in Europe as well as The Hague was lower than that of most

[1] This timeline is consistent with my estimation by the progressive wave. The reaction-diffusion equation is supported by the Schellekens and Poppel's research results.

> other religious groups (Van Popple, Schellekens, and Liefbroer 2002). Demographic transition theory would predict an early fertility decline among Jews in such a case, all else being equal. This is indeed what several authors claim happened. Livi Bacci (1986) describes Jews as forerunners in the fertility decline in Italy. An early fertility decline has also been ascribed to German Jews (Knodel 1974). However, Derosas (2003:114) finds no evidence of an early fertility decline among Venetian Jews. Furthermore, one of the few German family-reconstitution studies to compare Jews with other religious groups does not fit this characterization. For couples who married before 1830, the marital fertility of Protestants in the village of Nonnenweier already shows a large deviation from the Hutterite age pattern, while the first signs of family planning among local Jews only become visible among couples who married in the 1840s (Goldstein 1981: 132, Table 8). The result shown in the present study raises further doubts about the special character of Jews during the marital fertility transition. We did not find any evidence of higher levels of parity-dependent fertility control among Jews than among the rest of the population before or during the transition. **In some places, Jews may have been forerunners, but this was not a general European phenomenon.***
>
> *Boldface incorporated by the author of this book.
> J. Schellekens and F. van Poppel, Religious differentials in marital fertility in The Hague (Netherlands) 1860–1909, p. 80,
> in *Religion and the Decline of Fertility in the Western World* [68].

The marriage and birth functions extended to high-dimensional space can solve this apparently discrepant phenomenon. It is not sufficient to recognize some religious groups statistically, but it is necessary to grasp their dispersal in high-dimensional space.

9.1.1.1 Jewish Fertility Decline in 19th-Century Venice

R. Derosas reported significant research findings on the fertility decline among Jews in Venice in the 19th century. He originally considered intra-Jewish differential fertility.

> Its main concern is not–or not primarily–about fertility differentials between Jewish and non-Jewish populations, but rather about intra-Jewish differentials. Unfortunately, such a viewpoint is usually lacking in historical studies of Jewish fertility. ...
>
> R. Derosas, Between Identity and Assimilation: Jewish Fertility in Nineteenth-Century Venice, p. 177,
> in *Religion and the Decline of Fertility in the Western World* [18].

His analyses of Venetian Jews almost perfectly corresponds to "the Background Independence of Fertility Decline." Below is a quote from his Cox regression analysis of inter-birth intervals in Venice (1850–1869).

> All models show that the demographic factors heavily influence the hazard rates in the expected direction. They also remain remarkably constant, showing that the way of considering religion does not influence their effect at all.
>
> Socioeconomic conditions do not seem to play any significant role in shaping hazard rates. This is rather unexpected, since it clearly contrasts with the previous analysis of marital fertility rates.
>
> R. Derosas, Between Identity and Assimilation: Jewish Fertility in Nineteenth-Century Venice, p. 198. in *Religion and The Decline of Fertility in The Western World* [18].

During the process of fertility decline, background independence was also consistent in Venice. If some socioeconomic conditions appeared to have played a significant role, there might be an intense dispersal in high dimensional (geographical) space correlated with socioeconomic variables. Namely, the inference that socioeconomic conditions have some effect on birth behaviors is a perceptual illusion.

In this sense, Derosas' findings on the intra-Jewish differentials are an interesting case. Below is another quote from his analysis.

> We can now turn to the main interest of this analysis. Model 2 compares the hazard rates of Catholics and Jews. There is definitely no difference between the two groups. The difference in the maximum log-likelihood also shows that the covariate is not statistically significant. On the other hand, the coefficients estimated in model 1 are not modified once religion is added in model 2, suggesting that there is no interaction between socioeconomic conditions and religion. In fact, this is confirmed by a model including interaction terms, which is not shown here. In conclusion, this leads to the rejection of both the "particularized theology" and the "characteristics hypothesis." Neither religious affiliation in itself, nor socioeconomic conditions, nor any interaction between the two, have a significant effect on fertility, as it is expressed by the hazard rates of a conception.
>
> R. Derosas, Between Identity and Assimilation: Jewish Fertility in Nineteenth-Century Venice, p. 199, in *Religion and the Decline of Fertility in the Western World* [18].

As Derosas said, religious beliefs are unrelated to birth behaviors. I venture to say that any value has no relation to fertility. When some relation appears to exist, there must be a neighboring population with the same value.

9.1.1.2 True Nature of Difference Between a Ghetto and Elsewhere

The originality of Derosas' research is the comparison between the fertility of Jews inside and outside a ghetto. He analyzed the difference as follows.

> The situation changes dramatically once we distinguish between the Jews living inside the Ghetto and the Jews dwelling outside. As the maximum log-likelihood shows, the covariate is statistically significant. The coefficients for the two Jewish groups are both significant (p-value < 0.01). The relative risk of a new conception for the Jews of the Ghetto is 19.5 per cent higher than that of the Catholics. On the other hand, the other members of the Jewish community have a lower risk than the Catholics by 13.2 per cent. The relative difference between the two Jewish subgroups is therefore about 38 per cent. It should also be noticed again that there is no significant evidence of any interaction between religion and socioeconomic conditions.
>
> ... Thus far I have assumed that the residential choices of Venetian Jews reflected their different attitudes about the relationship with the Gentile society. ...
>
> R. Derosas, Between Identity and Assimilation: Jewish Fertility in Nineteenth-Century Venice, p. 199, in *Religion and the Decline of Fertility in the Western World* [18].

However, the intra-Jew difference in fertility should not be due to their attitudes. On this point, I cannot agree with Derosas' inference. It takes us back to where we started. The cause of the difference must be quoted in the space in which our birth behaviors occurred.

It must be noted that fertility decline in Venice began from the Eastern-South of this islet (probably the famous San Marco parish). The map is shown in Figure 9.1. His analysis was performed for the period 1850–1869. In the normal case, the progressive wave of fertility decline did not leave vestiges because its velocity is 10 km/year. However, the parishes of Venice islets are segregated by canals. Derosas selected five areas, as shown in Figure 9.1 (originally Map 2). As shown in Figure 9.1, all Jews analyzed outside the ghetto dwelt in "S. Luca." Therefore, it is natural that Jews outside the ghetto presented relatively low fertility compared to the Catholics, including having segregated parishes. On the other hand, the ghetto was unique for its exclusiveness, in fact, it is separated from "San Marco" by canals. It is already known that natural obstacles such as rivers, mountains, and valleys prevent or delay the diffusion process.

If Derosas compared the fertility within the ghetto with that within "S. Geremia," he would not find a significant difference. R. Derosas failed to explain the difference between the fertility of Jews outside the ghetto and that of the Catholics. The birth function in geographical dimensional space can provide a reasonable explanation. The geographical location of residents maps

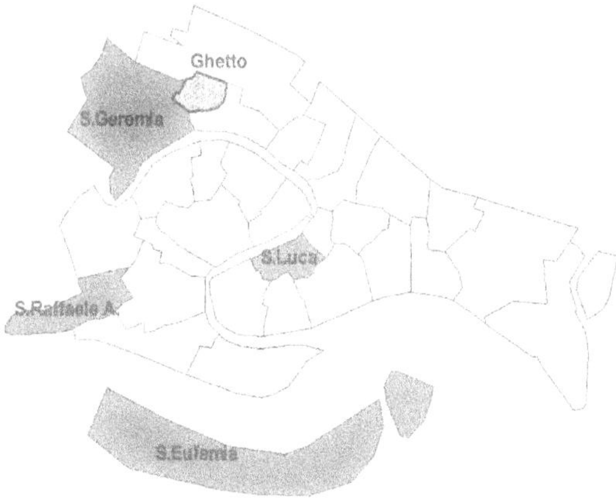

Figure 9.1: Areas analyzed by Derosas.

R. Derosas, Between Identity and Assimilation: Jewish Fertility in Nineteenth-Century Venice, p. 184, in *Religion and the Decline of Fertility in the Western World* [18].

Figure 9.2: Distribution of Jewish dwellings in Venice, 1869.

Source: 1869 census (my elaboration).

R. Derosas, Between Identity and Assimilation: Jewish Fertility in Nineteenth-Century Venice, p. 183, in *Religion and the Decline of Fertility in the Western World* [18].

their attitudes. Human neighboring behaviors map our consciousnesses. This is essentially assimilation.

9.1.2 End of American Catholic Fertility

According to Westoff and Jones [79], Catholic and non-Catholic fertility in United States tended to converge before World War II. This trend reversed during the baby boom, and restarted after it.

The historical record of Catholic and non-Catholic cohort fertility is displayed in Table 1[2]. The inter-cohort series for fertility completed through 15 to 19 years of marriage suggests some early movement toward convergence, followed by a sharp reversal and divergence for the cohorts whose early childbearing was during the baby boom, and then a return toward convergence after the baby boom. The same patterns are evident for fertility at each duration, with the most recent convergence apparent in the first years of marriage of the latest cohort (1971–1975).

Westoff & Jones, The End of "Catholic" Fertility [79], p. 210.

Table 9.1: Cumulative Number of Births by Duration of Marriage for Catholic and Non-Catholic Marriage Cohorts Based on Merged Data from the 1955 and 1960 GAF Studies and the 1965, 1970 and 1975 NFS.

Marriage Cohort	Comulative Births by Duration(Year)				Number of Women			
	2.5	7.5	12.5	17.5	2.5	7.5	12.5	17.5
				Catholic				
1936–40	.61	1.92	2.78	3.26	150	150	150	150
1941–45	.72	1.99	2.76	3.13	346	346	346	206
1946–50	.87	2.28	3.18	3.74	671	671	463	318
1951–55	.89	2.67	3.62	3.97	656	501	348	167
1956–60	1.15	2.28	3.36	3.57	512	392	213	155
1961–65	1.06	2.32	2.86		427	226	162	
1966–70	.92	2.00	2.86		277	200		
1971–75	.62				282			
				Non-Catholic				
1936–40	.55	1.66	2.34	2.72	440	440	440	440
1941–45	.68	1.89	2.55	2.91	992	992	992	616
1946–50	.77	2.04	2.72	3.03	1694	1694	1240	882
1951–55	.84	2.20	2.80	2.92	1529	1170	853	463
1956–60	.87	2.24	2.72	2.88	1233	956	521	390
1961–65	.83	1.93	2.36		1027	628	476	
1966–70	.73	1.70			733	505		
1971–75	.57				760			
				Difference(C-NC)				
1936–40	.06	.26	.44	.54				
1941–45	.04	.10	.21	.22				
1946–50	.10	.14	.46	.71				
1951–55	.05	.47	.82	1.05				
1956–60	.28	.58	.64	.69				
1961–65	.23	.39	.50					
1966–70	.19	.30						
1971–75	.05							

ibid, p. 212.

Readers of this book can easily understand that the reaction-diffusion process of the number of children brought about this convergence of the fertility of Catholic and non-Catholic cohorts. The reaction-diffusion equation can provide a reasonable explanation for the dynamic change in Catholic fertility before and after the baby boom.

[2]Corresponds to Table 9.1.

However, during the baby boom, a phenomenon that is beyond the capacities of the reaction-diffusion equation occurred. Westoff and Jones recorded this:

> One interesting finding that emerges from a comparison of the inter-cohort trend by religion is that the Catholics appear to have contributed disproportionately to that part of the baby boom which consisted of a rise in completed fertility (the minor part, according to Ryder, 1978). The rise in completed fertility among non-Catholics is rather modest by contrast.
>
> Westoff & Jones, The End of "Catholic" Fertility [79], p. 211.

During the baby boom, fertility of Catholics was relatively higher than that of non-Catholics. To explain this phenomenon, the birth function theory must be employed.

$$b(a_{i+1}, m_{j+1}) = \rho(a_{i+1})\, Fe(a_{i+1})\, \big(\psi + \kappa\, \mu(a^*)\big)\, b(a_i, m_j), \tag{9.1}$$

As mentioned earlier, the baby boom brought about an increase in marriages due to soldiers returning after World War II. While the same ratio of soldiers returned to the Catholic and non-Catholic communities[3], the relatively higher birth probabilities in the Catholic community remained, and the catch-up term $\kappa\, \mu(a^*)$ was effective in its chain reaction. The slightly higher birth probabilities caused a more significant increase in $\mu(a^*)$, and the $\mu(a^*)$ in turn raised the birth probabilities of Catholics during the baby boom, which lasted over 10 years. Namely, the exothermic reaction was more prominent in Catholic communities.

On finding the explanation for the differential fertility of Catholics and non-Catholics by the birth function, I became confident[4] that the birth function has a catch-up term in it and that the birth function itself is cross-culturally and longitudinally valid.

By constructing the birth function in the extended high-dimensional space, the intra-Jewish and the inter-religion differentials can be explained from a highly integrated theoretical perspective while naturally maintaining background independence. This is possible given that socioeconomic conditions have no locality or laterality in the space.

> <u>The combination of all of the social forces encouraging low fertility</u> in the United States coupled with the undermining of the authority of the church on the birth control issue seems to

[3] I assume that Catholics resident geographically locally densely in communities. This assumption is necessary and valid.

[4] This confidence was unfounded. The catch-up effect is eliminated by the neighboring effect in the extended high-dimensional space. The catch-up term is necessary in the frame of a single cohort because the neighboring effect is not sufficient in this low-dimensional space.

> be an adequate explanation of the disappearance of "Catholic" fertility in the United States.
>
> As plausible an explanation as this seems, however, it does not fully explain the whole twentieth century trend of Catholic fertility. First, there is the trend processes of assimilation that have caused toward convergence prior to the baby boom that occurred before the big debate about contraception, before the defection of masses of Catholics from official church teaching on contraception, and before the diffusion of new contraceptive technology. It may be that the pressures of the Depression and of World War II were sufficient to reduce Catholic fertility even with less reliance on more effective means. The much greater rise of Catholic than non-Catholic fertility during the baby boom is consistent with the notion that Catholic fertility, although embodied in the long-term historical decline of American fertility in general, was lower than it would have been during the 1930s and 1940s had it not been for cataclysmic events of those decades.
>
> Westoff & Jones, The End of "Catholic" Fertility [79], pp. 216–217.
> *Underlined by the author of this book.

The birth function and the reaction-diffusion system can concretely express the "combination of all of the social forces encouraging low fertility coupled with the undermining of the authority of the church on the birth control issue" which they were only able to metaphorically verbalize.

"The trend processes of assimilation" are the result of diffusion within the geographical and age dimensions. Essentially, this means that we must receive the effect of our neighbors. The birth function theory can dedicate a reasonable and integrated explanation of the phenomenon of "the much greater rise of Catholic" fertility during the baby boom. The alteration in high-dimensional space is governed by neighboring birth behaviors. These are the true cause of human behaviors, and our consciousnesses are mirrors that reflect them.

The alteration in high-dimensional space is the alteration of the attitude as a result. I reiterate that conscious motivations are unnecessary to explain fertility decline.

9.1.3 Inability of Explaining Fertility Undulations through Socioeconomic Differences

Social scientists often assume that the undulations of fertility can be reduced to socioeconomic differences. This type of conventional thinking must be changed.

It is impossible to reduce geographical differences to the aggregation of individual features. Several historical demographers are reminiscent of these impressions. Josef Ehmer, an Austrian historical demographer, made the following statement:

> Eindeutig ist nur, dass sich alle Theorien und Erklärungsansätze, die in einzelen Faktoren die Ursachen des Gegurutenrückgangs sahen, seinen sie ökonomisher oder kullttureller Art, als emprisch nicht haltbar erwiesen haben.
>
> (Translation: It is already evident that all theories and explanations seeking the causes of fertility decline in an individual economic or cultural factor are no longer supported empirically.[5]
>
> Josef Ehmer [24], p. 46.

Novel factors are necessary to explain demographic phenomena, not new statistical techniques. Common statistical techniques, such as multivariate analysis, only confirm subjectively reasoned conscious relationships.

9.1.3.1 Universality of Fertility Undulation

The industrial revolution is generally regarded as the main driving factor of fertility decline. However, fertility undulations had occurred before it.

> First, economic historians have too easily assumed that there was something exceptional about the rate of growth of population in England in the eighteenth century. It is implicit in much that has been written on this subject that the very slow growth of population in England in the first half of the eighteenth century was normal for Europe at that time and that the acceleration which took place during the second half of the century was not paralleled elsewhere. Table 5·1[6] shows how inaccurate such an assumption is. In a number of European countries, population growth in the eighteenth century was very swift by the general standards of pre-industrial societies, and was often at its swiftest in areas most remote from rapid economic change. Rates of growth as high as those in England occurred in many parts of Europe, and in many countries, there was an acceleration of the rate of growth about the middle of the century. Comparison with Table 5.12[7] shows that rates of growth did not in general change very much in the nineteenth century, contrary to what is often assumed.
>
> E.A. Wrigley [83], p. 152.

It can be inferred that demographic phenomena such as fertility increase and decline have an autonomy in themselves, that is, background independence. Their dynamics are solved only by computations using the equations of the function. In fact, reality is highly complex, and existent phenomena are so transcendental for human consciousness that they cannot be reduced to our conscious actions. Wrigley wrote about this problem as follows:

[5] I have translated this to English.
[6] Corresponds to Table 9.2
[7] Corresponds to Table 9.3.

Table 9.2: Population growth in 18th century Europe (in '000s).

	England & Wales	France	Italy	Sweden	Württem-berg	East Prussia
1700	5,826	19,000	11,500		340	400
1720				1,450		
1740		20,000			472	
1750	6,140			1,740		
1800	9,156	25,000	18,000	2,347	660	931

	Pomerania	Silesia	Austria	Bohemia	Hungary
1700	120	1,000			
1720					
1740	309	1,100			
1750			1,360	1,940	3,000
1800	500	2,000	1,888	2,922	8,500

Table 9.3: Percentage rates of growth per annum.

England	1701–1801 = 0.45	East Prussia	1700–1800 = 0.84
& Wales	1751–1810 = 0.80	Pomerania	1740–1800 = 0.80
France	1700–89 = 0.31	Selesia	1740–1804 = 0.94
	1740–89 = 0.45		
Italy	1700–1800 = 0.45	Austria	1754–89 = 0.94
Sweden	1974–1800 = 0.59	Bohemia	1954–89 = 1.18
Wüttemberg	1740–1800 = 0.56	Hungary	1754–89 = 3.01

E.A. Wrigley [83], p. 153.

Patterns of social behaviour and their associated demographic characteristics were often slow to change. The notion of leads and lags is helpful here, particularly for changes in age at marriage and fertility in general. A delayed response in say, age at marriage to changing economic conditions might itself in turn provoke further economic changes as well as being intimately bound up with other aspects of social behaviour. By introducing the idea of time lags into the model of relationships set out in Figure 4.1[8], for example, some interesting further possibilities appear.

The history of Colyton's population provides a striking instance both of the absence of quick response in marriage habits to changing circumstances, and of the possibility of lurches in either direction which this implies. During most of the sixteenth century and the first few decades of the seventeenth, population was rising rapidly in England, but thereafter for a further century it grew very little. Indeed, in many places, the earlier growth gave way to decline. During the period of rapid population growth, real income

[8]Figure 4.1 is presented in this Chapter Appendix as Figure 9.18.

> fell. Prices rose tremendously in the sixteenth century, but wages in general lagged well behind. When population growth had died away, however, real incomes stabilised, and in many areas probably began to rise slowly in the later decades of the seventeenth and the early decades of the eighteenth century. Direct evidence about the course of real wages in Colyton is lacking, but they are unlikely to have moved contrary to national trends over this long period.
>
> If marriage decisions in Colyton had reflected sensitively the changes in real income, the average age at first marriage of women should have tended to rise during the sixteenth century as population pressure grew. After the middle of the sixteenth century the average age should have tended to fall again. In fact, matters turned out quite differently, for over the long period from 1560 to the 1640s, the average age at first marriage of both men and women did not change at all but thereafter rose sharply (see table 3.5[9]) for women and slightly for men. Moreover, marital fertility was lower in this period than earlier and two changes together produced much lower general fertility rates than before. It is hard to resist the conclusion that this phase of Colyton's demographic history represents a lurch in the opposite direction to that of Le Vieuxbourg a hundred years later. If economic growth was being prejudiced, it was certainly not by an excessive press of numbers: and if real incomes were rising (which is probable) the fertility schedule was notably slow to respond in a 'Malthusian' way.
>
> E.A. Wrigley [83], pp. 140–142.

He pointed out that there were no consistencies in the relationship between fertility and socioeconomic variables in pre-industrial society in England. A meaningful theory of human birth behaviors must make no distinction between the pre-industrial and the industrial. When the birth theory is assumed for each classified case, it is no longer valid. There must be one single theory of birth behavior. As quoted previously, referring to the industrial society in England, Wrigley confessed that "it is evident that there is no simple general model of the relationship between demographic changes and economic variables." Even Wrigley's genius, in 1969, did not lead him to state that socioeconomic factors have a smaller effect on fertility. However, because I argue that neighboring birth behaviors are the principal factor affecting human birth behaviors, I assert that geographical differences occur, and (geographical) leads and lags occur. The extension of the birth function to geographical dimensions is equivalent to admitting that the effect of neighboring birth behaviors is principal. This was self-evident at all. Thus, past research suggests that this extension is unavoidable.

[9] table 3.5 is presented in the Chapter Appendix as Table 9.7.

Note the underlined part of the above quote. The notion of "leads and lags" (p. 252) are not appropriately modeled except for the function in a high-dimensional (time–space) space. The leads and lags of fertility can only be theoretically described on the time and space (geographical) dimensions.

> At the local scale of analysis, however, relationships between economic and demographic changes are no more easy to find. It may be plausible to attribute the sudden alteration in the balance of baptism and burial in, say, Worcestershire or Lancashire, to the changes in train in the local economies. It is more difficult to do so in a parish like Hartland, where the change was also marked and began about 1750, as early as in many industrial areas. Hartland was far from the small centres of new industrial growth. Its economic base did not begin to change for many years after the sudden appearance of substantial baptism surpluses.
>
> E.A. Wrigley [83], p. 156.

9.2 Equivalent Principle of Time and Distance

Following Wrigley's intellection, I conjecture an equivalent principle that for fertility-related phenomena, elapsed time is equal to distance. Time and distance are causally equivalent because our behaviors are instigated by neighboring behaviors.

Principle 9.2.1 (Equivalent Principle)

Elapsed time is equal to distance in fertility-related phenomena. Time and distance are causally equivalent because human behaviors are instigated by neighboring behaviors.

● represents Hartland; ★ represents Colyton

Figure 9.3: Hartland, Devon, England.

In one instance, I observed two parallel running trains beginning to depart at the junction, and rode one of them while watching another train depart. I noticed that their synchronicity was rapidly dissolving as they departed in different directions. This demonstrated that to share a neighboring space is to share time with others in society.

Causality must be defined as the proximity of time and space. Through his introspective speculation, Spinoza made a similar proposition:

> PROP. X. Towards something future, which we conceive as close at hand, we are affected more intensely, than if we conceive that its time for existence is separated from the present by a longer interval; so too by the remembrance of what we conceive to have not long passed away we are affected more intensely, than if we conceive that it has long passed away.

Spinoza, *Ethics (Ethica Ordine Geometrico Demonstrata)* [16], p. 161.

Here, "which we conceive as close at hand" can be equated to geographical proximity.

Julian Barbour said, "The flow of time and motion are illusions" [2]. In the social world, the simultaneity is to be with oneself. Strictly speaking, time and the temporal consciousness have slight differences when we are in close proximity to each other, and they can be quasi-shared.

9.2.1 Effect of High-Speed Mobility on the Equivalent Principle

Some may suspect that traveling on a highway, by a high-speed railway, or in an airplane would destroy this equivalent principle. Following the industrial revolution, modern people have gained a sense that humans have transcended geographical distance. High-speed transportation cannot eliminate this limitation. This is because during transit, we become unconnected individuals and lose interactions with neighboring people. Causality equals neighborhood alterations; in other words, without neighborhood alterations, there are no causal dynamics.

On reaching the destination, we are thrust into interactions there and experience significant neighborhood alterations.

9.2.1.1 High-Speed Mobility is not Destroying the Equivalent Principle

If high-speed mobility could remove the limitation of geographical distance, the difference in fertility between areas could not persist in modern society.

In modern Japan, a high-speed transport network has been constructed. In particular, Tokyo and Nagoya are closely connected by the "Shinkansen" and highway; if high-speed mobility could destroy the equivalent principle, the fertility of Tokyo and Nagoya would be closer than their neighboring districts.

Here, the fertility index for the local area was obtained by considering child–woman ratio (CWR)[10] of Tokyo, Aichi, Kanagawa, and Gifu Prefectures. It was confirmed that the CWR of Tokyo gradually changed to that of Kanagawa, and the CWR of Aichi resembles that of Gifu more than that of Tokyo.

Table 9.4: Child–woman ratio of Tokyo, Aichi, Kanagawa, and Gifu Prefectures.

	CWR		
Prefecture	2010	2015	2020*
Tokyo	0.15962	0.16755	0.16564
Aichi	0.21322	0.20725	0.20698
Kanagawa	0.19045	0.18479	0.18524
Gifu	0.20685	0.20223	0.20316

*Estimated by National Institute of Population and Social Security Research of Japan.

On observing the CWR on a more micro scale, it is evident that the CWR continuously changes based on the geographical dimensions, as shown in Figure 9.4 and Figure 9.5.

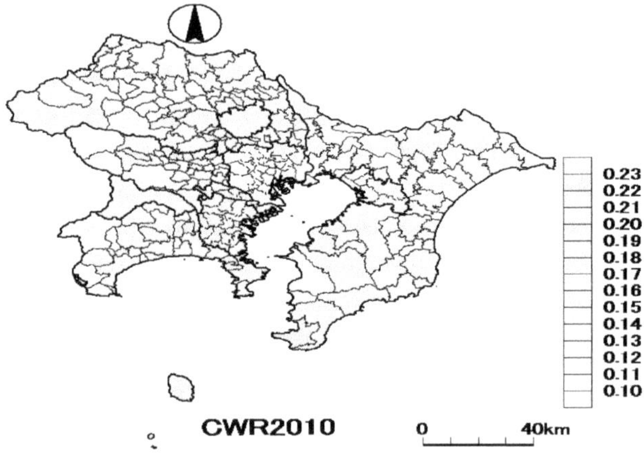

Figure 9.4: Regional CWR of Kanto Area in Japan in 2010.

Comparing Figure 9.4 and Figure 9.5, it can be seen that the Kanto area is relatively whiter than Aichi and Gifu. In the Kanto area, the low fertility of

[10]Child–woman ratio is a general index that indicates the fertility level of the local population and is commonly used to project the regional population.

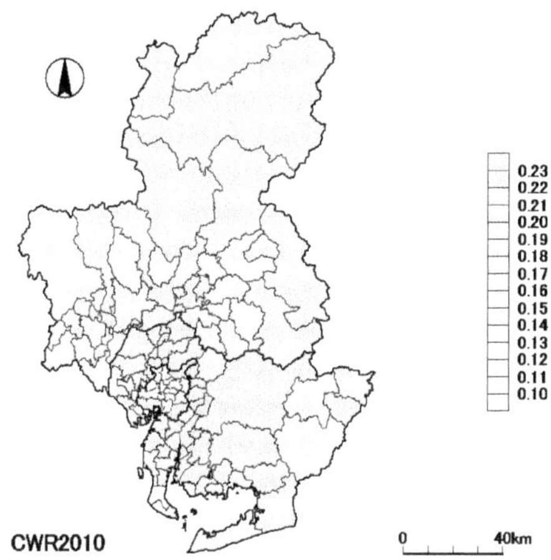

Figure 9.5: Regional CWR of Aichi and Gifu in Japan in 2010.

the center of Tokyo diffused to its peripherals. On the other hand, the Aichi and Gifu prefectures maintained relatively high fertility for all areas except in the very center of Nagoya city. The low fertility of the very center of Nagoya city was not a diffusion feature, as it did not diffuse when observed.

However, in this subsection, it is sufficient to state that, with sufficient verification, the equivalent principle can be protected from the hypothesis that high-speed mobility can transpose geographical distance. In this sense, the effect of modern traffic on human activity is not fundamental to demographic phenomena.

9.2.2 Effect of Network Communication on the Equivalent Principle

In the age of the Internet, worldwide communication transcending geographical distances is possible. Certainly, packets of information can be exchanged. However, there is little influence in terms of whether others are married or bear children. Similar to high-speed mobility, the information–interchange cannot affect the neighbor on the transmission path. Therefore, this may cause solely conscious change, and conscious change is always trivial and temporal. The alteration caused by symbols is always temporal.

In the field of sociology, it is generally accepted that information communicated by the media cannot change people's behaviors. As P. Lazarsfeld argued in his book *The People's Choice* [48], people vote based on their neighboring opinion leaders; the effect of mass media is secondary and indirect. The fundamentals of sociology caution against overestimating the effect of communication media such as hypertext space.

9.2.2.1 Present-Day No-acceleration of the Tempo of Fertility Undulation

If networks could eliminate geographical distance, the tempo of fertility undulation would be accelerated, whether in its decline or increase. On the other hand, the equivalent principle ensures a steady tempo of alterations.

The decline of total marital fertility occurred over about 50 years in Japan. In 1920, the mean number of children in a family was 5.0, and this number declined to around 2.0 in 1970. Thus, in the pre-Internet era, the rate of decline was estimated to be 0.06 per year.

In 2005, the TFR in Japan was recorded to be 1.26, its lowest thus far. This has been recovering slightly, having been recorded as 1.44 in 2017. Its rate of increase is estimated to be 0.015 in the Internet era. Information about the crisis and the fear of fertility decline can easily be found on the World Wide Web. However, the tempo of increase is slower than that of the pre-Internet era. Despite considering that the variation range of fertility is narrow at present, there is no empirical evidence that the network accelerates the alterations of demographic affairs.

9.2.2.2 High-Speed Mobility and Networks Transmit Conscious Information but cannot Alter Demographic Behaviors

It is always assumed that the network and high-speed mobility are two revolutionary and innovative powers. However, in terms of demographic phenomena, this assumption is illusory. Modern technological and mechanical innovations cannot change basic human behaviors.

The hypothesis tacitly postulates that network and high-speed mobility transmit symbols and information on the transmitters, and receivers consciously receive them and change their behaviors. This scheme is no longer valid or has always been false. Society is not constituted by consciousnesses.

9.2.3 Diffusion of Low Fertility in Spaces in Japan

The reaction-diffusion model is no longer valid in Japan strictly. It is merely a quasi-model, and it is valid on a macro scale. What is required is a genuine dynamic model that is more universal and describes on a micro scale.

See Figure from 9.6 to 9.8. As time passes, the maps become whiter. I speculate that this indicated decline in fertility in Japan will come to an end in around ten years. The reason for this speculation is explained further on. I argue that this slight fertility decline expressed in the figures is caused by the neighboring effect described by the birth function. However, the majority of people, including several sociologists, assume that the will of people may change owing to feminism or something similar. I call their assumption the "will alteration hypothesis."

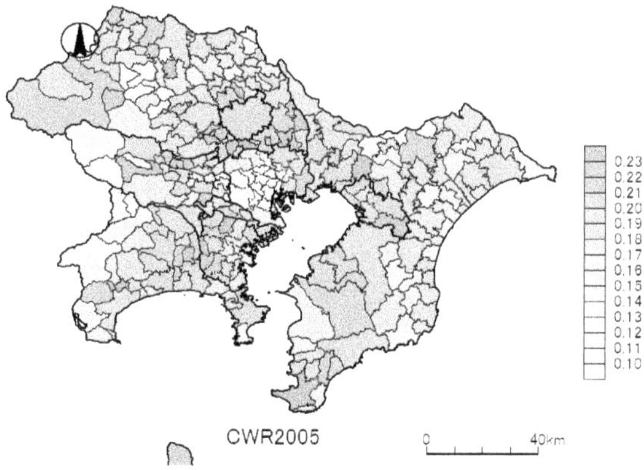

Figure 9.6: Regional CWR of Kanto Area in Japan in 2005.

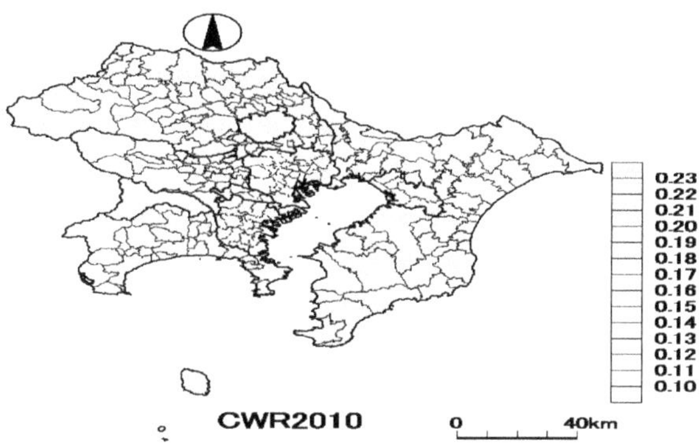

Figure 9.7: Regional CWR of Kanto Area in Japan in 2010 (reprint).

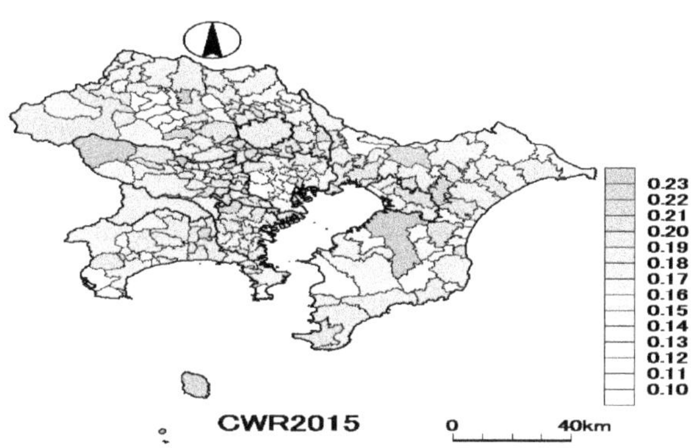

Figure 9.8: Regional CWR of Kanto Area in Japan in 2015.

9.2.3.1 Failure of the Will Alteration Hypothesis

The fertility decline in Japan at the end of the twentieth century was very slight (0.01–0.02 CWR decline over five years). It should be noted that all

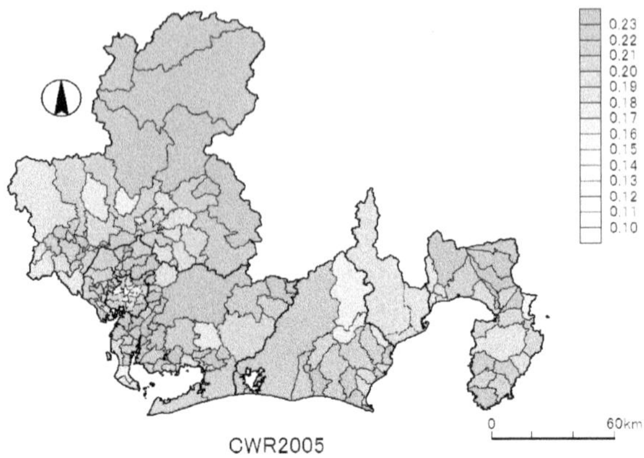

Figure 9.9: Regional CWR of Gifu, Aichi, and Shizuoka Prefectures in Japan in 2005.

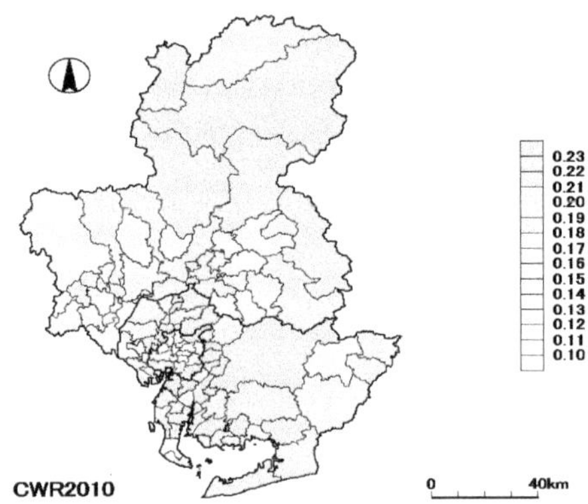

Figure 9.10: Regional CWR of Gifu, Aichi, and Shizuoka Prefectures in Japan in 2010 (reprint).

decline rates are and were small. However, what alteration in will corresponds to one-tenth of females changing their birth behaviors from two to one child in five years? It is difficult to assume that these alterations of will are really conscious.

The will alteration hypothesis is familiar to us but problematic for theoretical development. It is critical no note that this hypothesis cannot provide the grounds and conditions of will alteration. The reason for the minute nature of the alterations needs to be found, but within the hypothesis, there are no factors to adjust.

The will alteration hypothesis is comprehensible for humans, but it has no possibility of theoretical development. If it is valid, a theory explaining the lowest limit of fertility should be prepared. I expect that the extension of the birth function to high-dimensional space solves these problems. It is only possible for the birth function theory to solve these problems theoretically,

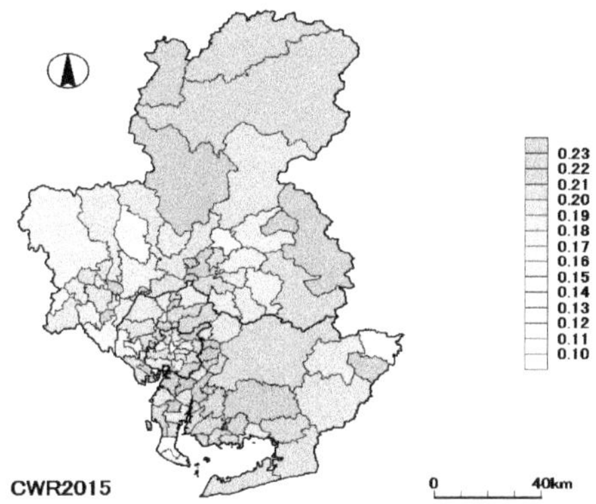

Figure 9.11: Regional CWR of Gifu, Aichi, and Shizuoka Prefectures in Japan in 2015.

because the will alteration hypothesis cannot solve them quantitatively and predictably. The will alteration hypothesis merely provides an *ex postfacto* explanation. In contrast to the failure of the will alteration hypothesis, the birth function can quantitatively predict the slight fertility decline and the possibility of will alteration.

9.2.4 Verification of the Principle by Historical Data

I try to verify the proposed principle by means of the historical demographic data.

9.2.4.1 Colyton

While Wrigley mentioned "as early as in many industrial areas," his sense of simultaneity was not rigid. Levine [50][11], who succeeded Wrigley reported the marriages, baptism, and burials in Colyton between 1550 to 1830. According to his report, baptisms began to increase beginning around 1780 (Figure 9.12). On the other hand, as shown in Figure 9.13, around 1750, baptisms began to increase in Shepshed. This does not indicate the growth of fertility in Shepshed. The constant immigration to Shepshed from the 17th century caused this increase. After 1750, this increase became drastic. The growth of the birth rate from 1750 onwards can be observed in Table 9.5. Essentially, there was a 20–30 year difference in time between Shepshed and Colyton.

Using the equivalent principle 9.2.1 and the velocity of the traveling wave—10 km per year, as estimated in Chapter 2—the distance from Shepshed to Colyton is about 200–300 km. As a Japanese researcher, I have not studied

[11]Levine firmly believed in the close correlation between demographic behavior and socioeconomic environment. Of course, I sympathize with Wrigley's doubt. Levine lacked the doubt that a researcher should have regarding their research objects and hypotheses. Nevertheless, I have availed Levine's data.

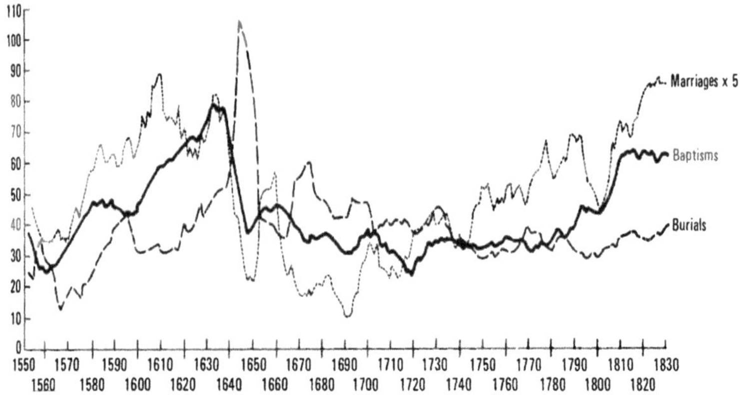

Figure 9.12: A nine-year weighted, moving average of baptisms, burials, and marriages (×5), Colyton.

*Source: D. Levine, Family Formation in an Age of Nascent Capitalism [50], p. 105.

the geography of England in detail. Measuring as per the metric system using Google Maps, the distance is found to be 257.95 km as the crow flies. The relationship between the distance and the time difference does not contradict the equivalent principle.

9.2.4.2 Bottesford

However, at first glance, Shepshed and Bottesford do not have the expected relationship. The distance from Bottesford to Shepshed was 38.12 km, but the number of baptisms actually began to increase in 1795. Does this disprove the equivalent principle? It must be noted that baptism data represent the number of cases depending on the population of adults who conduct baptisms, not the ratio. From the late 17th century to the end of the 18th century, the population of Bottesford declined owing to pastoral farming. Levine wrote about the circumstances of those days.

> ... In Leicestershire, "the movement towards the large-scale grazing received immense impetus from the 1660s onwards," and the eastern sections of the county began to phase out crop farming in favor of pastoral farming.[3] By Defoe's time, the beginning of the eighteenth century, eastern Leicestershire was "a vast magazine of wool for the rest of the nation."[4] Farmers in Bottesford were caught up in these changes: although the surviving inventories for the later seventeenth century show that the wealthiest farmers, those leaving estates valued at more than twice the average, owned more than 90 sheep, by the first half of the eighteenth century, members of this class had folks averaging 140 sheep.[5] This movement to pastoral farming produced a decline in the demand

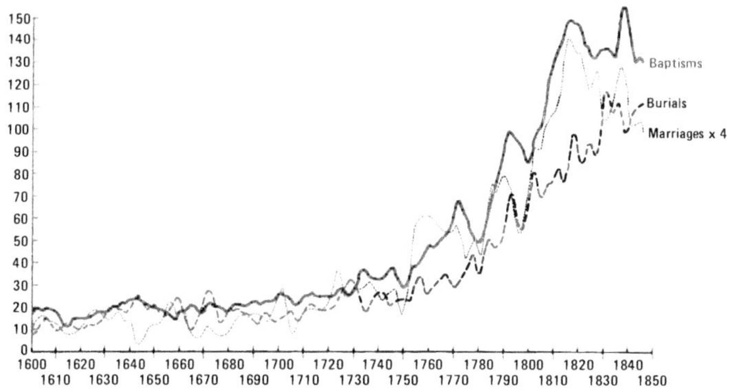

Figure 9.13: Baptisms, burials, and marriages of Shepshed.

*Source: D. Levine [50], p. 59.

Table 9.5: Age–specific birth rates (in Shepshed).

	Year at risk	Children	Rate/000
1600–1699			
Under 25	62	22	355
25–29	174	67	385
30–34	224	68	304
35–39	234	61	260
40–44	207	25	121
45–49	177	8	45
1700–1749			
Under 25	111	44	395
25–29	220	81	368
30–34	316	94	297
35–39	282	72	255
40–44	241	25	112
45–49	212	11	52
1750–1824			
Under 25	568	254	447
25–29	765	263	344
30–34	778	245	315
35–39	624	160	256
40–44	482	65	135
45–49	482	10	32
1825–1851			
Under 25	1129	486	430
25–29	1121	396	353
30–34	898	272	303
35–39	562	132	235
40–44	312	38	122
45–49	120	2	17

*Source: D. Levine [50], p. 65.

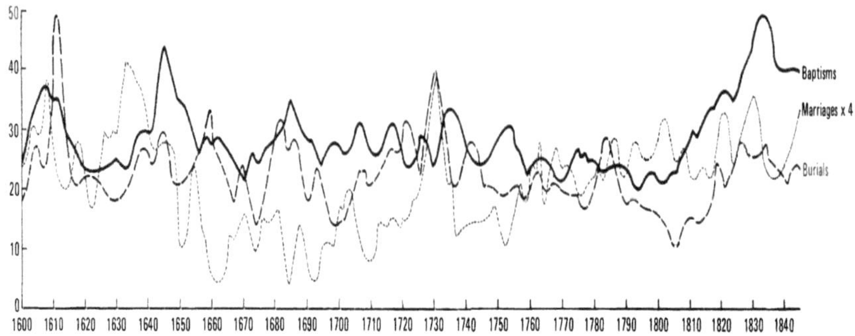

Figure 9.14: A nine-year weighted, moving average of baptisms, burials, and marriages (×4), Bottesford.

*Source: D. Levine [50], p. 89.

> for labor and led to a reduction in the village population from about 870 in 1676 to 772 in 1792.[6]
>
> *Source: D. Levine [50], p. 90.

Although the total population was decreasing, the number of baptisms was stagnant from 1760 to 1790, suggesting that the birth probabilities of parents increased during this period. Thus, the case of Bottesford does not contradict the equivalent principle.

9.2.4.3 Terling

As shown in Figure 9.15, around 1745, an increase in baptisms occurred in Terling, located in the county of Essex. This demonstrates that the time to increase its birth probabilities was the same as or faster than that of Shepshed. This does not contradict the equivalent principle; it indicates that the progressive wave came from the east. I estimate that fertility growth, which is the cause of the population increase, has been spreading from the European continent. As Wrigley said, there had been fertility fluctuations in pre-industrial Europe, and the progressive waves of fertility growth appear to have proceeded from Europe to the British Isles. The analysis by the equivalent principle suggests this possibility. The wave of the population increase did not come from the direction of Lancaster or Liverpool, which are thought to be the birthplace of the industrial revolution.

We must question the common belief that the industrial revolution prior to the demographic phenomena brought about the population increase in 18th century England. The reality is more complex. Wrigley stated the following:

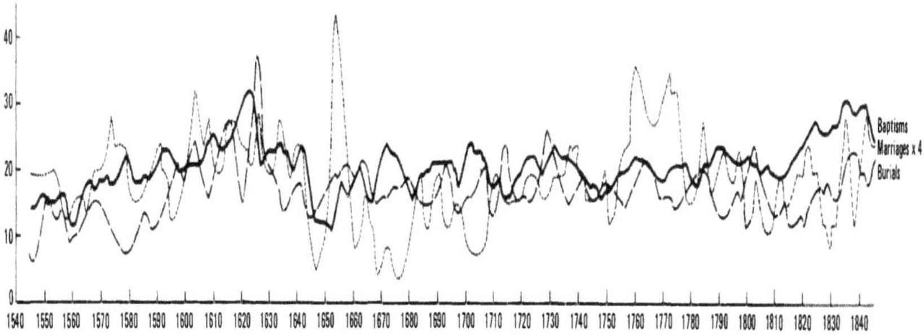

Figure 9.15: A nine-year weighted, moving average of baptisms, burials, and marriages (×4), Terling.

*Source: D. Levine [50], p. 117.

> The discussion of population totals in Chapter 2 will already have made it clear that the relationship between a society's demography and its economic and social structure are very intricate. This is true both of the functioning of the system of relationship at a given point in time, and still more of the process of change. We should, therefore, be wary about any explanation of these relationships, which is claimed to have a universal validity.
> (several sentences have been omitted).
> ... But it is wise to be suspicious of any treatment of these relationships, which purports to have solved matters simply by bringing them under a single general head of analysis, such as the view that all population changes occur in the long term in response to changes in the demand for labour.

E.A. Wrigley [83], p. 108.

For instance, modern China achieved massive economic development while their government rigorously enforced the one-child-per-family policy on their people. Is the one-child policy the result of their economic development? Without taking into account the chronological order of events, we tend to assume economic development to be the cause of Chinese fertility decline.

It should be noted that marriage function and birth function do not "solve matters simply by bringing them under a single general head of analysis" because both functions propose that time-space neighboring states entirely decide the following state. Because this formalization excludes general socioeconomic variables (conscious rational causes), it enables high flexibility of the theory.

We should cease to look for causes in human conscious will, because our consciousness is already condensed and misrepresented. Despite being for a social science, the theory that should be constructed is not one that we consciously agree with, but one that provides logical findings. Meaningful scientific sociology and demography must discard the commonplace understanding, despite it being a seemingly straightforward solution.

9.2.4.4 Falsifiable Theory

I believe that the only solution to building a falsifiable scientific theory of birth behaviors is to establish it in multi-dimensional space and define mappings from one point to the next.

The conventional research procedure that first formalizes conscious human motives as causes and then tests them statistically cannot describe the dynamics of birth probabilities. Essentially, this procedure can explain anything *a posteriori*, but it can make no predictions. This conventional procedure should be abandoned. By means of a mere statistical procedure, we may not reach a logical scientific theory. This makes numerical predictions through the only scientific approach.

9.3 Alteration of Values does not Cause Human Behaviors

A person autonomously intends to behave in a certain manner and subsequently acts out that behavior. This is the modern model of our behaviors, which is outdated and has been found to be false by phenomenology. Human will is decided within the given time–space. It is *ex post facto*. In the existing world–society, our senses act as a mirror of our surroundings. Some statistical analyses support this view. Hereinafter, those analyses that represent the optimal usage of the conventional research procedure will be introduced.

9.3.1 Alteration of Fertility Before Change in the Value of Childbearing

Mainstream sociology insists that human conscious values engender the alteration of our behaviors. This inconsistency is not supported by an empirical analysis. In the 1970s, the TFR of the United States was declining, and TFR $= 1.74$ was recorded in 1976. However, the attitude toward childbearing had begun to change following the decline of TFR. Rindfuss et al. demonstrated that the generation who raised children in the 1970s continued to maintain the traditional values of raising children. They noted the following:

> There is no evidence that this profound shift in behaviour was engendered by changing values or ideas concerning what is appropriate in raising children. Indeed, this shift occurred in the face of the prevailing belief that it was best for the child if the mother was a primary care giver and if the mother did not work outside the home while the child was young. Even though such a value position was not backed by rigorous scientific evidence, it

was widely held by the American public and promoted by many 'experts.'

Given that the behavioural change was already under way when our attitudinal time series began, we cannot definitively say that the behavioural change brought about attitudinal change through a societal process similar to that predicted at the individual level by theories of cognitive dissonance. Yet, all the evidence we could examine is consistent with an argument that, in the important area of women's roles and childrearing, behavioural change led and probably brought about attitudinal change. There was no widely reported scientific evidence that could account for the attitudinal change. During the same time period that a majority of mothers of pre-school children worked, a majority of the population still thought that pre-school children suffered if the mother worked. Even in 1991, close to half of the adult population thought that pre-school children suffered if the mother worked.

Ronald R. Rindfuss, Karin L. Brewster, and Andrew L. Kavee, Women, Work, and Children: Behavioral and Ideational Change in the United States, in *Dynamics of Values in Fertility Change* [64], p. 170.

They clearly stated "Yet, all the evidence we could examine is consistent with an argument that, in the important area of women's roles and childrearing, behavioral change led and probably brought about attitudinal change." Conscious changes always follow behavioral change. This is a solid fact that should be depended upon. The behavioral change is precisely the change in neighboring behavioral changes. Namely, when neighboring behaviors can be changed, conscious attitude and will can also be changed.

Rindfuss and his colleagues drew out their inference analyzing GSS data by the multinomial logit model. Their findings are important, and the birth function theory will allow readers to more profoundly understand the time lag between behaviors and attitudes. Therefore, their inferences will be introduced in detail.

Refer to "Age, Race, Gender" in Table 9.6 for both models[12]. The probability of answering "Agree" to the question *"Pre-schooler suffers if mother works"* is significantly larger than the reference categories. The group in which individuals are over 50 years of age and racially white, and the male gender tend to agree with *"Pre-schooler suffers if mother works"* because they did not bring up their children under their own care. Therefore, they selected the commonly accepted value judgment in those days.

In contrast, non-white females who were married and working, who had no choice but to work, had to positively evaluate their current situation. Therefore, they tended to disagree with *"Pre-schooler suffers if mother works."*

[12] The difference between the models is that Model 2 contains interaction terms. The interaction indicates the effect of the combinations of variables in statistical models.

Table 9.6: Logistic regression models of attitudes toward working women, total adult sample, United States (Question: Pre-schooler suffers if mother works. Agree/Disagree).

	Model 1	Model 2
Intercept	-0.24	-0.22
Year		
1985-86	-0.61***	-0.61***
1988-89	-0.74***	-0.73***
1990-91	-0.70***	-0.69***
Age		
Less than 30	-0.16*	-0.11
50 and older	0.73***	0.67***
Race		
White	0.57***	0.58***
Gender		
Male	0.66***	0.67***
Education		
Less than high school	0.28***	0.37**
Some college	-0.09	-0.10
College or professional	-0.25***	-0.34***
Religious affiliation		
Catholic	-0.16**	-0.16**
Other	-0.19*	-0.18*
Church attendance		
Occasional	-0.02	
Regular	0.47***	0.45***
Marital status		
Never married	0.04	0.03
Post-married	-0.16*	-0.17**
Respondent's mother's work status		
Worked and lived with mother	-0.25***	-0.24***
Did not live with biological mother	-0.11	-0.11
Education by age interaction terms		
Less than high school by < age 30		-0.47*
Less than high school by age 50+		0.03
Some college by < age 30		-0.05
Some college by age 50+		0.04
College by < age 30		0.30
College by age 50+		0.12
Model $\%^2$(degree of Freedom)	912.5(18)	929.4(24)
N	7,790	

* $p < 0.05$
** $p < 0.01$
*** $p < 0.001$

Ronald R. Rindfuss, Karin L. Brewster, and Andrew L. Kavee [64], p. 168.

Namely, the analysis by Rindfuss et al. shows that human values change depending on situational behaviors. In addition, situational behaviors are stochastically decided by their neighboring behaviors.

Another suggestion must be introduced: that these changes in value do not depend on socioeconomic differences. Rindfuss et al. wrote as follows:

> More importantly, our results also suggest the pervasiveness of the attitudinal shift. While the change is not monotonic, in later years respondents were less likely to believe that young children are harmed by having a working mother, even with individual characteristics held constant. That the effects of time remain net of individual characteristics suggests that this shift has occurred independently of changes in the distribution of the population across subgroups. Further, the attitudinal shift appears to have occurred more or less simultaneously across population subgroups.
>
> Time continues to have significant main effects: the more recent the survey, the more positive the orientation toward working mothers. Most important is the lack of time interactions, reinforcing our earlier finding that the structure of the determinants of attitudes toward working mothers has remained stable.
>
> The results of these analyses suggest two conclusions. First, although substantial group differences in orientation remain, the liberalization of attitudes toward working mothers has been pervasive. We have found scant evidence that the shift occurred earlier or more rapidly among any one sector of the population; rather, it appears to have occurred simultaneously across social groups. Second, our results suggest, albeit more tenuously, that the aggregate-level trends are due more to attitudinal change than to shifts in the composition of the population.
>
> Ronald R. Rindfuss, Karin L. Brewster, and Andrew L. Kavee [64], pp. 167–169.

Rindfuss et al. suggested that people's attitudinal shift had propagated independently from their socioeconomic attributions. Their suggestion supports the background independence of fertility decline. Unfortunately, they did not refer to the actual cause of these attitudinal shifts. As has been discussed, the true causes are the reaction-diffusion of the working mother, which are neighbor-dependent.

However effectively they planned their multinomial logit model to analyze birth behaviors, they were not able to find the true cause of the attitudinal shift. This is because each sample had no data on its neighboring behaviors. This is a limitation of social surveys by random sampling. The aggregation of random samples is not equal to that of the society. The dynamic alteration of society cannot be conceived without geographical space and time. The fact that the "Year" variable is the largest risk factor of the attitudinal shift suggests that the shift occurred in the form of neighbor-dependent diffusion.

The limitation of the social survey in measuring each individual as a research object must be recognized. The modern illusion that the individual consciously makes decisions and acts needs to be destroyed. Social research that uses the questionnaire tacitly assumes that each individual consciously determines their behavior. Therefore, this approach appears obsolescent. Individualism and the doctrine of prioritized self-interest must be abandoned.

9.4 Decline of the Number of Children by Unconscious Parents

I put forth the question of what conscious birth control French people in *Lot-et-Garonne* had toward the end of the 18th century. Further, at the beginning of the 19th century, French people did not have family planning, although they lost significant human resources in the French Revolution and Napoleonic wars. Le Play[13] pointed out that the impact of the Napoleonic Code (French Civil Code), as mentioned by Wrigley:

> Le Play, for example, believed that the great fall in fertility among the French peasantry occurred because under the provisions of the *Code civil*, the small proprietor obliged to divide his estate among his offspring equally. He supposed peasants to dislike the subdivision of their holdings sufficiently strongly to try to keep their families small.[30]
>
> E.A. Wrigley [83], p. 190.

However, the reaction-diffusion equation proposed herein estimated the beginning of fertility decline far before *Code civil.* It began around 1770–1780 in South-West France. Wrigley also criticized Le Play's hypothesis as follows.

> In general form, these arguments are difficult to resist. They must have much truth in them. In more particular settings, there are often difficulties. Events seldom fall out as neatly as models of social and individual action suggest. And even when events do fit the suggested patterns quite well, there is frequently some difficulty in understanding just what the trigger mechanism was, which started up a new pattern of behavior. For example, the argument from the *Code civil* through economic self-interest and a sense of the importance of undivided holdings to family limitation among the French peasantry does only rather rough justice to the facts, since family limitation was already beginning to appear

[13]Le Play was a renowned sociologist in the initial stage of sociology. He created the social survey and researched poverty. As his positivism, he inferred conscious motivations that could not be when behaviors were acted out.

> in some districts before the *Code* was promulgated. Some French areas, notably Brittany, though the population contained many peasants, did not show large falls in marital fertility until several generations later. Moreover, there were other countries with similar codes in which fertility fell only much later in the nineteenth century. Belgium is a particularly interesting instance of this for there the Walloon and Flemish area behaved quite differently.
>
> E.A. Wrigley [83], p. 192.

The reaction-diffusion equation intelligibly solves the Walloon and Flemish problem. I argue that when conscious self-interest is not considered, the problem of fertility decline can be solved.

9.5 Unconscious Birth Control

In the 19th century, it was difficult to find conscious birth control and family planning throughout Europe. Ehmer pointed out that conscious birth control and family planning began in the 1920s. At this time, the fertility of Europe had substantially declined. Namely, European people unconsciously declined their fertility for several decades.

> Auch in der Weimarer Republik nahm die Sterblichkeit noch weiter ab, aber Deutschland wurde nun stärker von einem schnellen Rückgang der Geburten geprägt. In ganz Europa fand in the dieser periode der Übergang zu bewusster Geburtenkontrolle und Familienplanung statt.
>
> (Translation in English by author)
>
> In the Weimar Republic also, the mortality was still further declining, but in Germany then, the stronger than the passed rapid decline of fertility emerged. During this period, the transition to conscious birth control and family planning took place throughout Europe.
>
> J. Ehmer, *Bevölkerungsgeschichte und Historische Demographie 1800–2010* [24], p. 10.

What must be confirmed is that the fertility decline in Germany had already begun at the end of the 19th century (about 1880). Namely, unconscious birth control had continued for over 20 years before conscious contraception began.

From Figure 9.16, it is evident that during the period of unconscious birth control (1880–1910), fertility in Germany declined more drastically and rapidly.

Abbildung 2: Der langfristige Abnahmetrend der Geburtenrate im Deutschen Reich und in der Bundesrepublik Deutschland von 1890 bis 2000

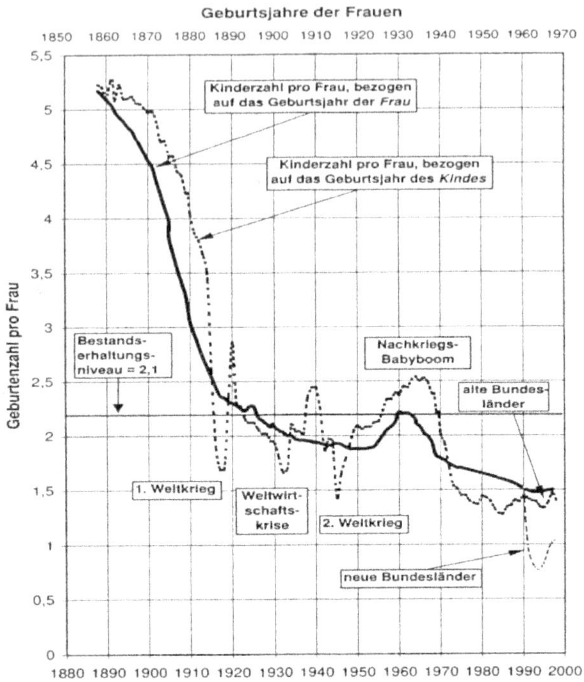

Dünne unterbrochene Linie: Fruchtbarkeitsrate, fette Linie: durchschnittliche Kinderzahl.
Quelle: H. BIRG/D. FILIP/E.-J. FLÖTHMANN, Paritätsspezifische Kohortenanalyse des generativen Verhaltens in der Bundesrepublik Deutschland nach dem Zweiten Weltkrieg. Bielefeld 1990.

Figure 9.16: Undulation of Fertility in Germany, 1880–2000.

*Source: J. Ehmer, *Bevölkerungsgeschichte und Historische Demographie 1800–2010* [24], p. 43.

9.5.1 *Failed Efforts of the Third Reich*

When Hitler acquired his dictatorship in 1933, based on racism and eugenics (both completely unscientific and false), Nazi Germany began a pronatalist policy.

It must be noted that the powerful pronatalist policy of Nazi Germany could not increase the overall fertility in Germany. Pronatalist policies are empirically incompetent. Propaganda cannot change birth behaviors, because it can only affect conscious attitudes.

See Figure 9.16. The dotted line in the figure is "Kinderzahl pro Frau, bezogen auf das Geburtsjahr des Kindes" which represents the TFR of the period. The TFR of Germany began to rise around 1933, when Hitler established his dictatorship. This increase in fertility continued until 1940. This rise in fertility was not evoked by the pronatalist policy of Nazi Germany; it was a compensation for and recovery from the previous time of crisis. The evidence for this inference is that the "Kinderzahl pro Frau, bezogen auf das Geburtsjahr der Frau," which represents the TFR of the cohort corresponding to that of the period, had been monotonically declining. Ehmer wrote the following:

> In the third Reich, the number of births increased until 1940, that is, the first year of World War II; however, it expressed none fundamental alteration of birth behaviors. The desire to have children that could not be realized from 1929 to 1933, during the Great Depression, was actualized during the first year of the national-socialist governance. This indicates that many people have positive expectations of the new regime. However, we cannot find successful pronatalist elements of the national-socialist demographic policy. For the 1920s and 1930s cohort, the average number of children born by a female throughout her life remained rather constant at around two. The moment of birth was carefully planned and completely dependent upon economic and political situations. However, the number of children that people want to have in all was influenced by neither economic trends nor demographic policies.
>
> J. Ehmer, *Bevölkerungsgeschichte und Historische Demographic 1800–2010* [24], p. 10. (translated by the author of this book.).

Thus, the pronatalist policy propelled by the overwhelming power of Nazi Germany was not able to reverse the decline in fertility in the country. Propaganda cannot alter the present number of children of parents, and thus, the next time, the number of children of parents does not change. To change birth behaviors, neighboring birth behaviors must be changed, and this is only possible to a small degree. It requires a long time to alter human birth behaviors because they are governed not by our conscious wills, but by our neighboring behaviors.

9.5.2 Unconscious Alteration of the Number of Children per Family in Japan

9.5.2.1 Childbearing and Increased Fertility Until the Defeat in World War II

Similar to Nazi Germany, Japanese militarism could not control fertility. Observe Figures 2.23–2.29 on p. 52. I believe that these are reaction-diffusion processes, and none of our conscious intentions provoked them. In fact, no Japanese individual was or is able to control their fertility intentionally.

The decline of the number of children in Japan began around 1920 and steadily continued throughout the 1930s and 1940s. Japanese parents of those days may have had no subjective and conscious will to decline their fertility.

I cite the part of the column of the 1941 Japanese newspaper titled "Should be five children per family—go for one hundred million Yamato folks!". The then-Director-General of the Ministry of Health and Welfare made the following statement:

> ... If these situations continue, we worry about the future Japanese population that must develop Asia. How, then, do we increase fertility?—the answers are the thought process behind the "Bear and Increase" campaign. We must cultivate the idea of pronatalism and patriotism. In contrast to the fertility of agricultural districts, that of metropolises is very low. What are the causes of this contrast? The urban characteristics of individualism and liberalism brought by Western civilization are guilty. It is most regrettable that many city-dwellers avoid bearing many children and prefer egoistic life. We must eliminate this idea in favor of our duty to develop Asia.
>
> Asahi Shinbun, 01.23.1941. Translated to English by the author of this book.

The then-Director-General of the Ministry of Health and Welfare supposed that the decline in fertility was caused by conscious egoism worsened by Western civilization, and it was temporally caused by the partial population. He believed that this trend could be reversed by a change in citizens' attitudes. I am astonished to find a similar explanation for the recent low fertility in Japan.

In the totalitarian regime in those days of Japan, the majority of Japanese people could not explicitly oppose the fascist slogan "Bear and Increase." However, just as in those days, Japanese people today have no evident will to decline their fertility. They (including analysts) were and are incompetent to assume nothing else to be the cause apart from their conscious will.

> PROP. XXIX. The idea of each modification of the human body does not involve adequate knowledge of the human mind.

Spinoza, *Ethics (Ethica Ordine Geometrico Demonstrata)* [16], p. 68.

As Spinoza mentioned above, humans have continued to completely misunderstand our own behaviors. The arbitrary explanations by our conscious memory are fantastical.

If the pronatalist policy of militarism Japan raised Japanese fertility, a fluctuation in the number of children would have been observed. However, the cohort's TFR rate has been monotonically declining since the middle of the 20th century in Japan. From Figure 9.17, it can be seen that there are no vestiges of "Bear and Increase," but rather, a logistic curve of decline. The commonplace idea that human behaviors are based on our motivations and affected by propaganda cannot explain reality.

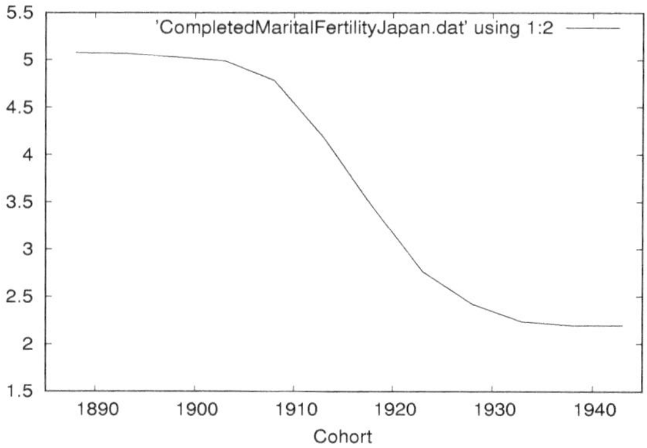

Figure 9.17: Total marital fertility in Japan, 1890–1947.
*Source: S. IKE, *Decline of Marital Fertility and Diffusion Hypothesis Reaction-diffusion model of completed marital fertility* [35], p. 171.

9.5.2.2 Antinatalist Policy in Japan Following World War II

Almost all Japanese people lack the realization that the fertility decline had already begun far before the beginning of World War II. Therefore, there is a misunderstanding that the antinatalist policy following the war, including the Eugenic Protection Law, caused the decline in fertility in Japan. I have read about this misunderstanding.

In particular, feminists tend to infer that the decriminalized induced abortion of the antinatalist policy caused the fertility decline of Japan. This inference is a perfect misunderstanding. It is evident that there is no undulation in the logistic curve of Figure 9.17, which might be caused by the antinatalist policy. From right-wing nationalists to feminists, they are only seeking conscious causes, which are not supported by empirical data. Both pronatalist and antinatalist policies and their evaluators made the same mistake of believing that humans responded to the policies. In reality, birth behaviors are insensitive to propaganda.

Unfortunately, we humans do not understand our own actions. Our "subjective intentions" were finally found at a final stage of fertility decline. This consciousness is not a real motive, but an adaptation to the neighboring necessity. We cannot have consciousness beyond one that corresponds to the neighboring situation. R.A. Fisher had two sons and six daughters, while the TFR of England was near 2.0. A genius like him, obsessed with the false idea of eugenics, was independent from neighboring birth behaviors. As a result, he utterly wasted his own resources.

9.6 Birth Function as a General Law

We have understood the flaws of considering human consciousness as the cause of fertility decline. We must cease to seek consciously plausible reasons for the decline in fertility. It is not sufficient to explain social phenomena to interpret our conscious motivations. As I have repeatedly mentioned, human consciousness is a vague correspondence to neighboring situations. It is evident that the law of the dynamics of proximity is far more significant than human consciousness.

Therefore, I infer that almost all human conscious thought regarding fertility is decided depending on the age · marriage-age space. Birth behaviors are only mechanical and active reactions to neighboring behaviors in "age · marriage-age space." Specifically, humans are also animals responding to external factors.

Thus, I advocate for the construction of the birth function in accordance with age · marriage-age space. This path is the only one that can explain the dynamics of fertility and human conscious will.

Chapter Appendix

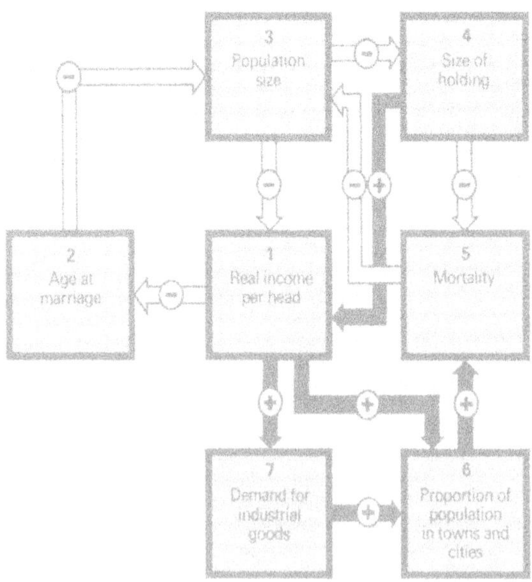

Figure 9.18: A hypothetical model of relationships between demographic, social and economic change.

*Source: E.A. Wrigley, *Population and History* [83], p. 109.

Table 9.7: The changing demography of Colyton.

	Age in years at first marriage M	F	Completed family size of women marring under 30		Expectation of life at birth in years (sexes combined).
1560–1646	27	27	6.4	1538–1624	43
1647–1719	28	30	4.2	1625–99	37
1720-69	26	27	4.4	1700–74	42
1770–1837	27	25	5.9		

*Source: E.A. Wrigley, *Population and History* [83], p. 87.

Chapter 10

Partial Constant Birth Probability

> PROP. XVII. God acts solely by the laws of his own nature, and is not constrained by anyone.

Spinoza, *Ethics (Ethica Ordine Geometrico Demonstrata)* [16], p. 20.

A study of the marriage and birth functions suggests that a small percentage of early teenagers must marry (including by common law) or cohabit and bear children as long as human reproduction continues. In terms of the reaction-diffusion theory, these singular marriages and births diffuse in the *marriage-age* · *age* space. Because marriages and births are stochastically decided by married early teenagers, the significance of them in human reproductive behaviors must be understood, even though it remains an estimation.

Human reproductive behaviors are chained and connected such that we live under common predetermined conditions. They are also stochastically chained, and thus, should not be forced by any other means.

10.1 Idiosyncratic Teenagers

There is a population of individuals who act solely according to the laws of their own nature in terms of birth behaviors. The population who marry between the ages of 15–18 behave their birth behaviors idiosyncraticly until they are 18 years old. Here, it is inferred that the development of marriage and birth functions depends on this quantum. To explore this further, this is considered in the context of the United States.

10.1.1 Teenage Pregnancy in The United States

The United States recorded a TFR of 1.87 in 2017. Compared with Japan, this is relatively high. The United States has no social insurance, and the public child support and public education systems are poor. Researchers who insist that the insufficiency of the public childcare support system in Japan is the cause of its low fertility must explain how fertility in the United States is higher than that in Japan, Germany, or the Mediterranean countries.

The relatively higher TFR of the United States can be traced to its higher teenage pregnancy rate. The central principle of the theory proposed herein is that the dynamics of a system is dependent on the variables in it. In the 1980s, over 2% of 15–17 year old females and 6% of 15–18 year old females married and cohabited in the United States. These values were obtained following the declining tendency, as shown in Figure 10.1.

The results of these relatively higher common law marriage rates were higher pregnancy and birth rates among females aged 15–17 (over 30/1000). This probability governs the birth function.

Table 10.1: Probability of married or cohabiting females of some countries in 1980.

Age	U.S.	Japan	Germany	Italy	Spain
15–19	0.093	0.01	0.023	0.046	0.056
20–24	0.533	0.222	0.24 (1987)	0.412 (1981)	0.408 (1981)

*Source: `https://esa.un.org/ffps/Index.html#/maritalStatusData`
E. Jones, *Teenage Pregnancy in Industrialized Countries* [27], p. 44.

The relatively higher fertility of the United States among developed countries in the 1980s and 1990s was dependent on the active marriage behaviors of young people. This movement, so to speak, occurred under the principle of neighborhood dependence. E. Jones described it as follows:

> Births to teenagers also became far more visible. Even though the teenage birthrate started to fall before 1960, the number of births to teenagers and the proportion of all births that occurred among adolescents continued to increase for 10–15 years after the total number of babies born in the United States began a precipitous decline (U.S. Bureau of the Census, 1984d, Table 82; NCHS, 1984b, Table 3). In addition, pregnant teenagers have become much more visible to their schoolmates, teachers, and the general community since, in 1972, it became illegal to expel a student from school because of pregnancy (Title IX, 1975, Educational Amendments of 1972).
>
> Early marriages during the 1950s, perhaps half of them to pregnant brides, served to screen high levels of teenage sexual

> activity from the public view (O'Connell and Rogers, 1984). As fewer and fewer teenagers married during the 1960s and the 1970s, and nonmarital birthrates began to soar, early sexual activity and premarital pregnancy became more apparent.
>
> E. Jones, *Teenage Pregnancy in Industrialized Countries* [27], pp. 37–38.

In the 1970s, marriage rates of American teenagers between the ages of 15–19 were about 10 percent. This is abnormal in a sense, because the "marriage boom" had already ended. The reason for this high teenage fertility cannot be determined. However, the fertility decline of the United States over a long period is inevitable. The young marriage tendency began to decline in the 1980s. It gradually fell to five percent in 1990 and two percent in the 2010s. However, the marriage rate of teenagers remains higher than that of Japan.

Except for the temporal decline, the TFR of the United States began to decline in 2007, and it has been monotonically declining since. This decline in TFR has a certain relation with the decline of teenagers' marriage rates and teenagers' pregnancy. I speculate that it took about 15 years after the teenager's pregnancy rate went below five percent for the TFR to go below the level of population replacement. I conjecture that, if "World Marriage Data 2015"[1] is accurate, the fertility decline of the United States will continue for the next 20 years.

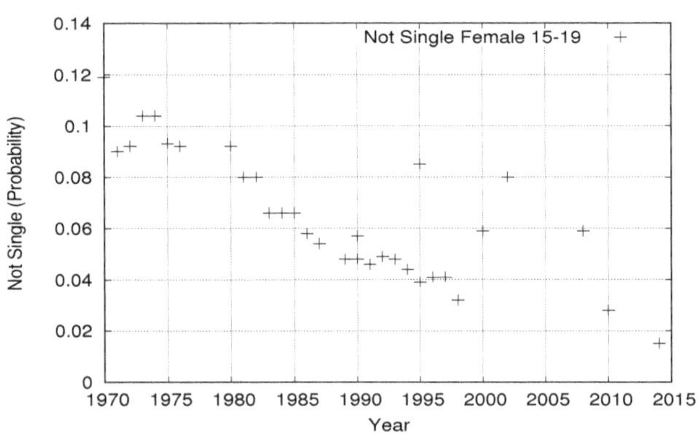

Figure 10.1: Married or cohabiting teenagers (aged 15–19) in the United States.

*Source: https://esa.un.org/ffps/Index.html♯/maritalStatusData.
The figure is drawn by the author.

From 2000 to 2015, available data values are discrepant depending on the type of data process (census, survey, or estimation). In general, the statistics on teenage fertility is not very accurate, and therefore, it is difficult to make

[1] https://esa.un.org/ffps/Index.html♯/maritalStatusData.

accurate judgements. However, at present, a monotonic decline in fertility suggests a low marriage probability among teenagers in the United States.

In the United States, the distinction between marriage and cohabitation is natively obscure. The decrease in marriage was not compensated by the increase in cohabitation.

10.1.2 Teenage Pregnancy in France

France is the first country that experienced the fertility decline nearly to the population replacement level. From the end of the 19th century to the first half of the 20th century, France had been captured by the fear of population decline. However, it has not yet experienced complete population decline. The reason for this is that teenage pregnancy has been changing steadily.

According to E. Jones, while not as high as the United States, mid-teenage French individuals had relatively higher birth rates during the 1970s. On comparing French data with those of Japan, more French teenagers married and bore children until the end of the 1980s than Japanese ones.

In the 1980s, marriages began to decrease and cohabitation began to increase in France and the United States. In 1986, E. Jones wrote the following:

> Cohabitation was beginning to become more common around this time in a number of countries, including the United States. Two percent of all U.S. women aged 15–19 were cohabiting in 1980 (Table 3.4), a marginally greater proportion than reported for France in 1981. In addition, the proportion of married is significantly higher in the United States; 8 percent of American 15–19-year-olds were married in 1980 compared to 5 percent of French teenagers. Thus overall exposure to the risk of pregnancy as a result of participation in a sexual union appears to be lower in France than in the United States, and may account for some of the difference in the pregnancy rates between the two countries.
>
> E. Jones, *Teenage Pregnancy in Industrialized Countries* [27], p. 131.

The development of cohabitation in France afterward was significantly different from that in the United States. Originally greater rate of cohabitation than the United States has developed to offset the decrease in marriages of teenagers in France.

Marriages also declined steeply in France. However, the birth rates of teenagers appear to have remained constant around 2010, as shown in Figure 10.2.

This flourishing state of affairs may not last long in France. After 2010, a decline in teenagers' birth rate can be observed. Although this decline may be temporal, the decline in fertility will proceed if cohabitation and pregnancy of teenagers does not compensate for the decrease in marriage. A decline in the population of France will approach.

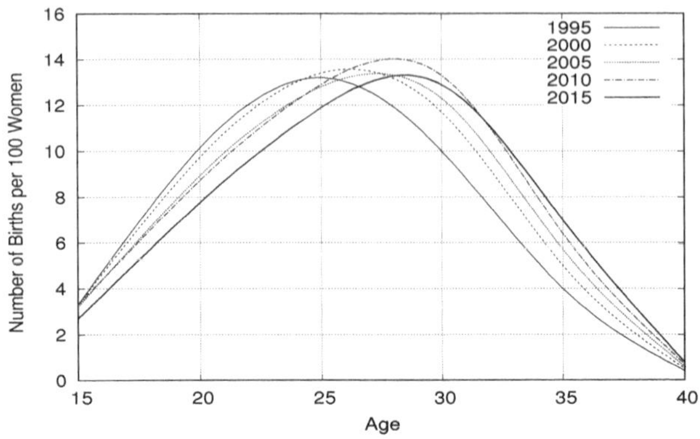

Figure 10.2: Fertility indicators in France from 1995 to 2015.

10.1.3 Teenage Pregnancy in Sweden

Sweden was contrasted with the United States in its low teenage pregnancy rate. In 1986, Jones and Westoff sought the model that declined teenage pregnancy in Sweden. In those days, the low fertility of developed countries was not a significant problem. It may be noted that despite the social problems associated with it, the population decline of developed countries positively impacts global issues such as global warming.

Ever since the beginning of the 20th century, the fertility of Sweden has fluctuated around the population replacement level. In 1970, the TFR of Sweden began to decline, and this continued until 1983. Accompanying this decline, or more precisely, causing it, teenage fertility began to decline.

> Teenage fertility, on the other hand, continued to decline at a rapid rate, from 35 per 1,000 women aged 15–19 in 1971, to 25 in 1976, to 14 by 1981 (Statistics Sweden, 1984; and Table 2.8). Since the mid-1970s, teenage fertility dropped from 5.3 percent to only 2.5 percent of the TFR. (In the United States in 1980, teenage fertility made up 14 percent of TFR [Appendix 1]).
>
> E. Jones, *Teenage Pregnancy in Industrialized Countries* [27], p. 181.

Jones noted that "Sweden has been in the avant-garde of the decline in marriage rates seen in most developed countries." Normally, like Japan, Sweden should have experienced a significant decline in fertility. However, Sweden maintained relatively high fertility at the end of the 20th century, equal to that of the United States. The reason for this relatively high fertility is that cohabitation offsets the decline of marriage.

> In theory, this evidence of the decline in the rate of marriage among young women is consistent with the hypothesis that exposure to the risk of pregnancy may have been reduced. However,

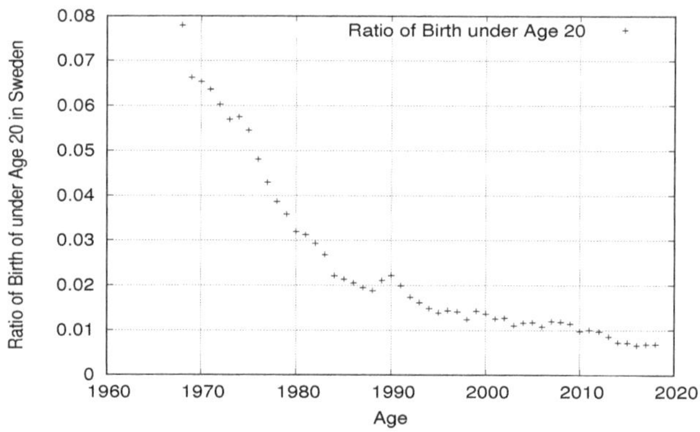

Figure 10.3: Rate of birth under age 20 in Sweden.

> there has been a more than compensating increase in informal cohabitation among young Swedish women. Thus, if women legally married are combined with those informally cohabiting, the proportion in a sexual union by age 20 has increased markedly. Result from the 1981 Swedish Fertility Survey indicate an increase in the percentage of women who by age 20 have ever lived with a man for at least one month from just under 30 percent for women born between 1956 and 1960 (International Statistical Institute, 1984).
>
> E. Jones, *Teenage Pregnancy in Industrialized Countries* [27], p. 185.

Namely, cohabitation has maintained the fertility of Sweden thus far. However, there are indications of a decline in Sweden's fertility. Human birth behaviors depend upon neighboring birth behaviors, and the decline of teenage fertility impedes the development of the birth function.

10.1.4 Teenage Pregnancy in Canada

A study by E. Jones et al. in 1980s reported that Canadian teenagers' fertility history was parallel to that of the United States. However, the teenagers' fertility was quite dissimilar.

> The U.S.–Canadian difference is large for both the younger and older teenagers, although it is slightly greater for the youngerones. For ages 15–17, the Canadian fertility rate is 56 percent lower than the U.S. rate (Tables 2.2 and 2.4). Among those aged 18–19, the difference is 46 percent.
>
> S.K. Henshaw, *Teenage Pregnancy in Industrialized Countries* [27], p. 67.

The teenage fertility rate in Canada being lower than that in the United States may be the cause of the subsequent lower overall fertility in Canada.

I believe the low fertility in developed countries is caused by the low fertility of their teenagers, because the most important factor in the development of the integral function is the initial value.

10.2 Effect of Teenage Marriage Rate on Birth

Initially, I had not meaningfully understood the relationship between the teenage marriage rate and the birth probabilities of all populations. The decline in marriage rates of teenagers does not bring about a direct decline in their birth probabilities, when they select cohabitant instead of marriages, because if they do not have an abortion, they are sure to be pregnant and bear a child. Therefore, married and cohabited teenage women have a constant birth probability. The birth function (probabilities) of teenagers is scarcely altered.

If teenage birth probabilities decline significantly, fertility declines more sharply in the United States and Sweden. Since the 1980s, it has taken a long time for the fertility of Sweden and the United States to gradually decline. This suggests that the decrease in teenage marriages has had a gradual effect on declining birth probabilities, that is, the declining birth function.

10.2.1 Turn in the Trend

The trends discussed above might have already turned. From Figure 10.1, it can be seen that the probability of married or cohabiting teenagers aged 15–19 began to decline at the end of the 1970s, and this tendency may continue.

The particularly high values from recent years have been recorded by social surveys. Therefore, it is possible that the values are not very low, as shown in Figure 10.1.

10.2.1.1 Reasons for the Fertility Decline of Teenagers

The causes of this change in the trend are unclear. A reason may exist; perhaps the popularization of secondary education suppressed the sexual activity of teenagers. It is strange that both politically left-wing and right-wing groups eagerly disagree with the marriage or cohabitation of teenagers. The modern educational system may be guilty of causing low fertility beyond the population replacement level. The public education system opposes eros.

In the spit of these types of persecution, married teenagers are genuine apostles of reproductive behaviors. Their birth probabilities are highly idiosyncratic compared to others.

10.2.2 Idiosyncrasy of Teenage Marriage

See Figure 10.4. Teenage marriage appears to be highly idiosyncratic. Their birth probabilities are abnormally high; however, they are not as idiosyncratic as their birth probabilities. The combination of the under-registering of

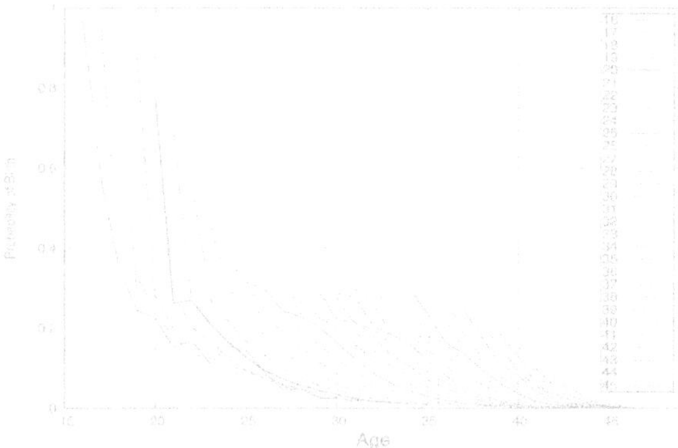

Figure 10.4: Birth Probabilities in 2000.

teenage marriages and their slight idiosyncrasy exaggerate the phenomenon. Considering the fecundability of their ages estimated by Woods & Weinstein, teenage birth probabilities are too high. Only those females who bear children legally register their marriages.

Individuals exhibiting such idiosyncratic behavior are rare. There are very few differences between them and other populations based on the perspective of the birth function. The ordinary population receives the effect of others' birth behaviors on marriageage dimension asymmetrically, but both sides.

However, another problem appears. What is the period of their idiosyncrasy? In other words, which age range is completely asymmetric? Youth is not everlasting; by their mid-twenties, their behaviors may have become normal. However, the exact age at which this change occurs cannot be deduced. For the time being, 18 is considered as the age up to which they are idiosyncratic in terms of their birth behaviors.

10.3 Choosing Initial Constant Values?

We should think separately about the birth function from the very high birth probabilities of very young married females. The morphology of birth probabilities of England presented by C. Clark (Figure 10.5) is much different from that of Japan. (for example, Figure 8.4 at p. 239)[2].

This is solely dependent on the initial values. The law governing the birth occurrences is the same, but the initial conditions are different. The indigenous difficulty of the birth function is that initial values cannot be conceived in a statistical manner, and they decide every subsequent probability.

Therefore, it must be assumed that the initial values yielding the most likely present (as futures for the past) birth probabilities are the optimum estimations of the initial values.

[2] I understood afterward this morphological difference was mainly dependent on the elapsed time from the beginning of the alteration.

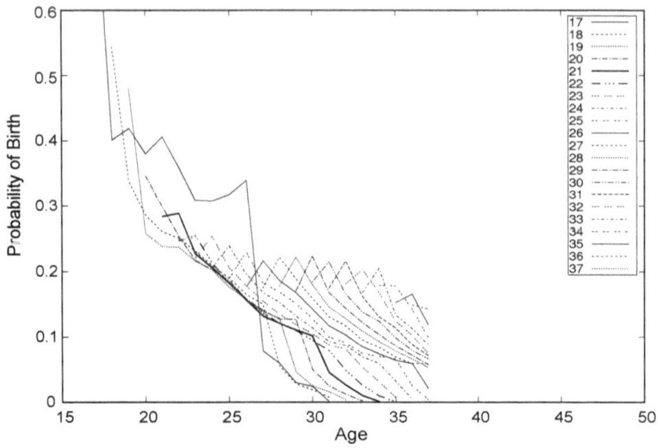

Figure 10.5: Birth Probabilities in England in 1939.

The data of the *marriage-age* · *age* specific probabilities of 1939 are available for the initial values. Data on the *marriage-age* · *age* specific probabilities from 1980 to 2010 are also available.

The cumulative probability of marriage from age 15–18 is under five percent at most. Therefore, the initial values themselves have little affect on fertility. However, they govern the subsequent birth probabilities.

10.3.1 Distinctive Features

Notice the first- and second-year birth probability of marriage for the different $marriage - age$ groups. It can be observed that these values are inherently dependent on their initial values of first married group. The birth probabilities of individuals married at age 15 cannot be estimated based on other birth probabilities. Their birth behaviors govern other birth probabilities as initial values.

Therefore, their probabilities were artificially set herein. In fact, they were unaffected by other *marriage-age* specific females.

10.3.2 End of Teenage Birth Probabilities

Another phenomenon that remains unexplained is that individuals married at age 15 form the lowest limits of the birth function. It is considered that their lowest limits are already known. Considering that C. Clark's data do not show the same morphology, physiological universality might not be assumed.

Clearly, the problems of initial values cannot be solved here. Therefore, the analysis is performed by setting the most likely values. One possible postulation is that there exists a lower limit of the birth function based on its fixed points. Similarly, the upper limit of the birth function can be postulated.

Chapter 11

Birth Function for High-Dimensional Space

> PROP. XVIII. If the human body has once been affected by two or more bodies at the same time, when the mind afterwards imagines any of them, it will straightway remember the others also.

Spinoza, *Ethics (Ethica Ordine Geometrico Demonstrata)* [16], p. 62.

In the frame of a single cohort, the possibility of theoretical development has arrived at its boundary, as mentioned in Chapter 8. Geographical extension of the birth function is inevitable, just as Wrigley imagined. This extension leads us to consider the causal relation between cohorts. First, the *age* and $marriage - age$ dimensions are extended to several cohorts.

11.1 Age and Marriage-Age Dimensions

For a given cohort, the period birth probabilities specific to age at marriage were regarded as the cohort birth probabilities. In this chapter, period birth probabilities specific to age at marriage are presented in order to discern the causal chain of births.

Consider Table 11.1. A 19-year-old female who married at the age of 18 bore a child with a probability of 0.3794 in 1985 in Japan. Now, consider the probability of females age 19 who married at the age of 18 bearing children

Table 11.1: Birth probabilities for females married between the ages of 16–25 in 1985 in Japan.

Age \ Marriage -age	16	17	18	19	20	21	22	23	24	25
16	0.7728	0.7882	0.5171	0.4123	0.3424	0.2845	0.2112	0.1496	0.1735	0.1544
17		0.8302	**0.5835**	0.3885	0.3463	0.2877	0.2127	0.1790	0.1615	0.1186
18			0.7449	**0.3794**	0.3526	0.2946	0.2210	0.1920	0.1427	0.1154
19				0.7384	0.3653	0.3398	0.2917	0.2183	0.1837	0.1417
20					0.6400	0.3470	0.3167	0.2863	0.2252	0.1815
21						0.5099	0.3557	0.3108	0.2931	0.2321
22							0.4353	0.3802	0.3085	0.3058
23								0.3843	0.3949	0.3059
24									0.3572	0.3959
25										0.3571

in 1986. The analysis presented in Chapter 8 suggests that the birth probability of an 18-year-old female who married at the age of 17 (0.5835) affects that of 19-year-old females who married at the age of 18 in 1986. The birth probabilities do not change over time owing to this effect. This also indicates that birth probabilities have no dynamics[1].

11.1.1 Causes of Dynamics are Birth Probabilities and Background Independence

As discussed earlier, the causes of the alteration of human behaviors are not our conscious wills, but neighboring birth behaviors themselves. The causes of the dynamics of birth probabilities are the undulations of probabilities themselves, which is another expression of background independence. Our birth behaviors, including our conscious wills, are responses to neighboring birth behaviors. Birth probabilities and their dynamics are unique objects which are falsifiable in the demographic study of births.

11.1.1.1 Sensitive Range of a Female

Birth probabilities can be interpreted as the frequencies of births in a neighborhood of females. A married woman is sensitive to the effect of the frequency of births in her neighborhood.

What is her neighborhood? It is not limited to females of the same age. She is also impacted by preceding birth behaviors. Further, younger females or those of the same age who married previously or simultaneously bore children affect her. Birth behaviors of significantly older females do not affect her. This limitation can be established because this assumption is in agreement with the causality and continuity of generations. If she was sensitive to her parents' generation, she might be sensitive to nearly twenty generations prior, because there are no reasons to deny the effect of intermediate generations. Thus, she could not decide her birth behavior, and we would become lost in the maze of causality. It has been empirically recognized that the number of children borne by a female is not significantly impacted by her parents'

[1] I believe that some dynamics caused by the alteration of the marriage function remain.

generation. If it was, there would be no difference in the mean number of children of the two generations, and a circulating wave of the mean number of children might be observed in our demographic history. Such a circulating wave has not been seen yet.

R.A. Easterlin [21] hypothesized that the American baby boom had ended based on the comparison of children's and parents' generations. However, this was called relative deprivation, and as a result of relative derivative comparison, the children's generation was observed to have relatively lower fertility compared to the parental one. With all due respect to Easterlin's work, his hypothesis is no longer supported.

The sensitivity range can be restricted to a very small age range (within one year). Additionally, the question of what type of sensitivity the female has should be answered. Does she have constant isotropic sensitivity? I confidentially answer this question in the negative. From Table 11.1, it can be seen that females are not sensitive to those married at an older age than them, i.e., to the elements below them in the matrix. Namely, in the *age and marriage-age* dimensions, the birth function is not isotropic.

A female who married at a young age is not sensitive to the birth probability of females who married at an older age. It is not observed that previously married females bear children in response to childbearing by females married at older ages.

Females' sensitivity range is limited to their closest predecessor who married at the same age or younger. This postulation naturally enables conjunction between cohorts. Sensitivity to the birth behaviors of a predecessor married at the same age or younger serves as the connection between the birth probabilities of successive cohorts. It connects successive time periods[2].

Using the probabilities in Table 11.1, the birth probability of 19-year-old females married at the age of 18 in 1986 in Japan is concretely computed as follows:

$$\begin{aligned} b(18,19,t_{1986}) & \\ = \quad & (rpc + \varphi\,\eta_{1985})\,Fe(19)\Big(\mu\, b(17,18,t_{1985}) + (1-\mu)\, b(18,19,t_{1985})\Big), \\ {*}\mu \leq 1 & \end{aligned} \tag{11.1}$$

$b(18,19,t_{1986})$ is the birth probability of marriages at 18 and 19 years of age in 1986. Let rpc be the reproductive coefficient. Let φ be the catch-up coefficient, η_{year} be the mean number of children in a society in that time, and μ be the diffusion coefficient.

The left and right sides of Equation (11.1) are regarded as the future and present, respectively.

[2] At this stage, I did not understand the true figure of the birth function. Please follow my history of trial and error. Before extending to higher-dimensional space (adding geographical dimensions), I had to consider the effect of birth probabilities within the same age group. I believe that this helps in gaining a strong understanding of the birth function theory.

Of course, the effects of other cells in Table 11.1, such as $b(17,19,t_{1985})$, were examined. I was unable to find substantial reasons to add other directional effects to the *marriage-age* and *age* dimensional space. According to the principle of Occam's razor, the simplest equation was selected.[3] In addition, more stable estimations of coefficients were expected.

It is known that $rpc \sim 1.47$. $Fe(19)$ is a constant estimated by Wood and Weinstein [81]. Now, φ, μ must be estimated. Let m be the age at marriage.

Equation (11.1) has been re-written in a more generic form. Let a be the age and μ be the influence of the neighboring $age \cdot marriage - age$ parameterized population. Of course, μ can be interpreted as the diffusion coefficient[4].

$$\begin{aligned} & b(m_i, a_j, t_{k+1}) \\ &= (rpc + \varphi\,\eta(t_k))Fe(a_j)\Big(\mu\, b(m_{i-1}, a_{j-1}, t_k) + (1-\mu)\, b(m_i, a_j, t_k)\Big), \end{aligned} \tag{11.2}$$

In Equation (11.2), $Fe(a_j)$ has already turned into a vector of constants.

As shown in Table 11.1, there are several real equations in the matrix and in longitudinal matrices. Thus, the coefficients can be estimated using iterative methods.

Equation (11.2) is compared with Equation (8.18) on page 227. It is presented again for convenience.

$$b(m+1, a+1) = \rho(a+1)\; Fe(a+1)\; (rpc + \varphi\,\mu(a))\; b(m, a),$$

The term $b(m,a)$ expands to $\Big(\mu\, b(m_{i-1}, a_{j-1}, t_k) + (1-\mu)\, b(m_i, a_j, t_t)\Big)$. With this expansion, Equation (11.2) acquires the dynamics of the function. It can essentially describe the birth functions from 1939 to the present. If it were not for significant turnover, this tendency would continue after 2019.

This simple extension to the neighboring *marriage-age* and *age* dimensional space enables the dynamics of the birth function. However, it is not sufficient.

Note that, to consider the $marriage - age$ of 15 in terms of the initial probabilities and to introduce other coefficients, another equation must be used:

$$\begin{aligned} & b(15, a_j, t_{k+1}) \\ &= (rpc + \varphi\,\eta(t_k))Fe(j)\Big(\mu_s\, b(15, j+1, t_t) + (1-\mu_s)\, b(i, j, t_t)\Big), \end{aligned} \tag{11.3}$$

They are singular as they do not obey the formula within the system. Unfortunately, their birth probabilities had to be controlled outside the system. They are not causally chained to other factors. Their birth behaviors must be artificially controlled to describe real birth dynamics.

---

[3]My razor did not cut sufficiently.

[4]This interpretation will be explained in the following section.

11.1.1.2 Self-Reconditioning is Ignored

It is somewhat strange that the term $b(18, 18, t_{1985})$ is lacking in Equation (11.1). Who is move to $b(18, 19, t_{1986})$. The homogeneous group is not the object that is recognized as comparable. Humans are affected by others, not themselves. This fact suggests that the subject "I" is the zero point that produces nothing by itself.

This omission is suitable for the theory because another equation need not be prepared for the elements in the principal diagonal in Table 11.1 (birth probability matrix). However, it is not necessary to omit some probable factors such that the omission is convenient for a theorist. There may be a profound reason for this.[5]

If we were only affected by ourselves, social causality would collapse. Society is full of egoists—"Bellum omnium contra omnes." I myself am conscious of myself as a determinant subject. However, "I" is a transducer transducing present, limited[6] neighboring situations to subsequent neighboring cells that cannot transduce themselves.

11.1.1.3 Diffusion Model

Equation (11.2) can be regarded as a diffusion model. The normal random walk model, which is the basis of particle diffusion, involves transportation to the neighboring horizontal cells. Diffusion is a phenomenon in which the state of a space is fundamentally altered by the influence of its neighboring space. It is not necessary physically transport substances or ideas to the neighboring cells.

The birth function in Equation (11.2) is, as it were, an asymmetrical transportation between diagonally adjacent cells, as shown in Figure 11.1. Concretely, this diagonality is superficial. From t_t to t_{t+1}, the effect occurs as

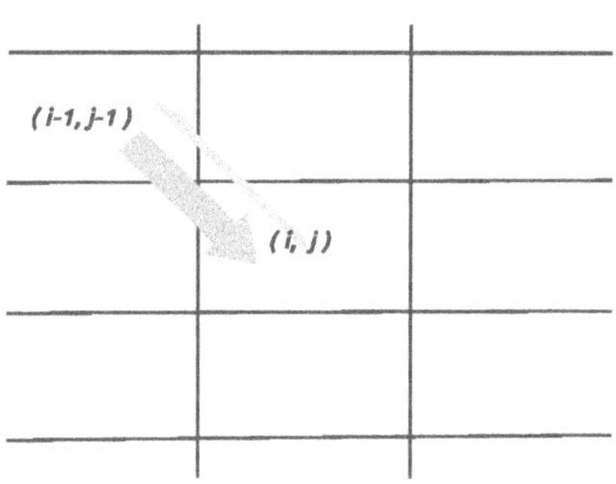

Figure 11.1: Asymmetrical diagonal diffusion.

[5] I did not realize this until I extended the birth function.

[6] Our lives are programmed as such.

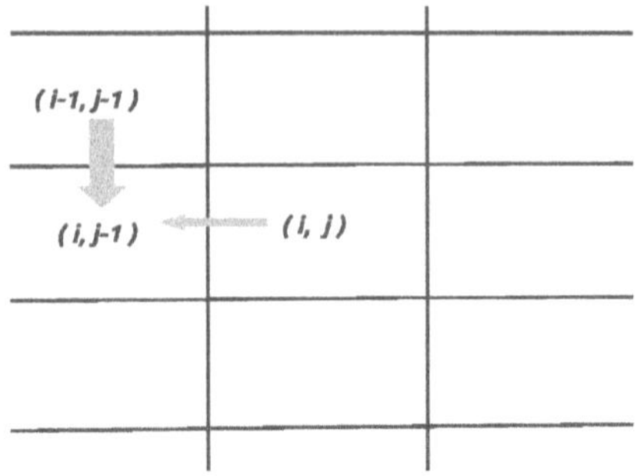

Figure 11.2: Concrete diffusion.

shown in Figure 11.2. After time t_{t+1}, the substantial population in the $i, j-1$ cell moves to the i, j cell. This anisotropic characteristic in the *marriage-age* and *age* dimensional space produces the peculiar dynamics in one direction.

11.2 Initial Values of the Birth Function

Thus, the birth function is a differential or difference function. When these types of formulas are used, their initial values are required. When birth probabilities are practically simulated, the initial values have two meanings. One is from when the simulation is started, such that the period data were selected for the initial birth probability of the *marriage-age* and *age* space. For these initial values, I have no options except for the 1939 data of Japan.

The other initial value is the selection of the first *marriage-age*. This selection has few options. There are no data on marriages at the ages of 13 and 14. Thus, the initial value of the *marriage-age* has to be set at 15. Earlier birth probabilities cannot be theoretically predicted.

Consequently, the real values must be empirically used as initial values of birth probabilities. It is extremely crucial that the birth probabilities of those very young married males and females stochastically decide every successive birth probability in a chain reaction. We are unconscious of this causality because it functions indirectly, leading to a large time difference. Human senses are not sufficient to conceive these theoretical causalities.

These teenagers' birth probabilities are required as initial values for the birth function. Without the birth probabilities of females married at the age of 15, probable dynamics of the birth function cannot be described concretely. It is no exaggeration to say that their birth behaviors govern all birth behaviors afterward.

I will explain the first meaning of the initial values of the birth function in detail. From which year should the numeric simulation begin? When computing the marriage function for Japan from the year 2000 ((Figure 10.4 in p. 322), the simulation is extremely easy to perform. This is because from 2000 to the present, the birth function in Japan has scarcely changed. Beginning from the year 1939 may result in a large number of dynamics. Japanese statistics from 1940 to 1979 lacked statistical data on age at marriage and marriage duration. Therefore, the simulation achieved by the marriage function is a direct evaluation of the validity of the birth function.

At first, I considered 1939 as the initial value. However, fertility had already begun to decline by this time. Therefore, an earlier year needed to be set as the initial value. I proposed to begin from 1930. This was because it is known that fertility decline in Japan began after the 1890 cohort. Females from this cohort would have been over the age of 40 in 1930, and it was coincidental that their birth probabilities began to decline just prior to their menopause.

11.3 Idiosyncratic Birth Behaviors

Social causality must be considered within the time–space where we reside. First, let us consider the *marriage-age* and *age* dimensional space. The geographical dimensions must now be considered.

For Japan, we must begin with the birth probabilities of females who married at the age of 15 in 1939 as initial values. In macroscale data, we cannot consider the probabilities of individual females who married before the age of 15. Who affects them? It can be supposed that neighboring preceding births by the same or slightly younger married females impacted them. This process forms an infinite loop. However, their real values cannot be theoretically decided. These are singularities of the difference Equation (11.2) and individuals who exhibit genuine idiosyncratic birth behaviors.

Consider Figure 11.3. As shown in the figure, in 1939 in Japan, the earliest birth behaviors were by 45-year-old females who married at the age of 15. They were the predecessors of all others within that time–space. To them, all others were passive in the causality. Those who pre-exist within a time–space necessarily make others passive. This is precisely the causality in the scientific

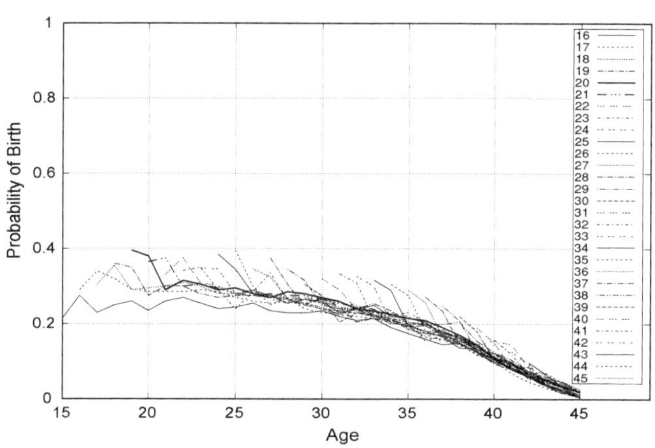

Figure 11.3: Birth probabilities specific to age and age at marriage in 1939.

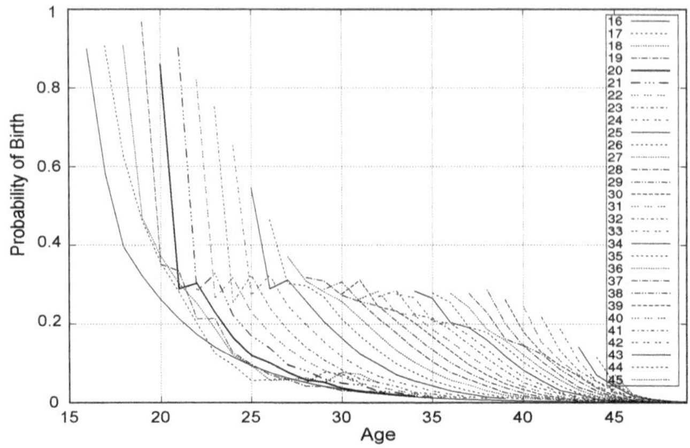

Figure 11.4: Birth probabilities of females married at the age of 15 in 2010.

and phenomenological definitions. The relationships between subjective states and behaviors established *ex post facto* are unscientific illusions.

Human birth behaviors are merely chain reactions to birth behaviors of the idiosyncratic population combined with biological fecundability Fe. Figure 11.3 is compared with Figure 11.4. I cannot help but assume that the same idiosyncratic population predominantly decides the chain actions. They are idiosyncratic not only because they have high birth probabilities at young ages, but also because they are scarcely affected by other populations. They have no preceding $married-age$ population to refer to. Their contemporaries married at just after in age dimension refer to them. It is astounding that the 0.1% population of a cohort have the maximum impact on determining the trend of the cohort and its successors.

This assumption implies exceptions to the difference Equation (11.2). These are manifest artificial conditions; however, they can effectively numerically describe the dynamics of birth probabilities. The features of birth functions caused by the idiosyncratic population must be carefully observed.

11.3.1 Features of Birth Functions

See Figure from 11.5 to 11.9. Since the 1980s, birth probabilities in Japan have been flattening. The birth function can describe a part of these dynamics.

However, it is necessary to make strong postulations regarding the birth probabilities of females married at the ages of 15–17. When the birth probabilities of females married at the ages of 15–18 are not controlled, the dynamics of the birth function deviate and do not trace the real dynamics. Married teenagers essentially control the birth function. Therefore, some adjustments must be introduced to control their probabilities and fit them to assumed realistic dynamics[7].

[7]Observed data are lacking.

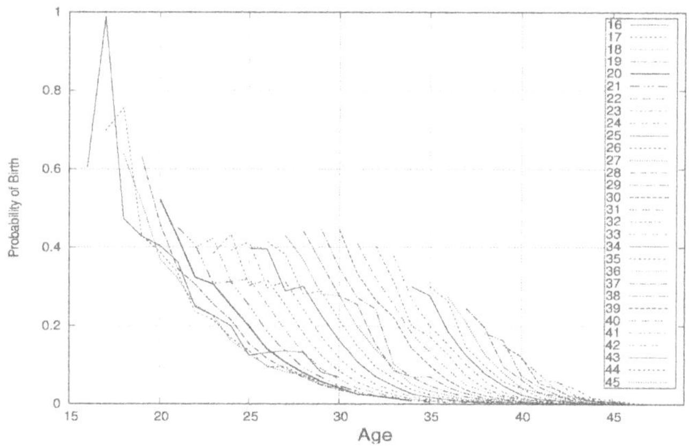

Figure 11.5: Birth probabilities in 1980.

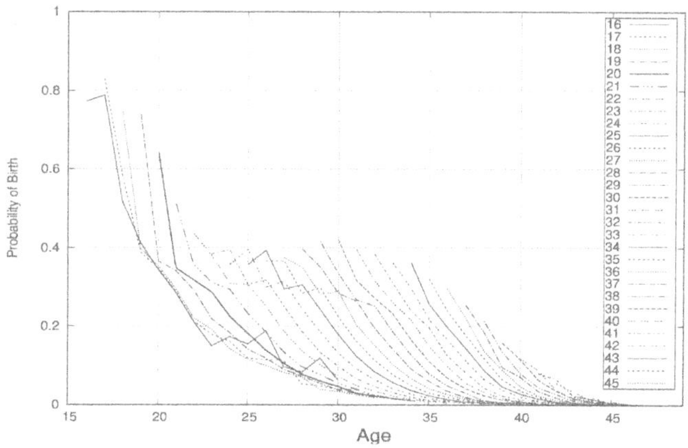

Figure 11.6: Birth probabilities in 1985.

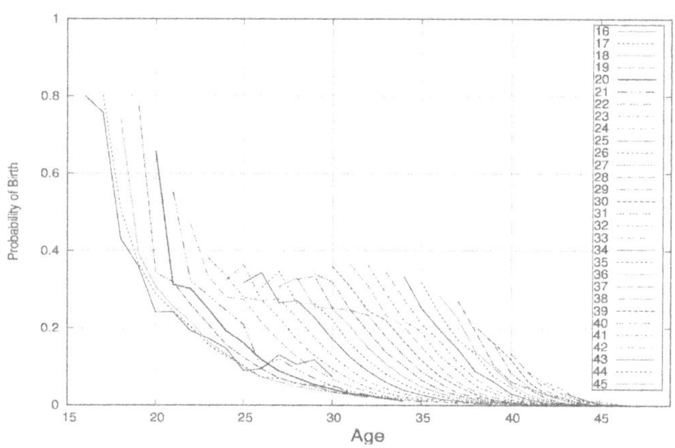

Figure 11.7: Birth probabilities in 1990.

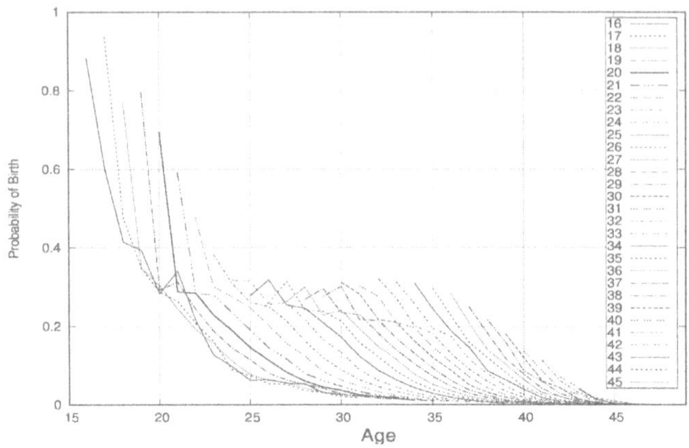

Figure 11.8: Birth probabilities in 1995.

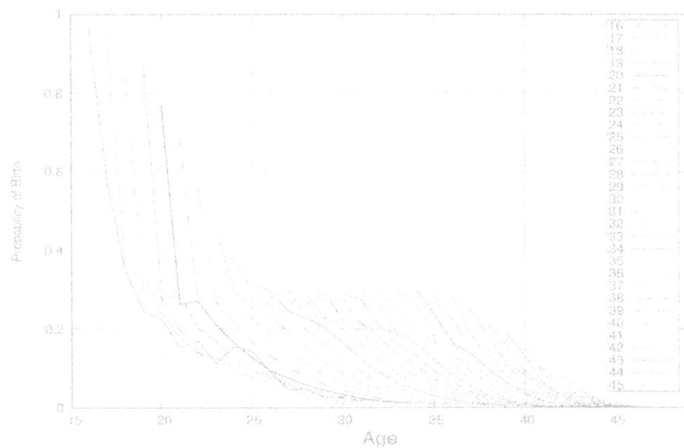

Figure 11.9: Birth probabilities in 2000.

This procedure is artificial, which is disheartening. However, over 90% of births are from females over the age of 20. Therefore, over 90% of births can be successfully explained.

11.3.2 Age of 18 as a Turning Point in the Birth Function for Japan

Compare Figures 11.4 and 11.3 (p. 293). A drastic conversion can be observed over about 70 years. Consider the intersection point of both the functions. I suppose that before this point, birth probabilities had increased and followed Equation (11.2).

I have prepared some scenarios regarding the decline of birth probabilities for this idiosyncratic population.

1. The birth probabilities are only idiosyncratic for the population of females married at the age of 15.
2. The birth probabilities is only idiosyncratic for the population of females married at the age of 15 and 16.
3. The birth probabilities of the idiosyncratic population of females married at the ages of 15–17 were natively high in 1985.
4. The birth probabilities of the idiosyncratic population of females married at the ages of 15–17 had gradually increased before 1980.

Each model has been examined numerically. Unfortunately, I am not certain of the precisely accurate model. I have not been able to solve the problem of the initial values. There is a lack of data on birth probabilities by *age* and *marriage-age* between 1940 and 1978 in Japan. This period might present the most dynamic alteration of birth probabilities.

However, the model begins from 1930 and continues from 1979 to 2010, describing the birth probabilities by age and *marriage-age* for each year.

11.3.2.1 Modification of Initial Values and Destinations

Figure 11.4 (p. 294) is unnatural. The birth probabilities of very young females are excessively high. Considering the fecundability, these values are impossible. The probabilities are so high that the denominator is very small. At present, marriages at very young ages are met with disdain and persecution just because they are a minority. Almost exclusively childbearing females legally register their marriages, and they were counted as denominators.

Therefore, their birth probabilities must be modified to attain reasonable values. I postulate that the behaviors of idiosyncratic groups are generally longitudinal. In 1939, marriages at very young ages were not hated and persecuted as they are today. Without these persecutions, couples in common-law marriages might register at the same rate as 1939.

11.3.3 Dynamics of the Birth Function for Japan

The dynamics of the birth function for Japan can be computed and illustrated using the difference Equation (11.2).

Because of the old format of Japanese vital statistics, for the years until 1978, birth probabilities cannot be computed by *age* and *marriage-age*. From 1979, the birth probabilities are computed by *age* and age at marriage and compared to the theoretical predictions using Equation (11.2).

Beginning with the knowledge of the birth probabilities by age and *marriage-age* in 1939[8], the birth function (11.2) computes each period's birth probability in a chain reaction. Of course, it presents the lacked periods by

[8]These probabilities are presented in *Marriages and Population* [59], pp. 232–233.

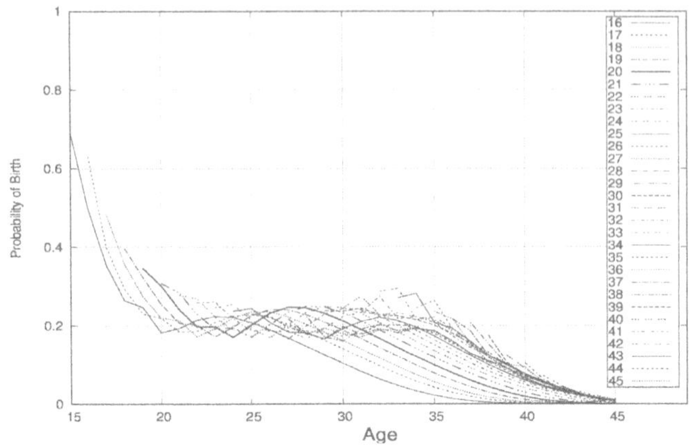

Figure 11.10: Birth probabilities computed for 1943.

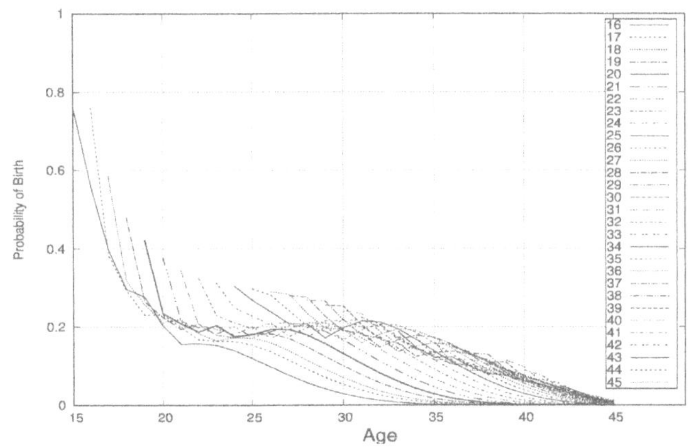

Figure 11.11: Birth probabilities computed for 1950.

World War II. It is described from 1985 to the unknown fixed points. Consider the dynamics of the birth function in Japan as you looking following figures.

Realistically, this period-based birth model in two-dimensional space is unstable. However, approximate estimations of μ, μ_s, φ can be obtained from it. Within this two-dimensional frame, the observed flattening feature of the birth function was not enough reproduced.

11.4 Addition of Geographical Dimensions

The *marriage-age* and *age* dimensions are intangible. However, we live in the tangible world, where the birth function must be considered. The physical world is thought to be three-dimensional; however, for the birth function, the *height* dimension can be temporarily omitted such that it can be reduced to two horizontal x, y dimensions.

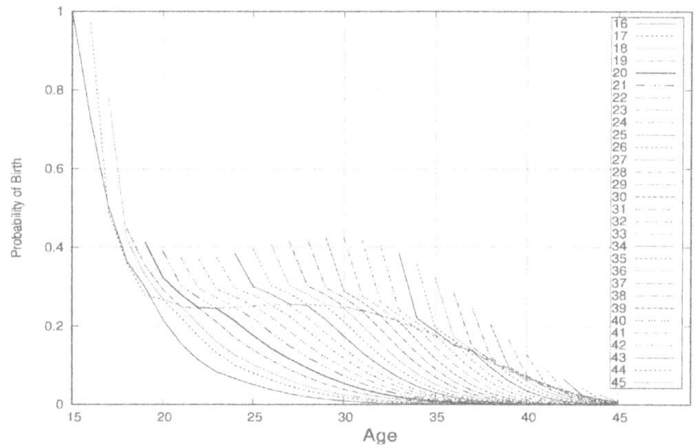

Figure 11.12: Birth probabilities computed for 1985.

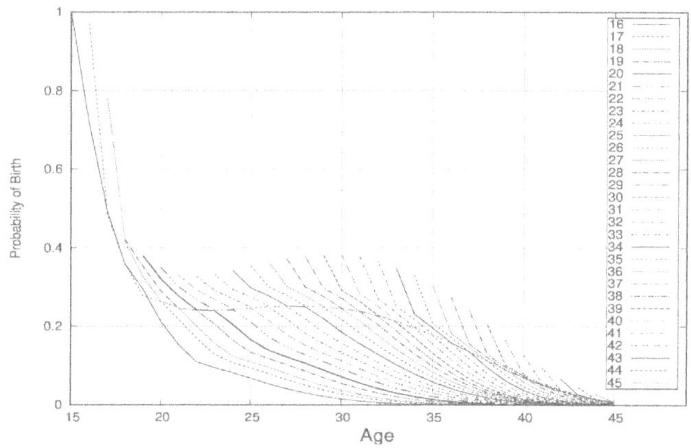

Figure 11.13: Birth probabilities computed for 1990.

For the birth function, *marriage-age* and *age* dimensions are added to the x, y dimensions. Namely, a four-dimensional space is built for the birth function, which has two symbolic physical *marriage-age* and *age* dimensions.

In this four-dimensional space, we are constantly affected by occurrences of neighboring births. In this setting, our behaviors are not individualistic, regardless of our beliefs. We are incessantly affected by others' behaviors, and our consciousness is developed *ex post facto*. The regularities of our behaviors are the result of this setting, and society is the state in this four-dimensional space. Thus, the birth function is a five-dimensional geometric object.

11.4.1 Extension in Lattice Space

Let us extend the model to the higher dimension in the lattice space, as in Equation (11.2).

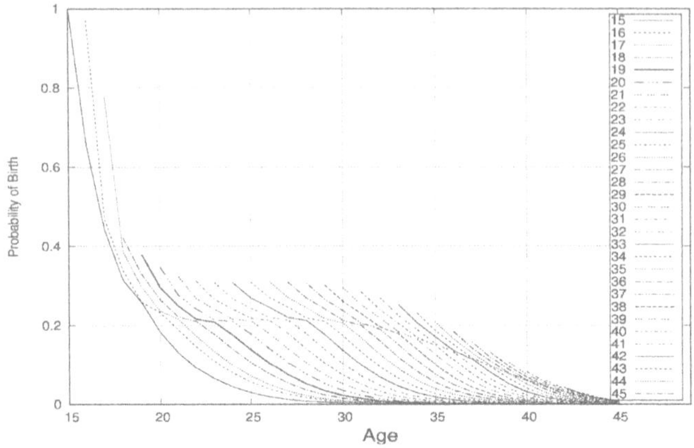

Figure 11.14: Birth probabilities computed for 2020.

However, I noticed that the catch-up effect should be eliminated from the birth function. Geographical extension to higher-dimensional space eliminates unnatural postulations of it.

11.4.1.1 Endogenously Computed $\eta(t_k)$–Expected Number of Children

As discussed earlier, during the first stage of this modeling of the five-dimensional birth function, the catch-up effect $\varphi\eta(t_k)$ was assumed. Namely, I postulated that females bear children catching-up based on the expected number of children in the space. For a single cohort in lower-dimensional space, this catch-up effect was necessary to correspond to the actual dynamics.

The average number of the children ever born, calculated by the National Fertility Survey of Japan[9], were applied exogenously for $\eta(t_k)$. It was surveyed and analyzed by the Japanese National Institute of Population and Social Security Research.

During the final stage of this research, I began to compute the birth function by employing the marriage function in the system, and endogenously computed the expected number of children. The birth function must be able to compute a valid series of the expected number of children in the space. Finally, I realized that the catch-up term $\varphi\eta(t_k)$ is not required. The coefficient φ computed an inaccurate series of the expected number of children. It was clear that it needed to be eliminated and another equation had to be formulated.

After extensive consideration, I set the singularity to the leading diagonal probabilities for the uni-cohort model. As there were no statistical data for them, I drew trajectories so as to correspond to reality. All enigmas are reduced to their singular birth behaviors. Owing to their singularity, $\varphi\eta(t_k)$ could be

[9](http://www.ipss.go.jp/site-ad/index_english/Survey-e.asp).

omitted, and consistent numbers of expected children could be computed in the system. On enhancing to the higher dimension, the catch-up effect revealed itself to be the neighboring effect.

Thus, I eliminated the catch-up effect. Interestingly, the catch-up effects (including the aging effect) result in inaccurate birth probabilities for the cleanup period of Japanese vital statistics, as shown in Figure 11.16 compared with Figure 11.15. Computed probabilities without the catch-up effect construct more suitable expected number of children.

Human birth behaviors are completely explained by the neighboring effect. Namely, it can be reasoned that we are not affected by the overall conditions of society. We behave only locally.

11.4.1.2 Difference Equation in the Extended Space

The structure of the neighboring lattice space is altered in the extended space. However, the same coefficient is temporarily adopted. Let x_i, y_j be the geographical position in it; m_ℓ be the marriage age; and a_k be the age.

$$
\begin{aligned}
b(x_i,&y_j,a_k,m_\ell,t_{t+1}) \\
&= rpc\, Fe(k+1) \\
&\quad \Big(\frac{\mu}{5}\, b(x_{i-1},y_j,a_{k-1},m_{\ell-1},t_t) + \frac{1-\mu}{5}\, b(x_{i-1},y_j,a_k,m_\ell,t_t) \\
&+ \frac{\mu}{5}\, b(x_{i+1},y_j,a_{k-1},m_{\ell-1},t_t) + \frac{1-\mu}{5}\, b(x_{i+1},y_j,a_k,m_\ell,t_t) \\
&+ \frac{\mu}{5}\, b(x_i,y_{j-1},a_{k-1},m_{\ell-1},t_t) + \frac{1-\mu}{5}\, b(x_i,y_{j-1},a_k,m_\ell,t_t) \\
&+ \frac{\mu}{5}\, b(x_i,y_{j+1},a_{k-1},m_{\ell-1},t_t) + \frac{1-\mu}{5}\, b(x_i,y_{j+1},a_k,m_\ell,t_t) \\
&+ \frac{\mu}{5}\, b(x_i,y_j,a_{k-1},m_{\ell-1},t_t) + \frac{1-\mu}{5}\, b(x_i,y_j,a_k,m_\ell,t_t)\Big)
\end{aligned}
\tag{11.4}
$$

However, $\mu \leq 1$.

The birth function is a five-dimensional anisotropic hypersurface. This is a fascinating object that deserves exploration.

11.5 Benefit of Extending the Birth Function

The fact that the catch-up effect is unnecessary is a significant benefit of extending the birth function. In the case of the two-dimensional—*marriage-age* and *age* space, the coefficients needed to be adjusted to a highly impractical extent. In the two-dimensional space, the many coefficients must be set to be anisotropic. However, the geographical extension naturally resolves these difficulties to some extent. It finally became clear that the anisotropic coefficients

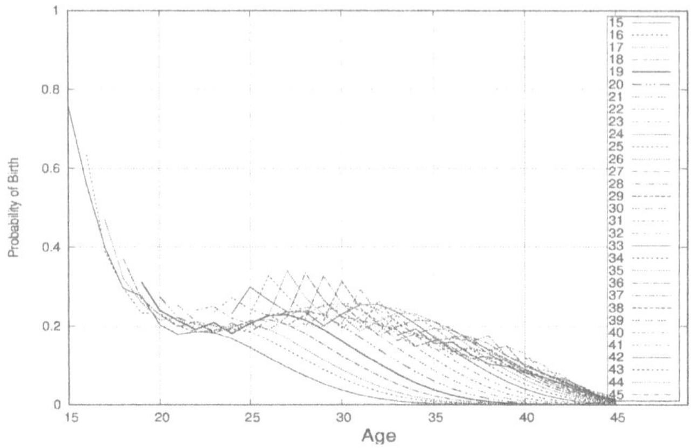

Figure 11.15: Birth probabilities computed for 1950 with catch-up effect.

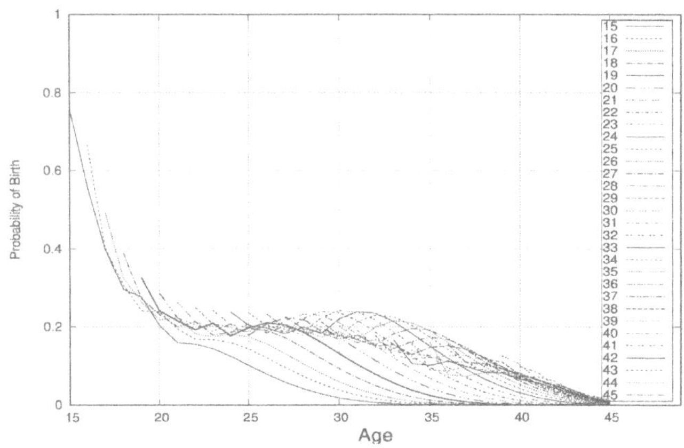

Figure 11.16: Birth probabilities computed for 1950 without catch-up effect.

of the two-dimensional space are superficial and caused by the asymmetrical phases of the diffusion process.

11.5.1 Significance of Four Dimensions

First, the *marriage-age* and *age* dimensions were added in order to explain birth probabilities. In other words they were required to trace the birth function. Now, I have realized that the most fundamental significance of the degree of dimensions is in the four-dimensional space, where the function itself can be mapped.

For the first married population of a cohort, from a space to its neighboring space, the birth functions deform based on the operator. Various operators may be naturally developed to describe diverse social phenomena, that is, to operate the functions corresponding to altering the frame of reference based on perspective. When a high symbolic dimensional space is constituted, modifications to our manner of thinking can be handled.

However, for human birth probabilities, I restrained myself from developing any dimensional space higher than the four-dimensional one. This is because I espouse *"Occam's razor"* and I cannot justify how further development would benefit the theory. I the most parsimonious theory to be the optimal one.

11.5.2 Original Values of Coefficients

In the case of the two-dimensional *marriage-age* and *age* space, no accurate estimations of the coefficients could be found, and unnatural and complicated effects of the specific birth probabilities were set. Using geographical extension, the number of coefficients could be reduced and more fundamental values could be set.

11.5.2.1 Elimination of Alternative Reference

For the first married population, $(1-\mu)$, which represents the ratio of the effects of the contemporary and immediately preceding populations who married at the same age in the lattice space, was eliminated. The only effect of the preceding population married at a slightly younger age must be supposed such that $\mu = 1$. This value is in accordance with the results of the uni-cohort, where only the effect of the contemporary population married at a slightly younger age was supposed.

Moreover, for the total population (other *married-age*, other *age*) in the space, $\mu = 1$ can be supposed before and after the fertility decline—constantly for almost all except for the first married population from the beginning of the fertility decline.

Thus, the Equation (11.4) changes to the following more elementary one.

$$\begin{aligned} & b(x_i, y_j, a_k, m_\ell, t_{t+1}) \\ &= rpc\, Fe(k+1) \\ &\quad \Big(\frac{\mu}{5}\, b(x_{i-1}, y_j, a_{k-1}, m_{\ell-1}, t_t) + \frac{\mu}{5}\, b(x_{i+1}, y_j, a_{k-1}, m_{\ell-1}, t_t) \\ &+ \frac{\mu}{5}\, b(x_i, y_{j-1}, a_{k-1}, m_{\ell-1}, t_t) + \frac{\mu}{5}\, b(x_i, y_{j+1}, a_{k-1}, m_{\ell-1}, t_t) \\ &+ \frac{\mu}{5}\, b(x_i, y_j, a_{k-1}, m_{\ell-1}, t_t)\Big) \end{aligned} \tag{11.5}$$

And, $\mu = 1$

The result of the computation with the coefficient $\mu = 1$ suggests that birth probabilities rapidly converge to fixed points. This beautiful simplicity evokes the law. The majority of humans are uncomplicated in their birth behaviors but complicated in their explanations depending on others' variations. It is likely that the first married population alone varied their manner of reference for childbearing. However, the modification was not drastic. It is a delicate and intangible alteration, but all subsequent behaviors are a consequence of it.

11.5.2.2 Infinite Transportation

The birth function can be conveniently imagined by converting the difference Equation (11.5) to a differential equation, as below.

$$\frac{\partial b}{\partial t} = rpc\,\frac{dFe}{da}\left(\frac{\partial^2 b}{\partial x^2} + \frac{\partial^2 b}{\partial y^2} + \frac{\partial b}{\underset{\partial m}{\swarrow}} + \frac{\partial b}{\underset{\partial a}{\swarrow}}\right). \tag{11.6}$$

$\swarrow$ means that m, a partial differentiated by descending neighborhood. The partial differentials of m, a are asymmetric and directional.

I feared that if $\mu = 1$, there may be no connection between the previous and present cohorts. This was my prejudice. The differential equation above expresses the connection between the cohorts. The elementary fact that present behaviors depend on immediately preceding neighboring behaviors guarantees the existence of a connection between generations, and consequently, the causality.

11.5.2.3 Considering $\mu = 1.0$ for the Diagonal Elements

For the uni-cohort model and the two-dimensional dynamic model, a value different from other elements has to be used for the diagonal elements. For low-dimensional space, coefficients need only be adjusted to fit the model to the real observed data in their decline. However, time can be gained when the spatial diffusion process is assumed. Another problem emerges wherein the model must be fitted to the velocity of diffusion; however, it is easy to adjust the velocity. Finally, the birth function theory is connected with the reaction-diffusion theory of fertility decline.

Regarding human behaviors, it should be noted that all phenomena occur in the geographically and spatially expanded space, and time must be considered in this process. The actual real value of the coefficient cannot be estimated depending only on the normal statistical data. Statistical data is the basis of empirical and falsifiable science, but its insuperable limitation is already evident, particularly since the advent of random sampling statistics. The boundary of social research must be recognized.

When deciding the coefficient for the diagonal elements, I finally had to adopt the fundamental theoretical policy of the simplest solution being the most optimal one. The diagonal elements represent females who are just married. For them, the same law of birth probability can be applied such that $\mu = 1.0$.

The extend to which humans are controlled by prior neighboring situations is astounding. This is a necessary law because neighbors cannot be eliminated. The diagonal elements are not exceptions.

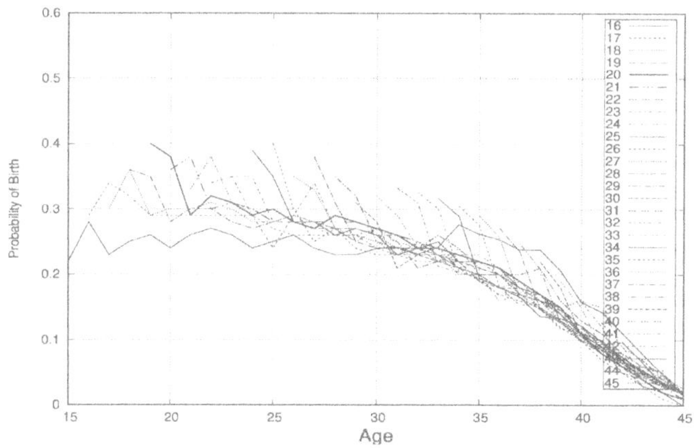

Figure 11.17: Initial birth function (correspond to 1930).

Other coefficients can be set to compute the most realistic trajectories. As mentioned earlier, the dynamics of the birth function are the dynamics of its geographic space. I believe that this is also a profound principle.

11.5.3 Changes in the Neighboring Function

Another significant feature of the extended four-dimensional birth function is that in each (lattice) space, the function itself alters and effects the neighbors. In the lower-dimensional diffusion process, the single value fluctuates. In the three-dimensional marriage function, the integrated value also fluctuates. However, in the birth function, the function itself fluctuates for the first married population. At first, this was difficult to conceive. In the four-dimensional birth function, functions interact within space. It takes time to become accustomed to this idea.

Corresponding to the real-world diffusion of the function is the interactions of the *marriage-age, age, and geographic dimensional* neighboring population. Thus, our interactions are symbolically—biologically—geographically structured, and this precisely reflects society. Society does not constitute what we consciously build, but it still exists. In order to formalize this matter, the four-dimensional space must be employed. An even higher-dimensional space may be built to explain other social phenomena; for instance, money may be added as a symbolic imaginary dimension. However, this possibility of the higher order of space is limited in modern physics.

11.5.3.1 How is the First Birth Function Altered?

It is difficult to imagine that the first married population alone fundamentally alters their birth behaviors — their birth functions — in space. However, they are the only population who can change their birth function. If not,

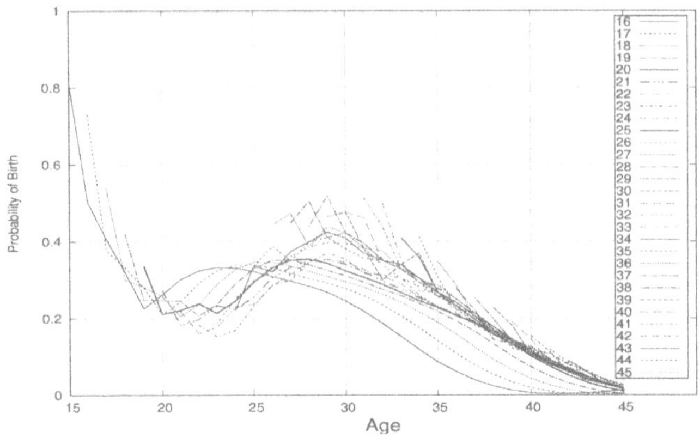

Figure 11.18: Birth function for 1938 at a hole in Japan by simulation.

no reasonable explanation for the dynamics of birth probabilities would be possible.

How does this population alter their birth function? How do they judge the state of neighboring fertility (birth behaviors)? As long as the simulation program was written based on their own declining birth probabilities, I was unable to reasonably reflect the decline. Only after writing the program to decline as per the neighboring decline, a reasonable decline was observed. Thus, although the singular population must be controlled by a neighboring singular population, they stochastically decide the birth behaviors of the remaining population. Through the challenge of seeking a valid model, I am confident that they are genuine leaders of our birth behaviors. However, they are not conscious of it.

11.5.3.2 Rapid Alteration by the First Married Population

Did the first married population ($\doteq$ married at age 15) rapidly or gradually alter their birth behaviors? By means of numerical simulation, I found that finer fits are achieved if they are considered to have rapidly altered their birth probabilities, as shown Figure 11.18. In reality, in the 1980s in Japan, the first married population appear to have reached the recent level, as shown in Figures 11.5 and 11.6.

They rapidly increased their initial (first married year) birth probabilities and quickly declined their later birth probabilities. We have been delayed by them.

Birth probabilities as an exponential function

What function is used to simulate the birth probabilities of the first married population? This question is not easy to answer. This process must be inferred to explain their decline in birth probabilities, and thus, my formulation is somewhat tautological.

I assume that their birth function was altered when their birth probabilities began to decline.

No simple function of the singular population for fertility decline could be found. Therefore, I empirically traced their birth probabilities in accordance with the observed data. I assumed that their initial birth probability (in their first married year) rapidly increased to around 0.9 and has remained at that value. Without this assumption, the dynamics of all birth probabilities cannot be traced. In fact, in 1985, the birth probability of females married at the age of 16 had already reached 0.773. Essentially, the singular population has been reducing the probabilities of elderly childbearing and increasing the probabilities of childbearing within the year of marriage. Whether a realistic birth function is produced is solely dependent on the ability to capture their behaviors graphically.

To achieve this, I adopted a time-dependent exponential function. I estimated that with the appropriate coefficients ϱ_i, this function effectively describes their birth probabilities.

$$(constant + \varrho_1 t)(age)^{-1-\varrho_2 t-\varrho_3 age^2} \tag{11.7}$$

I suppose that the shape of this time-dependent exponential function decides the distinctive pattern of the birth function. During the trial stage, the values estimated based on the observed values are included below the birth probabilities of females married at the age of 15. These models resemble the real data, but the rate of decline is not sufficient.

The problem was solved using this time-dependent exponential function. This solution suggests that the elderly birth probabilities of the first married population declined fairly rapidly as time passed. Nevertheless, I do not believe that I have gained a clear understanding of their behaviors. However, they may be the reaction term of the reaction-diffusion Equation (4.4).

11.5.4 Ground of β, γ

With this geographical extension, the basis of the reaction-diffusion of the number of children is obtained (Equation (2.1) on p. 23). I postulated γ as the minimum fertility in the equation, which was obscured owing to my subjective imagination.

However, by integrating Equation (11.5), the precise lower limit is confirmed. Its integration is always greater than the trivial 0. Moreover, the effect of smoothing the equation, which is naturally caused by the neighboring space, ensures approximately the same lower limit of the number of children. However, it should be mentioned that the probable birth probabilities of the initial age at marriage must be set in order to obtain reasonable birth probabilities. The lower limit of the number of children plays a critical role in setting γ.

Equation (11.5) similarly confirms the precise upper limit. Expectedly, its integration does not diverge. We can confidently postulate β and γ as the upper and lower limits of the number of children. Naturally, the birth function is more fundamental than the reaction-diffusion equation. Although the latter is useful, the former has more explanatory abilities in terms of the reality of demographic phenomena. Needless to say, the birth probabilities of the population with the initial age at marriage also play a significant role in the value of β.

In reality, as mentioned above, there remain slight geographical undulations of these upper and lower limits in space. There is probably a slight geographical undulation of these lower limits in Japan, as can be inferred from Figures 9.4, 9.5. The birth function can explain these undulations using the marriage function.

11.6 Origin of the Hole of Low Fertility

The origin of the hole of low fertility[10]—the decrease in the probabilities of the number of children—is unknown. The hole is four-dimensional, and thus, it cannot be observed like the reaction-diffusion of fertility decline. This problem is classified into two components.

The first is that the population decreased their birth probabilities. How old were they? Where did they live? I believe that the number of children borne by older females married as very young teenagers began to decline.

As this assumption caused the problem of the lack of predecessors in the practical numerical system, this population must be regarded as an exception in the iterative computation.

Predecessors must exist, and their behaviors must be approximately identical to those of very young married teenagers. Therefore, older females who married at the same age were substituted for their predecessors married at a younger age. Even with the geographical extension, an exception has to be made for females married at the age of 15. This means that their behaviors are independent of others in space, and they are unique and autonomous.

The second component of the problem is the magnitude of the decline in the birth probabilities of the population. Was it slight or significant? What was the area necessary for the beginning of the fertility decline? This problem will be solved to obtain likely estimations to fit the real data.

11.6.1 What Constitutes the Hole of Low Fertility?

My initial expectations regarding the hole of low fertility were false. Naturally, in real observed phenomena, the birth probabilities of older females who married as teenagers decreased. However, the theory of birth function transcends the reality observed from the data. As mentioned above, the hole is not the set of values, but the altered function.

[10]Which is equal to the singularity in case of the reaction-diffusion system.

The minimum area, the 1×1 lattice grid, is sufficient to start the fertility decline. This diffusion process does not depend on the amount of area as the initial value.

The birth function is fundamentally a mapping from the present high-dimensional surface to the next. The alteration of this hypersurface is more naturally created by the alteration of the mapping itself.

The first (teenage) married population varied their mapping at the beginning of fertility decline. This varied mapping diffuses exclusively to the first married population. This appears to be a peculiar phenomenon. At first, I believed that this harmed the integrity and universality of the theory. However, only the altered mapping diffuses in the four-dimensional space for the first married population. They precisely correspond to the $(1-\mu)$ population in the difference reaction-diffusion equation, which is extraordinary in the diffusion process.

Clearly, my work integrating them into the equation system was in vain. Therefore, I consider the original idea of how their birth probabilities declined.

The birth function cannot be developed without this assumption. The first married population are singular, meaning that they alone changed their childbearing behaviors. They are a genuinely sensitive population. The theory of the birth function has several unsolved problems. However, this suggests that the theory is not false but has certain limitations.

11.6.2 Peculiar Pattern of the Birth Function

From Figures 11.5 on p. 295 to 11.9 on p. 296, the peculiar pattern of each birth function can be observed. In fact, these peculiar patterns prove that human birth behaviors are controlled by the birth function. This is entirely the result of being directed by neighboring situations.

Unless human birth behaviors are controlled by the birth function, they cannot be represented as shown. Our distinguished ability to be followed by predecessors presents these analogous and parallel translations.

The emergence of these patterns depends on the longitudinal performance of the birth probabilities of the first married population. These patterns can be represented or erased by subtly manipulating these birth probabilities. Thus, the first married population governs all succeeding married populations.

11.7 Magnificent Metamorphosis

With these rather complex assumptions of coefficients[11] and with the singularity of the first married population, the fantastic dynamics of the marriage function can be drawn as Figure 11.19 to Figure 11.20.

[11]The number of coefficients is less than 10, including the setting of singularities. This is significantly less than the standard theories of quantum physics.

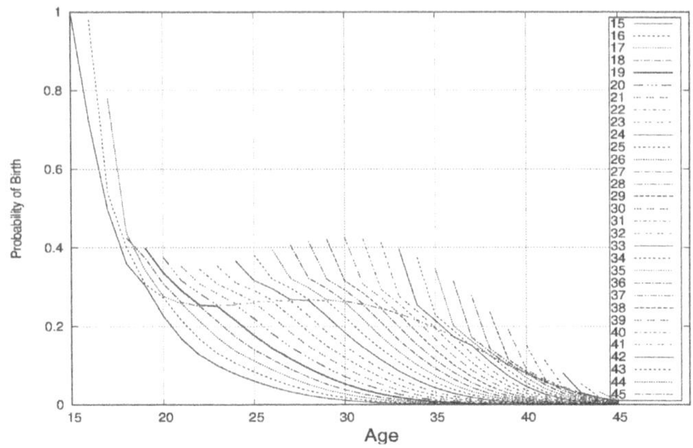

Figure 11.19: Birth function computed for the year 1985 in Japan.

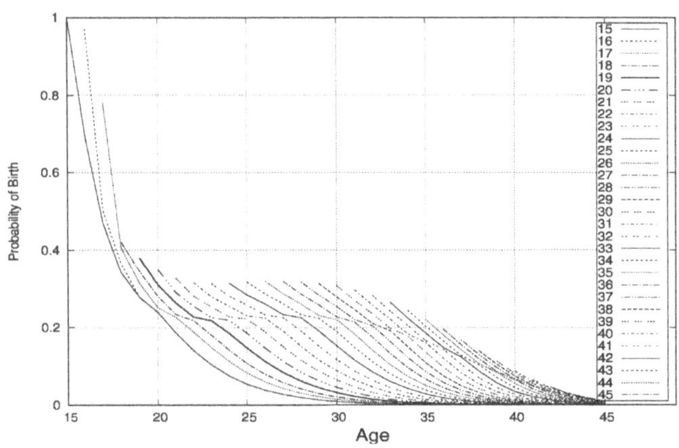

Figure 11.20: Birth function computed for the year 2015 in Japan.

An astounding metamorphosis can be observed from Figure 11.17 (1930) to Figure 11.14 (2020). The majority of the population simply follows the previous neighboring activity. We merely react to our environment, contrary to our assumptions. Our behaviors are unconscious, despite what we may think.

I anticipate that this metamorphosis is the result of the collective construction of the group, not of our subjective selections. Theories based on our subjective consciousness can never draw such dynamics. They are neither falsifiable nor impotent.

11.7.1 Schema of the Dynamics of the Birth Function

It is impossible to observe visibly the dynamics of the four-dimensional hypersurface. It is necessary to use our imagination for this.

For females of all ages in the first married population, the decline in the number of children (fertility decline) occurred as a slight decline in their birth probabilities. This marked the beginning of fertility decline. Although we may be inclined to suppose that the upper class or intelligent population began to decrease the number of children they bore, the leaders of fertility decline are none other than the first married population. The birth probabilities of older individuals tend to decline rapidly, and those populations tend to marry later. This coincidence causes us to misconceive the leader of this phenomenon.

Here, the above hypothesis was adopted for two principal reasons. One is that a lower probability of birth diffusing in space cannot be simulated when the hole of low fertility is supposed to be caused by the population of females who married later. Permanent declines in birth probability cannot be demonstrated with the diffusion process alone[12]. The other reason is that in the real progression of geographical diffusion, there was little differential fertility between different socioeconomic populations on a somewhat macro scale. The primary factor of fertility decline is the movement of the first married population of a cohort.

The first married population are most sensitive to the alteration of their neighboring contemporary homogeneous population (population married at the same age). This is possibly because they are a minority in their living space at any rate.

This initial local decline of the birth probabilities diffuses in the *geographical dimensions, marriage-age, and age* dimensions. The delay in the decline of the first married population constitutes the delay of local fertility decline.

11.7.1.1 Resemblance with the Birth Function of England-Wales

C. Clark presented birth probabilities based on age and age at marriage for England-Wales in 1939, as shown in Figure 11.21. Hereinafter, this state of the birth function is referred to as Clark's state.

As shown in Figure 11.17 (p. 305), it differs significantly from the Japanese birth function at the same time. However, during the process of metamorphosis, resembling patterns emerged in Japan. Around t = 18 (Figure 11.22), the birth function somehow resembles that of Clark's state. As Japan lacked data for those periods (t = 18 is correspond to 1948), the resemblance between the theoretically computed values and the observed data is a weak but interesting example.

In the 1880s–90s, fertility decline may begin in England-Wales. By then, over forty years had passed since the starting point of fertility decline in England. In Japan, about twenty years had passed since the beginning of

[12]This is already suggested by the reaction-diffusion equation of the number of children, especially by the reaction term.

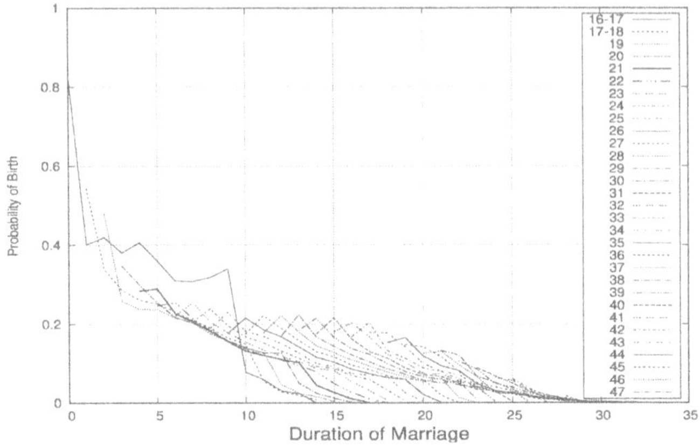

Figure 11.21: Birth function computed for 1939 in England-Wales.

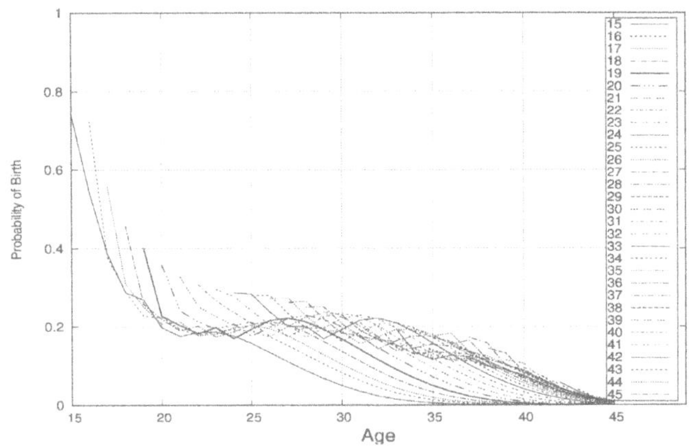

Figure 11.22: Birth function computed for t = 18 (1948) in Japan.

fertility decline. Regardless, the birth function of England-Wales and of Japan coincide in their shapes. This suggests the universality of the birth function[13].

11.8 Strange Ripples Following the Development of the Hole of Low Fertility

As this chapter nears its conclusion, it is a pleasure to report the discovery of some strange and unexpected birth function patterns. At first glance, the birth function appears to consistently alter itself; however, tiny ripples always occur in the latter half.

[13]The speed of the metamorphosis is quite different yet. This should be cleared in the future.

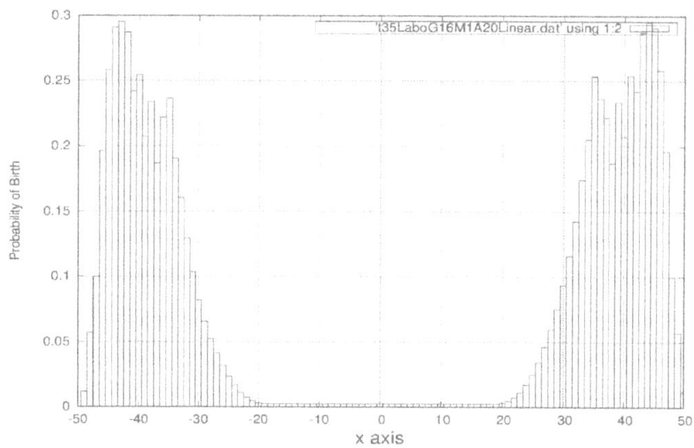

Figure 11.23: Birth probabilities of a 35-year-old female married at the age of 15, observed on a line passing through the hole $t = 35$.

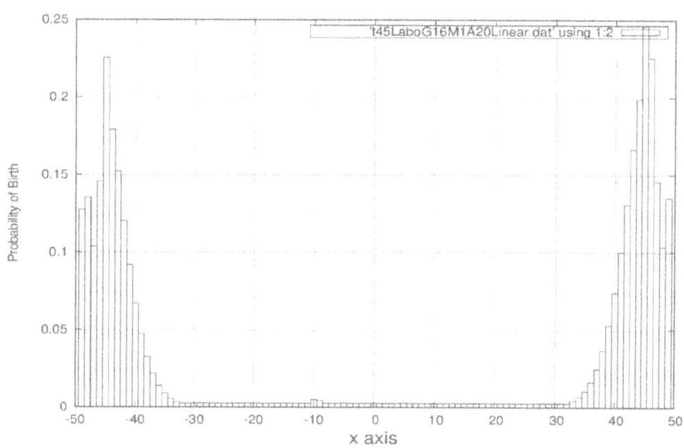

Figure 11.24: Birth probabilities of a 35-year-old female married at the age of 15, observed on a line passing through the hole $t = 45$.

It is impossible to observe the development of the hole of low fertility, because the birth function is a five-dimensional object. Therefore, the geographical undulation of the birth probabilities of a 35-year-old female married at the age of 15 was longitudinally monitored as a three-dimensional surface on a line passing through the hole.

The area of the hole gradually expanded to some point in time, as shown in Figure 11.23 and 11.24. This process is as expected.

However, following a certain point in time, strange, small undulations irregularly occurred at some points. These undulations were irregularly developed or dismissed. They are referred to as "ripples" herein.

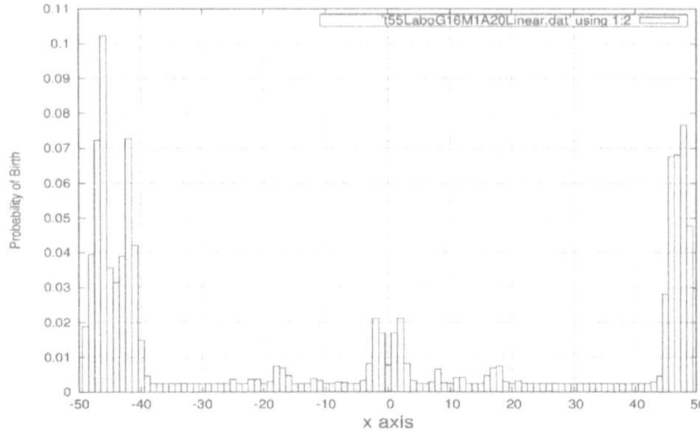

Figure 11.25: Birth probabilities of a 35-year-old female married at the age of 15, observed on a line passing through the hole $t = 55$.

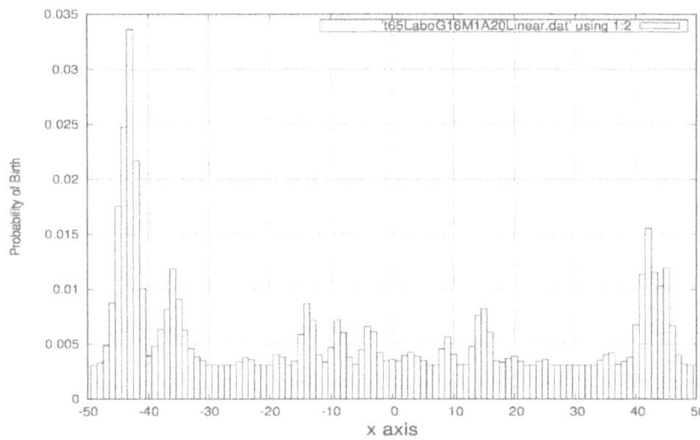

Figure 11.26: Birth probabilities of a 35-year-old female married at the age of 15, observed on a line passing through the hole $t = 65$.

These phenomena are surprising. Over time, these irregular undulations altered as if they were autonomous systems. When the line measured was changed to one that did not pass through the original hole, entirely different undulations appeared. I drew a three-dimensional surface of birth probabilities of a 35-year-old female who married at the age of 15.

These ripples suggest that there are irregular, instantaneous increases in births at some geographical points. The heights of the ripples ranged from 0.035 to 0.005. These values are small but not impossible to observe. When they are practically observed, the birth function can be verified.

Existing measures for birth probabilities are not sufficiently sophisticated to observe these ripples. Social scientists have been researching objects without recognizing what the theory should predict.

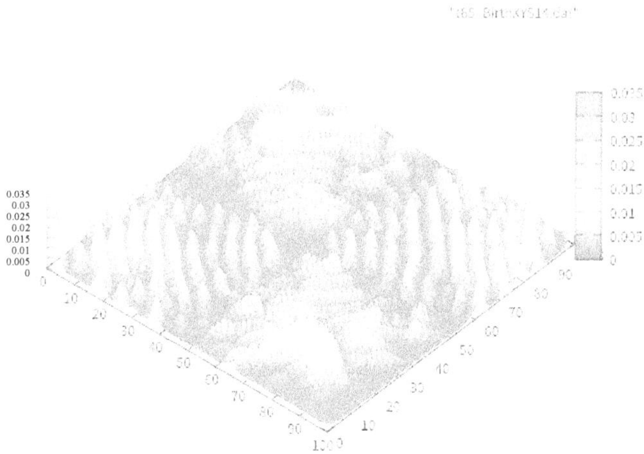

Figure 11.27: Birth probabilities of a 35-year-old female married at the age of 15, observed on a line passing through the hole $t = 85$.

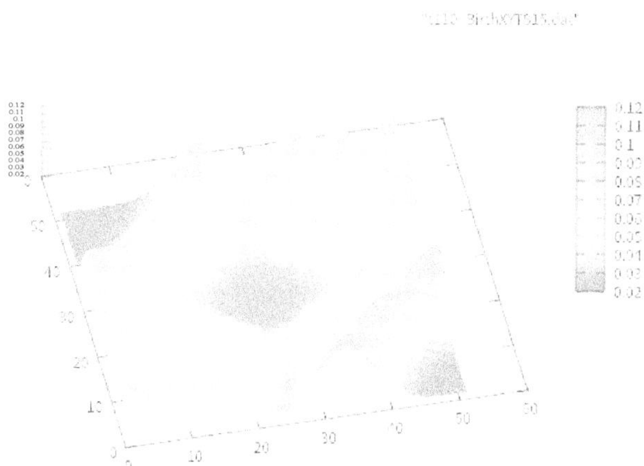

Figure 11.28: Birth probabilities of a 35-year-old female married at the age of 15, observed on a line passing through the hole $t = 110$.

11.8.1 Birth Function as a Reaction-Diffusion System

Various patterns are observed for each *marriage-age* dimension. They are bewitchingly beautiful. Provided that every ripple corresponds to each birth behavior, our choice to bear children is completely controlled by other births and is also unique.

I had hoped that such strange patterns might occur. The birth function is at least a fundamental diffusion equation. The reaction-diffusion system is the repository of pattern formation in time–space. However, the birth function has no evident reaction terms.

A. Turing's original idea of reaction-diffusion was expressed in the following simultaneous differential equations.

$$\begin{cases} \dfrac{\partial u}{\partial t} = \Delta u + f(u,v) \\ \dfrac{\partial v}{\partial t} = d\Delta v + g(u,v) \end{cases} \tag{11.8}$$

His intention was to explain the morphogenesis by the interaction of the active factor u and inhibiting factor v in the diffusion process. I was under the impression that the diffusion-driven instability requires two different materials in space. The birth function contains only one material: birth probabilities. I could not imagine the mathematical mechanism of reaction-diffusion in the birth function.

Therefore, I did not predict that the birth function exhibits these chaotic phenomena, as shown in Figure 11.27. I had been attempting to compute the macro birth function to fit the observed data. Therefore, I was not able to understand the birth function in terms of the reaction-diffusion equations. The birth function fundamentally lacks simultaneity at the first starting point. There appears to be only one variable b, and no interaction in it. I was lost in the maze of the high-dimensional birth function.

Finally, I realized that I should regard the females not bear children in a space as inhibiting factor. The birth function in a space is a kind of intraction system of bear and not bear females. Let be b the ratio of females bear children. Let be s(sterile) the ratio of females not bear child.

$$\begin{cases} \dfrac{\partial b}{\partial t} = rpc\,Fe\,(\Delta b + f(b,m,a)) \\ \dfrac{\partial s}{\partial t} = d\,(\Delta s + g(s,m,a)) \\ \dfrac{\partial b_s}{\partial t} = h(b_s) \end{cases} \tag{11.9}$$

In addition to, the initial married teenagers bear their children as separate from others. The birth function is an interaction between their singular minor birth probabilities b_s and other major birth probabilities.

$h(b_s)$ is a function that singular population obey after the childbearing behavioral shift.

On rearranging the birth function in view of the reaction-diffusion system, it shows the potential to form strange patterns in time–space. Human births occur depending upon others' birth occurrences such that the wonderful patterns appear in time space. However, we are incapable of conceiving them.

11.8.1.1 Mechanism of the Strange Patterns

The birth function postulates anisotropic diffusion. Namely, the diffusion of the birth function has directivity in the age dimension. During the initial stage of the diffusion of the birth function, owing to the geographical space extension, the undulations of birth probabilities did not completely disappear. They survived the smoothing process and developed in the latter half of the stage.

11.8.1.2 Reason for the Preservation of Convexoconcave Undulations

The principal problem is the mechanism of how the observed convexoconcave undulations are preserved against the flattening effect of diffusion.

In the high-dimensional space, the birth function immediately following the beginning of fertility decline instantly exhibits high birth probabilities. This difference in elevation results in very slight convexoconcavity of the birth probabilities.

As the diffusion process proceeds, the effect of the diffusion term weakens, and consequently, the flattering effect becomes weak. The preserved convexoconcave undulations begin to develop into ripples.

Disturbance may occur owing to various factors. I have not excluded socioeconomic causes; for example, the recession slightly delays births, causing a slight undulation in birth probabilities. However, its development into peculiar patterns is evidence of the birth function.

In the research on reaction-diffusion systems, the "Gray-Scott model" [29], [62] is known to exhibit various time–space patterns. According to Ueyama and Nishiura [74], it has the following six patterns:

1. Standing pulse solution
2. Traveling pulse solution
3. Self-replication pattern
4. Spatio-temporal chaos
5. Frontal wave solution
6. Traveling pulse solution (pair annihilation)

I conjecture that the birth function exhibits the frontal wave solution to some point in time on a relatively large scale compared to a later point. It also exhibits the traveling pulse solution on a relatively small scale after it in time.

Similar to other reaction-diffusion models, when coefficients are operated upon, various patterns may appear. However, the coefficients of the birth function should not be altered easily. Nevertheless, the birth function bears various patterns for each *marriage-age* and *age* dimension.

11.8.1.3 Individual Cut Sections

When the high-dimensional space is cut by the initial year of *marriage-age* and *age*, no undulations are observed on the surface. This is because the model

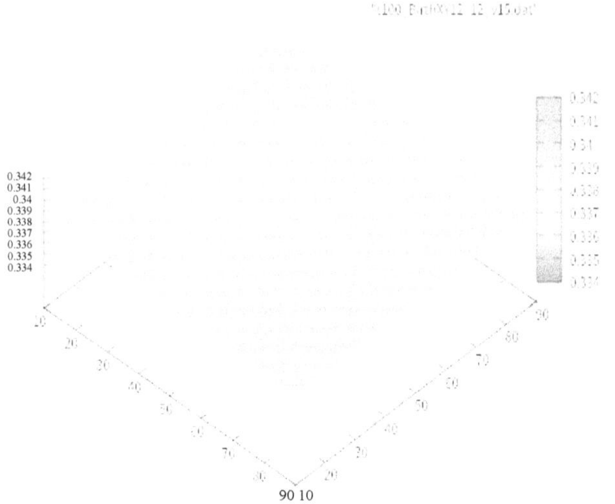

Figure 11.29: Birth probabilities of 27-year-old females who married at age 27, observed on a line passing through the hole $t = 100$.

has been programmed as such. I do not know how to combine the singular group with regular populations. Because the law determining the initial birth probabilities of this singular group is not yet understood, they can only be artificially decided. As time passes and age increases, for the *marriage-age* and *age*, cut surfaces form peculiar patterns. I believe that this is a natural phenomenon.

With the passage of time, the initial group married at the age of 15 exhibited unknown patterns, such as the "forest of ice pillars." This forest grows as if it contains wills. However, this growth is an inevitable process controlled by the simultaneous Equation (11.9). The development is in fact the result of neighbor interactions.

Human birth behaviors and conscious wills related to them result from our neighbors birth behaviors. As stated by Spinoza, "They are, as I have often said already, conscious of their own actions and appetites, but ignorant of the causes whereby they are determined to any particular desire[14]." The desire to bear children is determined by Equation (11.9).

For the *marriage age*, *age*, and *geographic* space, peculiar and astounding patterns emerge via the reaction-diffusion Equation (11.9). Our conscious wills are not the final causes but transforming shadows. Almost all patterns represent spatio-temporal chaos with different designs growing over time. The genuine complexity of chaos gives us the illusion that our wills are unique.

These undulations continue up and down cyclically. These patterns determining our behaviors are far more complex and abundant than we believe. We are not complex, but our interactions with others are. I consider that society is highly complex and cannot be computed and conjectured. The interesting

[14]Spinoza [16], p. 152.

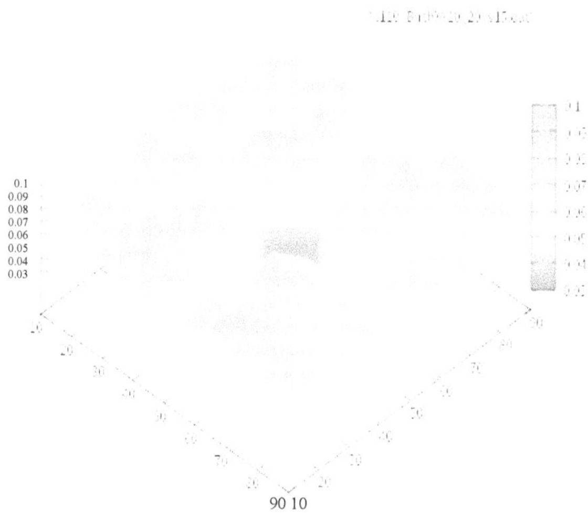

Figure 11.30: Birth probabilities of 35-year-old females who married at the age of 20, observed on a line passing through the hole $t = 110$.

peculiar patterns on the four-dimensional surface suggest that our conscious subjectivity cannot reliably predict our behavioral patterns.

I am not equipped to make a more profound mathematical analysis.

11.8.1.4 Transcendental Others

If these irregular undulations are observed, the birth function proposed herein shall be proved to be true in a single falsification. I believe that these undulations have a novel, notable significance in our understanding of human behavior.

Childbearing is the result of the dynamic pattern formation of the birth function, not conscious will. The behaviors of others decide our behaviors. Behaviors differing from neighboring ones are also caused by them, according to Equation (11.5). Essentially, human behaviors are nothing but reactions to neighboring behaviors. Even the consciousnesses to differ from others are realized owing to others. This is precisely the philosophy of *"In der Welt Sein"*.

Now, behaviors of others alter themselves autonomously. The idea that voluntary motivations cause behaviors is false. Humans are like machines that react to other machines within the parameters of environmental conditions. The only difference is that the human reaction approach is fairly complex.

I consider that we are merely organisms reacting to others and our environment[15]. However, the reaction mechanisms are too complex to be understood. From fifty years ago to the present, we have taken it for granted that human behaviors are based on a combination of reasonable self-selection, socioeconomic conditions, and some social costumes. This view is hackneyed and false. Our behaviors vary depending on those of others. Infinite combinations

[15] The organic conception of society is a fairy tale at all.

of our behaviors may exist in high-dimensional space. Now, we must consider this novel perspective on behaviors and discard outdated conceptions.

11.9 Birth Behaviors are not Individualistic

Equation (11.4) indicates that our birth behaviors are causally chained and result from neighboring birth behaviors. Evidently, birth behaviors are not the actions of independent individuals. The validity of the equation is not only in that it fits well with the data, but also that there is no other formalization of birth behaviors. We have no concealed motivations; our motivations are imagined and found in the neighboring chain behaviors. Free will merely exists in probability.

I reiterate that marriage and childbearing are not determined by individual conscious will. It is foolish to hope to bring about fertility behaviors in the future by some pronatalist policies. With the exception of initial singularities, all fertility behaviors are stochastically determined by neighboring fertility behaviors.

Macro trends of fertility can be predicted by recognizing marriage trends and patterns in the birth function. Individuals expect that their consciousness plays a significant role in the fertility process. However, conscious desire is a illusion programmed into our minds. We are merely animals reacting to neighboring animals. Our consciousness is not a conductor of behaviors, but a junk recorder.

Chapter 12

Distribution of the Numbers of Children

> After all, the self is a mirror to reflect others. Each mirror reflects slightly different others. This is the reason that personalities are different.

S. Ike

In the previous chapter, I made feeble complaints. However, is it true that there is no evidence of the birth function? The birth function is the only metrical theory that can yield quantitative predictions. On making full use of it, decisive evidence can be obtained. A useful theory appeals to logic, while a meaningless one appeals to emotion.

12.1 Validity of the Extended Birth Function

The birth function has already been built on the transcendental basis of human consciousness. Namely, the effects experienced in high-dimensional space cannot be adequately represented. Conscious reflection cannot be used to judge the validity of the function.

For instance, despite maintaining our positions, we cannot recognize the collisions of numerous particles from all directions. Boltzmann had to invent the kinetic theory of gases in order to explain phenomena of gases. Therefore, another logical method must be applied to judge the validity of the birth function. Another method that can prove the function is finding whether or not it can derive a numerical prediction and whether it matches the real observed value.

In the previous section, it was made relatively clear that it is difficult to make micro or regional predictions about birth occurrences. There is an advantage of the birth function in macroscale dynamics: it is the only theory that can predict macro dynamics over several decades, even though its prediction is exclusively dependent on adjusted values.

Recall the background independence of fertility decline. As its definition states, the birth function is also background independent. Birth behaviors are stochastically determined by neighboring birth behaviors. The distribution of the number of children can be quantitatively predicted. No other theory can achieve this; this is the exceptional ability of the birth function.

12.2 Distribution of the Number of Children

If birth probability is a single value p, the distribution of the number of children of a female (parents) might be binomial. When the means and variances of the number of children of a cohort are calculated based on age at marriage, where the duration of marriage in years is y_i, the binomial distribution is obtained with $\mu = y_i p$, $\sigma = y_i p(1-p)$.

Naturally, the birth probabilities derived by the birth function are variable. The distribution of the number of children is not a binomial distribution. Namely, the comparison with the binomial distribution is an effective test for the validity of the birth function, because their means, variances, and frequencies can be compared. This is verification of the birth function.

12.2.1 Number of Children Based on the Birth Function

The variables calculated by the birth function, such as Equation (11.5), are not constant. Therefore, the expected number of children is computed using the following simple, straightforward approach.

Let p_i be the birth probability and q_i be no birth probability at during year i. The number of children per during marriage year is the product of p_i and q_i.

$$\begin{aligned}
&p_1, q_1\\
&p_1p_2, q_1p_1, p_1q_2, q_1q_2\\
&p_1p_2p_3, q_1p_1p_3, p_1q_2p_3, q_1q_2p_3, p_1p_2q_3, q_1p_1q_3, p_1q_2q_3, q_1q_2q_3\\
&\vdots
\end{aligned} \tag{12.1}$$

In each term, the frequency of p_i represents the number of children. For example,

$$\begin{aligned}
&1.\ p_1p_2p_3 \qquad \text{Birth} = 3\\
&2.\ q_1p_2p_3 \qquad \text{Birth} = 2\\
&3.\ p_1q_2p_3 \qquad \text{Birth} = 2\\
&4.\ q_1q_2p_3 \qquad \text{Birth} = 1
\end{aligned} \tag{12.2}$$

Clearly, $p_1 p_2 p_3 \neq p^3$ and $q_1 \neq q_3$; each term cannot be arranged in order as the binomial distribution. Comprehensive calculations must be performed for each individual term. In addition, the number of p_i in each term is distinguished. Based on these frequencies, the terms are sorted, and probabilities of the same frequency are summed.

The number of terms is large. For the duration of marriage y_i, the number of terms is 2^{y_i}. A woman who married at age 15 (and has remained married) has a birth probability composed of $2^{34} = 17,179,869,184$ terms, and her expected birth is computed by summing up these terms. It is a challenging task even for my personal computers to compute it. In particular, for a longer marriage duration, it is impossible to allocate a sufficiently large memory space corresponding to the long array to be computed. It is impossible for a normal processing system to define such a long array. For this reason, a relatively small y_i that could be easily defined on my processing system was used. Namely, the objects needed to be confined to analyze the number of children of females in relatively late marriages.

12.2.2 Distribution of the Number of Children at the Beginning of Fertility Decline—Japan, 1939

As can be seen from Figure 11.17 (p. 305), birth probabilities at 1939 in Japan were relatively quasi-constant during the beginnings of fertility decline. Thus, the distribution of the number of children may resemble a binomial distribution. First, females who were married at age 30 in 1939 are analyzed.

The mean number of children born is 2.64. In addition, their duration of marriage is set to 16 years. Therefore, the constant birth probability p assuming the binomial distribution of the number of children can be estimated as follows:

$$y_{46-30} \cdot p = \mu_{30}, \quad 16 \cdot p = 2.64, \quad p = 0.165 \tag{12.3}$$

On obtaining p, the binomial distribution of the number of children is easily computed using the formula.

In contrast, the distribution of the birth function for newborn children is complicated but computable. Thus, two similar and different distributions are obtained. It is fairly self-evident that birth probabilities in Japan gradually declined in 1939. Thus, it is also self-evident that the distribution computed by the birth function is valid. See Table 12.1, which presents the probabilities of the number of children. If a corresponding sample size greater than 5000 is obtained, the difference can be decisively tested.

Namely, the validity of the birth function can be verified by means of the distribution of the number of births. The future birth function can be computed/predicted using Equation (11.5), and thus, the future distribution of the number of children can be predicted.

Alternatively, the future regional birth function can be computed using Equation (11.5), and it can be tested using the distribution of the number of regional births. The birth function theory is completely falsifiable.

Table 12.1: Probabilities of number of children for females married at the age of 30 in 1939 in Japan.

Number of Children	Binomial	Birth Function
0	0.0558449116	0.0601714000
1	0.1765635528	0.1934544000
2	0.2616735289	0.2808687000
3	0.2413037332	0.2440325000
4	0.1549690143	0.1416620000
5	0.0734942870	0.0580517000
6	0.0266251758	0.0172840000
7	0.0075160890	0.0037868000
8	0.0016708671	0.0006107000
9	0.0002934856	0.0000716000
10	0.0000405959	0.0000059000
11	0.0000043756	0.0000003000
12	0.0000003603	0.0000000000
13	0.0000000219	0.0000000000
14	0.0000000009	0.0000000000

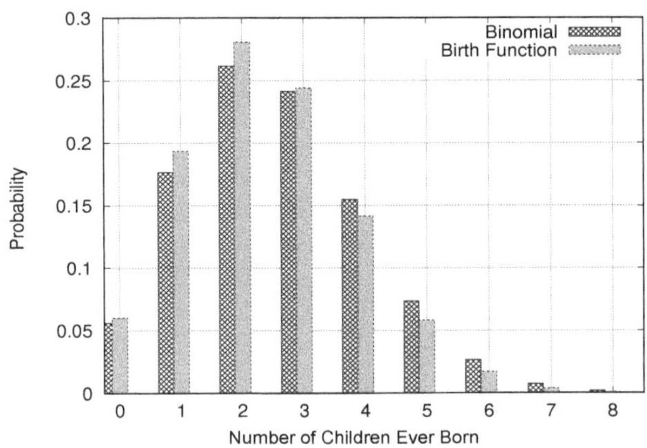

Figure 12.1: Probabilities of number of children for females married at the age of 30 in 1939 in Japan.

12.2.3 Distribution of the Number of Children During the Process of Fertility Decline—Japan, 1985

The difference between the binomial distribution and that computed by the birth function was greater during the process of fertility decline. Data on females married at the age of 25 in the year 1985 were taken (see Table 8.4, p. 204).

The expected number of children is 2.14150. Thus, the p value for the binomial distribution is calculated as follows:

$$y_{25-46} \cdot p = 2.1450, \;\; p = 2.145/21 = 0.1019762, \tag{12.4}$$

Table 12.2: Probabilities of number of children for females married at the age of 25 in 1985 in Japan.

Number of Children	Binomial	Birth Function Observed	Birth Function Computed
0	0.1044828998	0.0809011000	0.0747001000
1	0.2491582585	0.2426960000	0.2216320000
2	0.2829346005	0.3122304000	0.2931854000
3	0.2034834920	0.2260851000	0.2301434000
4	0.1039806362	0.1017775000	0.1204081000
5	0.0401459891	0.0297556000	0.0447020000
6	0.0121568671	0.0057507000	0.0122376000
7	0.0029581873	0.0007375000	0.0025331000
8	0.0005878608	0.0000625000	0.0004032000
9	0.0000964242	0.0000035000	0.0000499000
10	0.0000131395	0.0000001000	0.0000048000
11	0.0000014921	0.0000000000	0.0000004000
12	0.0000001412	0.0000000000	0.0000000000
13	0.0000000111	0.0000000000	0.0000000000
14	0.0000000007	0.0000000000	0.0000000000
σ^2	1.923117	1.560150	1.821877

Table 12.2 presents the respective distributions of the number of children in terms of the binomial distribution, the observed birth function, and the calculated birth function. The proposed test is highly rigorous and does not reject the birth function theoretically calculated by the equation. If there is an appropriate size sample, the birth function can be tested by the χ^2 test.

12.2.3.1 Test by Variance and Non-parametric Values

It is recognized that the sample's variance s^2 rapidly approaches the population's variance σ^2 as the sample size n increases. As Table 12.2 shows, the binomial, observed, and theoretical σ^2 differ. On measuring 5000 samples, the validities of these σ^2 values can be judged. The binomial value can easily be eliminated, and the problem lies in deciding whether the observed or theoretical value is superior.

Under normal conditions, the theoretical value would be likely to be eliminated. However, the observed data must neglect the undeclared marriages and their births. Thus, the theoretical value has a chance of being selected.

Comparing the ranks of zero and one child for corresponding females may serve as a simple and decisive test. In this case, if the rank is approximately 0.0747, the birth function theory is accurate. If the rank is larger than 0.080, the birth function is not as effective as the observed data.

12.2.4 Prediction and Test by Non-Parametric Statistics

Because the birth function theory is a true falsifiable and metric theory, it can predict future numbers of children. Here, the birth function is computed until 2099 in Japan in virtual space. A prediction is provided based on these computed data.

Comparing the ranks of zero and one child for corresponding females serves as a simple and decisive test. In this case, if the rank is approximately 0.0747, the birth function theory is accurate. If the rank is larger than 0.080, the birth function is not as effective as the observed data.

12.2.4.1 Gradual Variation of Birth Function Between Fixed Points

On reviewing the computed data, I found that the birth function will gradually vary to fixed points in 2025, as shown in Figure 12.3. These movements cannot be visually distinguished easily.

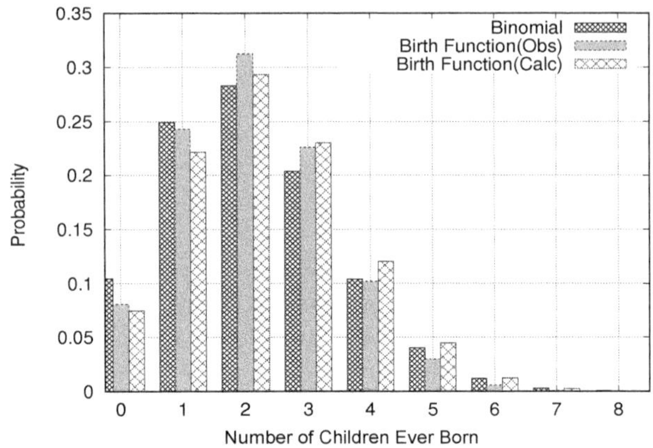

Figure 12.2: Probabilities of number of children for females married at the age of 25 in 1985 in Japan.

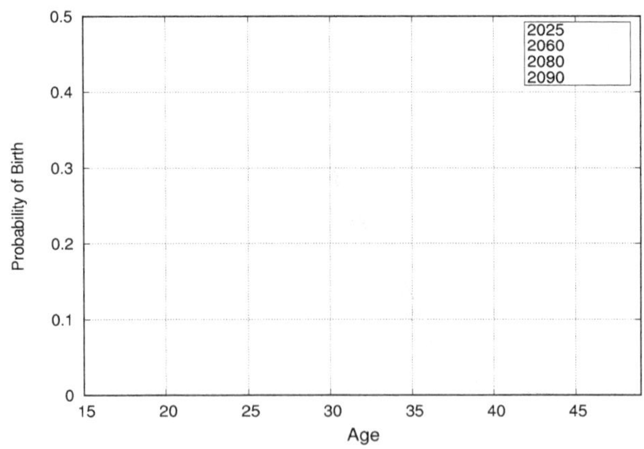

Figure 12.3: Birth probabilities of females married at the age of 30 between 2025–2090.

These movements depend on the space where the birth function compute the birth probabilities. Therefore, the space of the representative point is arbitrarily chosen.

Urban—30 km from center

First, birth probabilities are selected at a point 30 km from the center. The birth probabilities of the 1995 female cohort married at age 30 are reconstructed, and the distribution of the number of children is computed. The results are presented in Table 12.3 and showed graphically in Figure 12.4.

Table 12.3: Predicted probabilities of the number of children of females from the 1995 cohort, married at the age of 30, at an urban point in Japan.

Number of Children	Birth Function Computed
0	0.3508477
1	0.2883580
2	0.1328189
3	0.0378478
4	0.0069563
5	0.0008356
6	6.53E-05
7	3.3E-06
8	1E-07

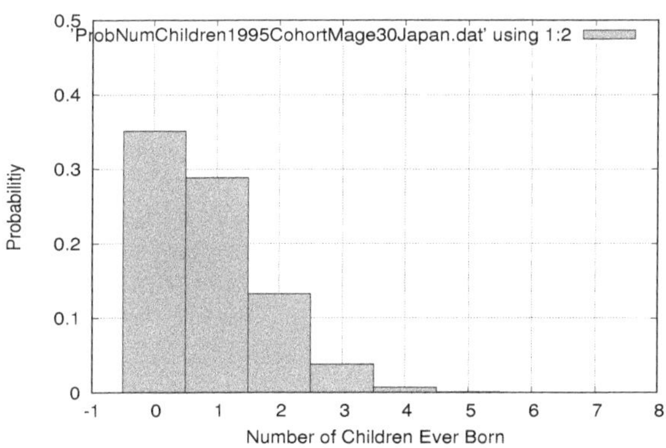

Figure 12.4: Probabilities of the number of children for females married at the age of 30 (between 2025–2040) at an urban area in Japan.

In Japan, a female who was born in 1995 and married at the age of 30 will bear 0.7 children in urban area.[1] The 35th percentile is the boundary between zero children and one child at this area. When we measure 35 percent zero children of them in 30 km from center around 2030, the birth function verified.

[1]This is a distressing prediction. However, we should be wary of bringing forth our own extinction.

Larger area

Next, birth probabilities over a larger area are selected and reconstructed in the same manner as before. The results are presented in Table 12.3 and showed graphically in Figure 12.5.

Table 12.4: Predicted probabilities of the number of children of females from the 1995 cohort, married at the age of 30, over a larger area in Japan.

Number of Children	Birth Function Computed
0	0.330506
1	0.295484
2	0.151403
3	0.049397
4	0.010788
5	0.001615
6	0.000167
7	1.19E-05
8	6E-07

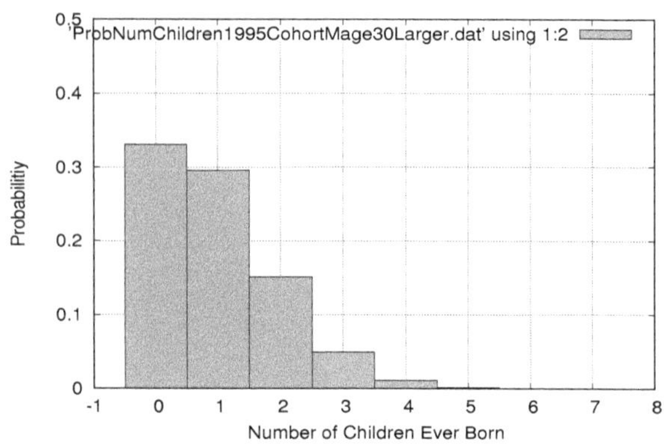

Figure 12.5: Probabilities of the number of children married at the age of 30 between 2025–2040.

Over the larger area of Japan, a female who was born in 1995 and married at the age of 30 will bear 0.8 children. The 33rd percentile is the boundary between zero children and one child for this larger area. If the birth function is valid, the frequency of one child of them will be not more than 30 percent in Japan.

Someone who is alive in 2040 will test my prediction. If individual married teenagers do not alter their birth behaviors, my prediction will be accurate. Despite it indicating a bleak future, the birth function can *prévenir*.

Chapter 13

Conclusion—to End This Book

> Sociology must enable the realization of human beings in time–space.

S. IKE

13.1 End of Eclecticism

Since the 1980s, social scientists have been inclined to believe that social phenomena are aggregates of rational individual behaviors. On encountering phenomena that could not be explained in this manner, we brought out cultural factors to explain them *ex post facto*.

This strategy has completely failed; it has not produced a single falsifiable prediction. Theorists who insisted on the second demographic transition did away with this strategy and indulged themselves in the cultural feminist revolution. After all, the assumption of rational individual behaviors was meaningless, particularly in terms of demography. This gave us a possible explanation for fertility decline, but it has remained just an explanation.

Using this idea of rational individual choice, the mean number of children born or the mean age of first marriage cannot be calculated. It is demonstrated that it is not a scientific theory.

Rational individualism assumes that the alteration of human birth behaviors was brought about by the alteration of human conscious motivation. This argument is tautological and says nothing about the causal relationship. Human consciousness being the cause of our behavior is not a scientific, falsifiable theory. The actual causes exist solely in space.

It is necessary to develop a scientific theory in space from one point in time to the next. Rational individual and social norms are completely static and *ad hoc*, and they are clearly unable to produce meaningful predictions. The only gateway to falsifiable sociology is to investigate cohabitation in the space in which we interact with each other.

Humans are not "rational fools." We are simply fools to believe ourselves to be independent individuals, not knowing that we are controlled by others. In such a situation, the eclecticism of rational individuals and cultural factors becomes essential. However, this eclecticism is conventional and infertile, and it does not cultivate scientific inferences. The binary opposition of rational and untilitarian individuals to the ethical norm is hackneyed. This opposition may itself be an illusion; it is time to abandon it.

Human behaviors must be studied in high-dimensional space. The patterns of our behaviors in it must be recognized. Space and those residing in it have been greatly ignored. Rather than abstractly considering them in terms of our thought processes, they must be considered as actual physical phenomena. This book is intended to mark the beginning of this approach. I believe that the formalization of humans living in space is the only scientific approach to sociology.

The idea of free will is also an illusion. First, "free" cannot be defined. At best, the chains of behaviors (i.e., causes) can be traced. However, investigating these chains in high-dimensional space is a valuable endeavor. The manner in which symbolic dimensions interact with actual geographic dimensions and control us in certain aspects should be understood.

Humans (as one animal species happens to call itself) depend on neighboring situations. Their thought processes reflect neighboring ones. Thus, consistent individual conscious thoughts do not exist.

13.2 Time and Space

As I noted Julian Barbour second in References, I do not place absolute trust in the existence of "Time" and Space. In fact, marriage function and birth function are formulated as a mapping from a instant to next instant receiving neighboring effect. Of course, my theory naturally assumes neighboring existences and space for them. Society may be the patchwork of small lattices. Events do not transmit instantly to remote lattices. Strictly speaking, simultaneity makes no sense in it. For sociology, fortunately, I think, we are allowed to use a standard time–space system as GPS or GNSS for measurements without deep consideration.

In order to think as a reaction-diffusion system, I introduced differential equations. However, I expect that the difference equation is more fundamental. I recommend that social scientists pay much more attention to space (it inevitably causes to consider time). We have just begun to explore the mappings of the relation with others.

Appendix

Convergence of the Velocity of the Progressive Wave to its Lowest Point

By means of simulation, it is confirmed that the progressive (FKPP) wave travels relatively rapidly immediately following the singularity. Its velocity then reduces and converges to a constant. Kolmogorov-Petrovsky-Piskunov proved this as a theorem [46]. The velocity will be explained herein.

Front

How does the front of a progressive wave form? When a concave surface is formed on the hypersurface, its differs from the next hypersurface in terms of the front.

$$\frac{\partial C}{\partial t} = \mu \left(\frac{\partial^2 C}{\partial x^2} + \frac{\partial^2 C}{\partial y^2} \right) + (1-\mu)\,\alpha \sqrt{(\beta - C)^2 (\gamma - C)} \tag{1}$$

Immediately following the singularity, its neighborhood declines owing to the overwhelming decline of the logistic term. After the diffusion has progressed, the C value of the logistic term and the Laplacian become smaller as their limits. In particular, they are small because of the second-order partial derivative and C^2. Therfore, the size of the concave mainly depends on μ, α. The limit of concave is the ultimate cause of the certain finite velocity of FKPP-wave.

Thus, the alterations in time t and $C(\acute{x}, \acute{y})$ immediately preceding the front do not covariate. Following a slight delay, $C(x, y)$ declines. At the front, the neighboring hypersurfaces are almost all $C(x, y) = \beta$. It must be noted that at the micro scale, the reaction-diffusion process is a stochastic interaction process. In these processes, a substantial decline occurs following a time course. The constant velocity of the progressive (FKPP) wave is the result of these micro-interactions.

For the above-mentioned reasons, the product of μ and α relate to the velocity of diffusion. However, the definition

$$\mu\left(\frac{\partial^2 C}{\partial x^2}+\frac{\partial^2 C}{\partial y^2}\right)+(1-\mu)\,\alpha\sqrt{(\beta-C)^2(\gamma-C)} \tag{2}$$

implies that an increasing μ value is equal to a decreasing α value. On the contrary, an increasing α value strongly induces the unique instability of the reaction-diffusion. In addition, for the fertility decline process, when the scale of space is constant, the high α value practically implies a large interval of time. For these reasons, the velocity cannot be made as large as desired.

Bibliography

[1] *World Almanac and Book of Facts, 2007*. World Almanac, 2006.

[2] Julian Barbour. 2003. *The End of Time*. Weidenfeld & Nicolson. ISBN-13: 978-0753810200.

[3] Gary S. Becker. 1981. *A Treatise on the Family*. Harvard University Press.

[4] Ernst Benz. 2006. Family limitation among political catholics in baden in 1869. In Renzo Derosas and Frans van Poppel, editors, *Religion and The Decline of Fertility in The Western World*, chapter 9, pages 107–132. Springer, 2006.

[5] Gérard Calot. 1998. *Two Centuries of Swiss Demographic History—Graphic Album of the 1860–2050 Period*. Swiss Federal Statistical Office/ Observatoire démographque européen, Paris, 1998 with assistance from Alan Confession and Jean-Paul Sardon (Observatoire démographque européen, Paris), Erminio Baranzini, Stephane Cotter and Philippe Wanner (Swiss Federal Statistical Office), ISBN 3-303-01095-1.

[6] G. Carlsson. 1966. The decline of fertility: Innovation or adjustment process. *Population Studies*, 20(2): 149–174.

[7] John B. Casterline. 2001. *Diffusion Processes and Fertility Transition*. National Academy Press, Washington, D.C. National Research Council.

[8] Gregory Chaitin. 2005. Algorithmic irreducibility in a cellular automata Universe. *Journal of Universal Computer Science*, 11(12): 1901–1903.

[9] C.Y. Cyrus Chu. 1998. *Population Dynamics—A New Economic Approach*. Oxford University Press, Inc.

[10] Colin Clark. 1949. Age at marriage and marital fertility. *Population Studies*, 2(4): 413–426, 3 1949.

[11] A.J. Coale and D.R. McNeil. 1972. The distribution by age of the frequency of first marriage in a female cohort. *Journal of the American Statistical Association*, 67(340): 743–749.

[12] Ansley J. Coale. 1971. Age patterns of marriage. *Population Studies*, 25(2): 193–214.

[13] Ansley J. Coale. 1986. The decline of fertility in Europe since the eighteenth century as a chapter in human demographic history. *In*: Ansley J. Coale and Susan Cotts Watkins (eds.). *The Decline of Fertility in Europe—The Revised Proceedings of a Conference on the Princeton European Fertility Project*, Princeton, 1986. Conference on the Princeton European Fertility Project (1979), Princeton University Press.

[14] Ansley J. Coale and Roy Treadway. 1986. A summary of the changing distribution of overall fertility, marital fertility, and the proportion married in provinces of Europe. *In*: Ansley J. Coale and Susan Cotts Watkins (eds.). *The Decline of Fertility in Europe—The Revised Proceedings of a Conference on the Princeton European Fertility Project*, Princeton, 1986. Conference on the Princeton European Fertility Project (1979), Princeton University Press.

[15] Ansley J. Coale and Susan Cotts Watkins (eds.). 1986. *The Decline of Fertility in Europe—The Revised Proceedings of a Conference on the Princeton European Fertility Project*. Princeton University Press, 1986.

[16] Benedict de Spinoza. *The Ethics (Ethica Ordine Geometrico Demonstrata)*. The Project Gutenberg EBook of The Ethics, by Benedict de Spinoza, http://www.gutenberg.org/ebooks/3800, 1677. Translated from the Latin by R.H.M. Elwes.

[17] J.E. Dennis Jr. and Daniel J. Woods. 1985. Optimization on Microcomputers The Nelder-Mead Simplex Algorithm. Technical report, Rice University, 1985. available as Math. Science TR 85-9.

[18] Renzo Derosas. 2006. Between identity and assimilation: Jewish fertility in nineteenth-century venice. pp. 177–205. *In*: Renzo Derosas and Frans van Poppel (eds.). *Religion and the Decline of Fertility in the Western World*, Chapter 9. Springer.

[19] Renzo Derosas and Frans van Poppel. (eds.). 2006. *Religion and the Declineof Fertility in the Western World*. Springer.

[20] Andreas Diekmann and Peter Mitter. 1983. The 'sickle hypothesis': A time dependent poisson model with applications to deviant behavior and occupational mobility. *Journal of Mathematical Sociology*, (9): 85–101.

[21] Richard A. Easterlin. 1961. The American baby boom in historical perspective. *American Economic Review*, 51(5): 869–911.

[22] Richard A. Easterlin. 1968. *Population, Labor Force and Long Swing in Economic Growth: The American Experience*. Columbia University Press.

[23] Richard A. Easterlin. 1987. *Birth and Fortune: The Impact of Numbers on Personal Welfare*. The University of Chicago Press.

[24] Josef Ehmer. 2013. *Bevölkerungsgeshichite und historishe Demographie 1800–2010*. Oldenbourg Wissenshaftsverlag GmbH, Múnchen.

[25] Patrick Festy. 1984. *Fertility in Western Countries from 1870 to 1970*. Population Studies Translation Series No. 7. The Population Devision Economic and Social Commission for Asia and the Pacific. UNITED NATIONS, Bangkok.

[26] Ronald A. Fisher. 1937. The wave of advance of advantageous genes. *Ann, Eugen*, (7): 355–369.

[27] Elise F. Jones, Jacqueline Darroch Forrest, Noreen Goldman, Stanley Henshaw, Richard Lincoin, Jeannie I. Rosoff, Charles F. Westoff and Deirdre Wulf. 1986. *Teenage Pregnancy of Industrialized Countries*. Yale University Press, New Haven and London.

[28] Guilherme Quaresma Gonçalve, Reinaldo Onofre dos Santos, Laura Lídia Rodríguez Wong, and Josė Alberto Magno de Carvalho. 2017. Time and space of the fertility transition in brazil: An exploratory analysis of brazilian municipalities in 1991, 2000 and 2010. Research Gate. See discussions, stats, and author profiles for this publication at: https://www.researchgate.net/publication/320888241.

[29] Peter Gray and Stephen K. Scott. 1984. Autocatalytic reactions in the system A + 2B → 3B, B to C. *Chem. Eng. Sci.*, (39): 1087–1097.

[30] Torsten Hlagerstrand. 1968. *Innovation Diffusion as a Spatial Process*. University of Chicago Press. Translation and postscript by Allan Pred.

[31] Louis Henry. 1961. Some data on natural fertility. *Eugenics Quarterly*, 8(2): 81–91, Jun 1961.

[32] Louis Henry. 1976. *Population: Analysis and Models*. Edward Arnold, London. Dėmographie: analysis et modėles, translated by Etienne van de Walle and Elise F. Jones.

[33] Gudmund Hernes. 1972. The process of entry into first marriage. *America Sociological Review*, 37: 173–182, April 1972.

[34] Alan L. Hodgkin and Andrew F. Huxley. 1952. A quantitative description of membrane current and its application to conduction and excitation in nerve. *Journal of Physiology*, (117): 500–544.

[35] Shuichirou Ike. 2009. *Decline of Marital Fertility and Diffusion Hypothesis—Reaction-diffusion Model of Completed Marital Fertility*. Kokon Shyoin, Bunkyou-ku, Japan. The Japanese title is translated into English by the author of this book.

[36] Shuichirou Ike. 2012. A singularity of fertility decline is Lot-et-Garonne. *Journal of Mathematical Sociology*, (36): 137–155. DOI:10.1080/002225 0X.2011.556917.

[37] Shuichirou Ike. 2015. *The Mathematics of Marriage Function—Its Dynamics and Predictions*. Kokon Shyoin. Grants-in-Aid for Scientific Research No. 265181.

[38] Shuichirou Ike. 2016. *Fertility Decline and Background Independence—Applying a Reaction-Diffusion System as a Stochastic Process*. Population Studies of Japan. Springer.

[39] Hisashi Inaba. 2002 *Mathematical Demography*. University of Tokyo Press.

[40] Tatuya Ito and Eriko Bando. 1989. The trend of marital fertility in the first half of 1980s of Japan (translated). *Journal of Population Problems (Jinko Mondai Kenkyu)*, (189): 51–69, 1 1989. ISSN 0387-2793, http://www.ipss.go.jp/syoushika/bunken/sakuin/jinko/jsakuin1.htm.

[41] Koremichi Kametaka. 1977. *Hi-senkei Bibun Houteishiki*, volume 19 of *Surikaiseki to sono Shuuhen*. Sangyoutosho Kabusikigaisya.

[42] Yoshinori Kametaka. 1976. On the nonlinear diffusion equation of Kolmogorov-Petrovskii-Piskunov type. *Osaka Journal of Mathematics*, 13(1): 11–66. DOI: 10.18910/9093, translated into English by Gülbeyaz, Abdurrahman, Osaka University Knowledge Archive: OUKA, https://ir.library.osaka-u.ac.jp/repo/ouka/all/.

[43] Michael C. Keely. 1979. An analysis of age pattern of first marriage. *International Economic Review*, 20(2): 527–544.

[44] John Knodel and Etienne van de Walle. 1979. Lessons from the past: Policy implications of historical fertility studies. *Populations and Development Review*, 5: 217–245.

[45] Hiroshi Kojima, Makoto Atoh, Tatsuya Itoh and Masako Ikenoue. 1985. An application of the nypitiality model developed be Coale and McNeil. *The Journal of Population Problems*, (175): 36–47.

[46] A. Kolmogroff, I. Petrovsky and N. Piscounoff. 1937. étude de l'equation de la diffusion avec croissance de la quantité de matière et son application à un problème bibliogique. *Bull. Univ. d'État à Moscou*, Ser, Internat. Sect. A: 1–25.

[47] Karl U. Mayer and Nancy B. Tuma. 1990. *Event History Analysis in Life Course Research*. The University of Wisconsin Press.

[48] Paul F. Lazarsfeld. 1968. *People's Choice: How the Voter Makes Up His Mind in a Presidential Campaign*. Columbia University Press, Third Edition.

[49] Ron Lesthaeghe. 1991. The Second Demographic Transition in Western Countries: An Interpretation. In *Gender and Family Change in Industrialized Countries*.

[50] David Levine. 1977. *Family Formation in an Age of Nascent Capitalism*. Academic Press, Inc.

[51] Benjamin Libet. 2005. *Mind Time: The Temporal Factor in Consciousness*. Perspectives in Cognitive Neuroscience. Harvard University Press, 2004. First Harvard University Press paperback edition.

[52] Massimo Livi-Bacci. 1986. Social-Group Forerunners of Fertility Control in Europe. pp. 182–200. *In*: Ansley J. Coale and Susan Cotts Watkins (eds.). *The Decline of Fertility in Europe—The Revised Proceedings of a Conference on the Princeton European Fertility Project*, Princeton, 1986. Conference on the Princeton European Fertility Project (1979), Princeton University Press.

[53] Nancy Locklin. 2016. *Women's Work and Identity in Eighteen-Century Brittany*. Routledge. First published 2007 by Ashgate Publishing.

[54] Hector Malot. 1983. *Nobody's Girl (En Famille)*. CUPPLES & LEON COMPANY (Ernest Flammarion), 1922. Translator: Florence Crewe-Jones, The Project Gutenberg EBook of Nobody's Girl, by Hector Malot, Produced by Juliet Sutherland, Jen Haines and the Online Distributed Proofreading Team at http://www.pgdp.net, http://www.gutenberg.org/ebooks/27690.

[55] Labour Ministry of Health and Welfare. Vital statistics. https://www.mhlw.go.jp/english/database/db-hw/index.html.

[56] Mark R. Montgomery and John B. Casterline. 1993. The diffusion of fertility control in Taiwan: Evidence from pooled cross-section time-series models. *Population Studies*, 47(3): 457–479.

[57] Mark R. Montgomery and John B. Casterline. 1996. Social learning, social influence, and new models of fertility. *Population and Development Review*, 22 (Supplement): 151–175.

[58] G.L. Nemhauser, A.H.G. Rinnooy Kan and M.J. Todd. (ed.). 1989. *Optimization*, volume 1 of *Handbooks in Operations Research and Management Science*. Elsevier Science Publishers.

[59] Ayanori Okazaki. *Marriages and Population*. Chikura syobou, 1941. Kuress Press co. reprinted as *Family, Marriage* No. 12 in 1999.

[60] Kenji Otani. 1991. Fitting mathematical functions to the distribution of first marriage age of a cohort. *Population Association of Japan*, (14): 67–73. The Japanese title is translated into English by the author of this book.

[61] Talcott Parsons. 1968. *The Structure of Social Action*, volume 1. The Free Press, 1937.

[62] John E. Pearson. Complex patterns in a simple system. *Science*, (261): 189–192.

[63] Ross L. Prentice. 1974. A log gamma model and its maximum likelihood estimation. *Biometrika*, 61(3): 539–544.

[64] Ronald R. Rindfuss, Karin L. Brewster and Andrew L. Kavee. 1999. Women, Work, and Children: Behavioural and Ideational Change in the United States. In Richard Leete, editor, *Dynamics of Value in Fertility Change*, Chapter 7, pp. 148–175. Oxford University Press.

[65] Everett M. Rogers. 1973. *Communication Strategies for Family Planning*. The Free Press, New York.

[66] Luis Rosero-Bixby and John B. Casterline. 1994. Interaction Diffusion and fertility transition in Costa Rica. *Social Forces*, 73(2): 453–462.

[67] Carlo Rovelli. 2017. *L' ordine del tempo*. ADELPHI EDIZIONI S.P.A. Mi-LANO. Japanese Transtation.

[68] Jona Schellekens and Frans van Poppel. 2006. Religious differentials in marital Fertility in the Hague (Netherlands), 1860–1909. In Renzo Derosas and Frans van Poppel, editors, *Religion and the Decline of Fertility in the Western World*, Chapter 4, pp. 59–81. Springer.

[69] John G. Skellam. 1951. Random dispersal in theoretical populations. *Biometrika*, (38): 196–218.

[70] John G. Skellam. 1952. Studies in statistical ecology. *Biometrika*, (39): 346–362.

[71] Michael S. Teitelbaum and Jay M. Winter. 1985. *The Fear of Population Decline*. Academic Press, Inc.

[72] Teizo Toda. 1937. *The Constitution of Family*. Kobundo-shobo. The Japanese title is translated into English by the author of this book.

[73] Alan Mathison Turing. 1952. The chemical basis of morphogenesis. *Philosophical Transactions of the Royal Society of London*, 237: 37–72.

[74] Daishin Ueyama and Yasumasa Nishiura. 2006. *Pattern Formation anf its Dynamics*. Chapter 2, pp. 21–62. University of Tokyo Press. ISBN: 4-13-064094-1.

[75] Dirk J. van de Kaa. 1987. Europe's Second Demographic Transition. *Population Bulletin*, 42(1).

[76] Francine van de Walle. 1986. Infant mortality and the european demographic transition. pp. 201–233. *In*: Ansley J. Coale and Susan Cotts Watkins (eds.). *The Decline of Fertility in Europe—The Revised Proceedings of a Conference*

on the Princeton European Fertility Project, Princeton, 1986. Conference on the Princeton European Fertility Project (1979), Princeton University Press.

[77] Thorstein Veblen. 2008. *The Theory of the Leisure Class*. Project Gutenberg, 8 2008. https://www.gutenberg.org/ebooks/833, Project Gutenberg's The Theory of the Leisure Class, by Thorstein Veblen.

[78] Ben J. Wattenberg. 2004. *Fewer —How the New Demography of Depopulation Will Shape Our Future*. Ivan R. Dee, Chicago.

[79] Charles F. Westoff and Elise F. Jones. 1979. The end of "catholic" fertility. *Demography*, 16(2): 209–217, 5 1979.

[80] James W. Wood. 1989. *Fecundity and Natural Fertility in Humans*. Chapter 2, pp. 61–109. Oxford Reviews of Reproductive Biology.

[81] James W. Wood and Maxine Weinstein. 1988. A model of age-specific fecundability. *Population Studies*, (42): 85–113.

[82] Daniel J. Woods. 1985. *An Interactive Approach for Solving Multi-Objective Optimization Problems*. PhD thesis, Rice University, 1985. Available as Math. Science TR 85-5.

[83] Edward A. Wrigley. 1969. *Population and History*. World University Library. George Weidenfeld and Neclson Ltd., London.

[84] Eiji Yanagida. 2015. *Reaction-Diffusion Equation*. University of Tokyo Press.

Index

For Product Safety Concerns and Information please contact our EU
representative GPSR@taylorandfrancis.com
Taylor & Francis Verlag GmbH, Kaufingerstraße 24, 80331 München, Germany

www.ingramcontent.com/pod-product-compliance
Ingram Content Group UK Ltd.
Pitfield, Milton Keynes, MK11 3LW, UK
UKHW020021200926
475702UK00014B/1998
* 9 7 8 1 0 3 2 1 3 4 7 1 0 *